The Martian Climate Revisited

Atmosphere and Environment of a Desert Planet

Springer
Berlin
Heidelberg
New York
Hong Kong
London
Milan
Paris
Tokyo

Peter L. Read and Stephen R. Lewis

The Martian Climate Revisited

Atmosphere and Environment of a Desert Planet

Published in association with
Praxis Publishing
Chichester, UK

Professor Peter L. Read
Department of Physics
University of Oxford
Oxford
UK

Dr Stephen R. Lewis
Department of Physics
University of Oxford
Oxford
UK

SPRINGER–PRAXIS BOOKS IN GEOPHYSICAL SCIENCES
SUBJECT *ADVISORY EDITORS*: Dr Philippe Blondel, C.Geol., F.G.S., Ph.D., M.Sc., Senior Scientist, Department of Physics, University of Bath, UK; John Mason, B.Sc., M.Sc., Ph.D.

ISBN 3-540-40743-X Springer-Verlag Berlin Heidelberg New York

Springer-Verlag is a part of Springer Science+Business Media (springeronline.com)
Bibliographic information published by Die Deutsche Bibliothek
Die Deutsche Bibliothek lists this publication in the Deutsche Nationalbibliografie; detailed bibliographic data are available from the Internet at http://dnb.ddb.de

Library of Congress Cataloging-in-Publication Data
A catalogue record for this book is available from the Library of Congress

Printed in Germany

Cover design: Jim Wilkie
Project Management: Originator Publishing Services, Gt Yarmouth, Norfolk, UK

Printed on acid-free paper

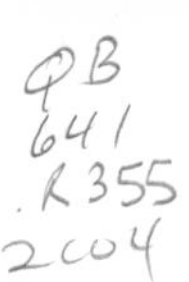

Contents

Preface

The past 5–6 years have seen a remarkable renaissance in the exploration of Mars from space, following a prolonged hiatus of more than 15 years since the Viking missions of the 1970s and 1980s. Despite the failure of the Mars Observer mission in 1993, which was supposed to herald a new age of Martian exploration by NASA, the Mars science community has shown considerable resilience in promoting this renewal of the mission to explore our nearest planetary neighbour. Although this new programme of exploration is still under way at the time of writing this book, the spectacular success of the Mars Global Surveyor (MGS) mission, which was the first major successor to Mars Observer and enabled several of the main instruments which formed the payload of that mission finally to reach Mars orbit, has largely completed its primary goal of mapping and surveying the entire planet in unprecedented detail. With the hiatus following the loss of both the Mars Climate Orbiter and Mars Polar Lander spacecraft in 1999, it seemed timely to take stock of what had been learned so far about some aspects of Mars.

Of course much has been written about Mars since the Viking missions themselves, either in the form of popular science articles or books, or as learned reviews on all aspects of the planet from its deep interior to its magnetic field and external particle environment. The immediate post-Viking position in the scientific literature is perhaps best represented by the monumental book on Mars published by the University of Arizona Press in 1992 (Kieffer *et al.*, 1992). But this tome was primarily aimed at the experts in the subject, and gives a 'no holds barred' discussion of every topic in terms which only professional researchers can fully appreciate. In our case, we felt that there was a clear need in the literature for a text which was aimed more towards the beginning student or scientifically literate lay person, and in particular to concentrate on a proper treatment of the climate and meteorology of Mars, which seems in the past to have been relatively neglected compared with other aspects of the planet. The question of life, past or present, tends to grip the imagination of the media whenever Mars is mentioned. And yet there is much more to this fascinating planetary system than whether or not microbial life (or something similar) was ever able to establish itself on a planet other than the Earth. One of the delights of physical science is trying to predict from first principles the behaviour of a complex system, and the many similarities between Mars and the Earth pose some particularly intriguing challenges to the terrestrial scientist to see if our understanding of the Earth is sufficient to predict what the environment of Mars might be like. Richard P. Feynman, as quoted at the end of our book, summed up this fascination in more powerful words than we could manage.

In deciding how to write this book, we felt it necessary to confine the limits of our task mainly to the present climate of Mars. Much has been written elsewhere on the possible conditions in Mars' early history, when the climate may have been relatively warm and wet. But so much is now known about the present state of the Martian environment that there is plenty of new science to discuss at the level of detail previously reserved only for studies of the Earth's circulation and climate. Indeed there is a strong argument that, unless we have a clear, quantitative understanding of the present climate of Mars, any ventures into speculation concerning its past and likely future states are likely to be ill-grounded. Only in the latter chapters, therefore, have we strayed into talking about the climate in other epochs, concentrating on relatively recent (and ongoing) climate change in an effort to place the present climate of Mars into some sort of broader context. We have also felt it appropriate to limit ourselves to a discussion of the neutral atmosphere of Mars, largely (though not completely) excluding the thermosphere and ionosphere. This is because, from a scientific viewpoint, there is a natural watershed at the base of the thermosphere, beyond which the range of physical and chemical processes which need to be considered changes markedly to take into account the ionisation and photodissociation which takes place at altitudes greater than about 100 km. Despite this elimination of a detailed discussion of aeronomy on Mars, there is much material to discuss and to fill a substantial number of pages!

As indicated above, our primary target readership is intended to be university students (both undergraduate and postgraduate) of physics, chemistry and space science. It is increasingly common these days to include courses on various aspects of astronomy and planetary science within mainstream science degrees. A wide range of textbooks on planetary science have appeared in recent years to cater for these students, but almost without exception, their treatment of atmospheric science (especially dynamics and circulation) tends to be relatively superficial. Of course this is often an expedient, in order to ensure that a broad range of topics can be covered in a realistic number of lectures and classes. But there is, in our view, a clear need for a text which does go into greater detail in various aspects of planetary atmosphere dynamics, and Mars provides an ideal context within which to introduce a range of basic ideas of dynamical meteorology. Accordingly, our approach here has been not simply to produce an up-to-the-minute review of the latest Mars science – such a text would be expected to date rather quickly as the subject moves forward. Rather, we have tried to orient the discussion more towards the introduction of a range of basic ideas in atmospheric physics and dynamics using simple theoretical models and Mars as the principal illustrative context. Being reasonably fundamental to all atmospheric sciences, these ideas are unlikely to go out of date, and we hope thereby to have produced a text which will continue to help new students and young scientists to gain a full and in-depth appreciation of Martian atmospheric science for some time to come. Although we have included a certain amount of mathematics in our development, this is mostly at a fairly basic undergraduate level. We would hope, therefore, that this text might also appeal to the more ambitious lay person, hoping to gain a thorough understanding of at least some of the major scientific issues which the Mars scientific community is wrestling with at present.

In producing such a text, we are very grateful to a number of people who have helped with proof-reading and commenting on our efforts. These include Claire Newman and Henning Böttger in AOPP[1], and Dr John Mason and Clive Horwood at Praxis Publishing Ltd. We are also grateful for encouragement and help from our colleagues and collaborators in the 'EuroMars consortium', especially Jean-Paul Huot of ESTEC[2] and François Forget at LMD[3] in Paris. We thank Dr Claire Newman, Henning Böttger and Christopher Lee (AOPP), Drs François Forget and Monica Angelats i Coll (LMD), Dr Michael D. Smith, Dr Tim Schofield, Dr Maria Zuber, David Blanchard, David Hardy, Michael Carroll, Dr Candace Gudmundson, Dr Manoj Joshi and Dr Jacques Laskar for help in providing illustrations and diagrams. Peter Read also wishes to send his appreciation of the hospitality of the Institut de Recherche sur les Phénomènes Hors Equilibre, Marseille, France in the Spring of 2003, which provided a much valued opportunity to bring this project to completion. We are sensitive to the possibility of errors and omissions in the text contained herein. While we have endeavoured to produce a treatment which is reasonably comprehensive and balanced, we certainly cannot claim infallibility in this matter and would welcome notification from any reader of any glaring omissions or errors.

Peter L. Read
Stephen R. Lewis
Oxford, November 2003

[1] Atmospheric, Oceanic and Planetary Physics Sub-department, University of Oxford, UK.
[2] European Space Technology and Engineering Centre, Noordwijk, The Netherlands.
[3] Laboratoire de Météorologie Dynamique du CNRS, Paris, France.

Abbreviations

AU	Astronomical Unit
CAT	clear air turbulence
CFC	chlorofluorocarbon
CISK	Convective Instability of the Second Kind
CNES	Centre National d'Etudes Spatiales
CNRS	Centre Nationale de Recherche Scientifique
EOF	empirical orthogonal function
ESA	European Space Agency
GCM	general circulation model
GFDL	Geophysical Fluid Dynamics Laboratory (Princeton, NJ)
GISS	Goddard Institute for Space Science
GRS	Gamma Ray Spectrometer
HST	Hubble Space Telescope
IRIS	InfraRed Interferometer Spectrometer
IRTM	InfraRed Thermal Mapper
L_S	Areocentric longitude (see Appendix A)
LT	local time
LTE	local thermodynamic equilibrium
MarsGRAM	Mars Global-Reference Atmosphere Model
MAWD	Mars Atmospheric Water Detector
MCD	Mars Climate Database
MCO	Mars Climate Orbiter
MGCM	Mars general circulation model
MGS	Mars Global Surveyor
MOC	Mars Orbiter Camera
MOLA	Mars Orbiter Laser Altimeter
MPF	Mars Pathfinder
MPL	Mars Polar Lander

PFC	perflourocarbon
PMIRR	Pressure Modulator Infrared Radiometer
PV	potential vorticity
QG	quasi-geostrophic
Ri	Richardson number
rms	root mean square
SSW	stratospheric sudden warming
TES	Thermal Emission Spectrometer
THEMIS	THermal Emission Imaging System (also the Greek goddess of justice)
UT	universal time
VL1 (2)	Viking Lander 1 (2)
WBC	western boundary current
WMO	World Meteorological Organization

Figures

Tables

Colour plates (between pages 150–151)

1

An introduction to Mars

The Solar System presents us with a wide variety of alien planetary environments for possible exploration, ranging from the fierce heat, and harsh, intense sunlight of the Sun-facing hemisphere of Mercury or the oppressive heat of the surface of Venus, to the intense cold of the outer planets and their satellites. At the furthest extremity of the Solar System, Pluto, volatiles comprising the atmosphere all but freeze out onto the ground. From our human perspective, however, interest naturally tends to focus on more temperate environments within which life (at least in a form we would recognize) could potentially survive or even flourish. For this reason alone, and its comparative closeness in distance from the Earth, it is hardly surprising that Mars has become a major focus in recent years for intensive exploration.

The location of Mars, at a comparable distance from the Sun to that of the Earth (see Table 1.1), puts it in a similar radiative environment to that of the Earth, so that the range of surface temperatures which might be expected to prevail ought to bring Mars close to the conditions which terrestrial organisms could find tolerable. The notion that Mars might provide habitable conditions conducive to the development of life gains further support on discovering that Mars also possesses a number of intriguing landforms which appear to suggest that liquid water may have existed and flowed in large quantities on the Martian surface at some stage in the planet's history.

This emphasis in the popular literature, however, on the possible existence, past or present, of indigenous life on Mars, in our view at least, should not form the sole (or even necessarily the most important) motivation for exploring Mars as a planetary environment. Our understanding of the physics and chemistry of Earth's environment and climate is currently going through a period of active advance, in which it is becoming increasingly important to take full account of the complexity and interdependency of many different kinds of process in determining the state of the climate – to the extent that such studies are now seen as part of a whole new

Table 1.1. Key planetary and atmospheric parameters for Earth and Mars.

	Earth	Mars
Mean orbital radius (10^{11} m)	1.50	2.28
Distance from Sun (AU)	0.98–1.02	1.38–1.67
Orbital eccentricity	0.017	0.093
L_s of perihelion	281°	251°
Planetary obliquity	23.93°	25.19°
Rotation rate, $\Omega\,(10^{-5}\,\mathrm{s}^{-1})$	7.294	7.088
Solar day, sol (s)	86,400	88,775
Year length (sol)	365.24	668.6
Year length (Earth days)	365.24	686.98
Equatorial radius (10^6 m)	6.378	3.396
Surface gravity, $g\,(\mathrm{m\,s^{-2}})$	9.81	3.72
Surface pressure (Pa)	101,300	600 (variable)
Atmospheric constituents (molar ratio)	N_2 (77%)	CO_2 (95%)
	O_2 (21%)	N_2 (2.7%)
	H_2O (1%)	Ar (1.6%)
	Ar (0.9%)	O_2 (0.13%)
Gas constant, $R\,(\mathrm{m^2\,s^{-2}\,K^{-1}})$	287	192
c_p/R	3.5	4.4
Mean Solar Constant ($\mathrm{W\,m^{-2}}$)	1367	589
Bond Albedo	0.306	0.25
Equilibrium temperature, T_e (K)	256	210
Scale height, $H_p = RT_e/g$ (km)	7.5	10.8
Surface temperature (K)	230–315	140–300
Dry adiabatic lapse rate ($\mathrm{K\,km^{-1}}$)	9.8	4.5
Buoyancy frequency, N ($10^{-2}\,\mathrm{s}^{-1}$)	1.1	0.6
Deformation radius, $L = NH_p/\Omega$ (km)	1100	920

discipline of Earth System Science. In this context, it is important to distinguish which features of the climate system are unique to Earth and which are generic.

1.1 COMPARATIVE ATMOSPHERIC SCIENCE

Because of the intrinsic complexity of the environmental systems of a planet such as the Earth, it is often very difficult to untangle the web of non-linear feedbacks and interactions between different components of the system, and even to begin to address ultimate questions of physical cause and effect from observations of the environment alone. Quantitative, predictive models of the Earth System are crucial in enabling scientists to perform 'experiments' to test various concepts and hypotheses. Such models are typically extremely complex numerical codes which reflect many aspects of the complexity of the Earth itself, but of necessity make many approximations and assumptions in order to produce a closed dynamical system. The validation and verification of such models and their various components are critical issues, even for simulating the present climate (let alone predicting future climates far removed from present conditions), but for which observations and measurements of the actual Earth are often insufficient by themselves. In this context, it may be extremely useful to be able to apply the same modelling methodology to more than one planetary system, especially if there is enough 'common ground' between them to enable meaningful comparisons to be made with essentially the same model.

Such comparisons might be expected to lead to new insights concerning the possible future state of the Earth's climate. A famous example of this kind of cross-fertilization of ideas between two distinct planetary environments actually emerged several years ago during the 'Cold War' between the West and the communist powers, when atmospheric scientists Richard Turco, Brian Toon, Tom Ackerman, Jim Pollack and Carl Sagan (1983) were evaluating the possible consequences of a large-scale 'nuclear exchange' between the major superpowers. Observations and other theoretical studies of large-scale dust storms on Mars at the time had indicated the possibility of large amounts of mineral dust accumulating in the Martian atmosphere, which could lead to the entire planet being shrouded in an aerosol cloud for months at a time. As we will see later, such a cloud on Mars has a dramatic effect on the radiative balance of the atmosphere, substantially reducing the amount of solar radiation reaching the surface and strongly modifying the climate near the ground. It was then realized that the likely association of nuclear detonations in urban regions on Earth with widespread firestorms could also lead to the injection of large amounts of smoke and ash high into the atmosphere. From a suggested analogy with Martian planet-encircling dust storms, supported also by observations of the effects of large-scale volcanic eruptions on Earth, it was suggested by Turco *et al.* (1983) that, if sufficient smoke and ash were to reach the Earth's stratosphere, the cloud might also envelop much of the planet and persist for many months before settling out. The effects of such a large and thick cloud surrounding the Earth for timescales comparable with the seasonal cycle were thought to include the possibility of a major cooling of the entire global climate system, leading to widespread crop failures throughout the world.

Whether this so-called *nuclear winter* is an inevitable consequence of nuclear war was (and still is) regarded as controversial, but it did inform the political debate at the time, reinforcing perceptions of the potentially dire consequences of nuclear war, not only for the likely direct target locations of 'nuclear exchanges', but on the entire planet.

Comparative insights between the Earth and Mars are more typically much less dramatic and extreme. But given that the surface environments of both planets are, or have been in the past, influenced by a range of common geochemical and geophysical processes, including erosion by wind and water, we might expect the emerging discipline of comparative planetology to be particularly valuable in seeking to clarify the origin, history and potential future of the Earth itself. It is in anticipation of a wide variety of comparative insights which might emerge between the climate and meteorology of the Earth and Mars that the present book will be concerned.

1.1.1 Planetary features

Mars has many aspects in common with the Earth, and much can be deduced from comparisons of even the most basic planetary parameters on Mars with equivalent aspects on Earth. Table 1.1 lists a selection of useful facts and figures on the two planets, including their orbital and rotational parameters, atmospheric composition and surface conditions.

Even from this simple table of statistics alone, it is clear that Mars is a rocky body, roughly half the linear size of Earth, and rotates about its axis with a period approximately 40 min longer than the length of day on Earth. Surface gravity on Mars is around a third that on Earth, which ought to make moving around on the surface somewhat easier than on Earth for prospective Martian astronauts. Like the Earth, the circulation of the Martian atmosphere is driven by differential equator–pole heating from the Sun, balanced by infrared cooling from its surface and atmosphere. The obliquity of both planets (the angle between their equatorial plane and the plane of their orbit) is also approximately equal at the present time (though both vary slowly with time), which would imply (a) that their tropics are generally heated more intensely by the Sun than their polar regions, and (b) that they both exhibit strong seasonal variations at mid- and high-latitudes, with each hemisphere alternating from summer to winter during the course of a year. The atmospheric circulation on both planets thus seeks to transport heat from the warm tropics to the cold poles (albeit at somewhat different rates relative to the strength of direct solar heating), and motions are similarly affected by the rapid rotation of the planet. This leads to the development of baroclinic instabilities at mid-latitudes which result in intense cyclone–anticyclone weather systems on a horizontal scale of a few thousand kilometres, comparable with the deformation radius L (see Table 1.1).

Mars is roughly 50% further away from the Sun than the Earth, so the intensity of solar heating (the solar constant in $\mathrm{W\,m^{-2}}$) is roughly half that found at the Earth's orbit. Mars' greater distance from the Sun also leads to the Martian year being almost twice as long as on Earth. Mars' orbit is significantly more elliptical than the Earth's, which leads to some significant asymmetries between the northern and southern hemisphere seasons. At present, perihelion occurs during southern summer, so southern summer (and northern winter) are relatively short but intense, while southern winter and northern summer are long but relatively mild.

1.1.2 Atmospheric properties

Mars has a thin atmosphere composed largely of CO_2 with small amounts of nitrogen and argon, with a total surface pressure around 0.5–1% that of the Earth. The Martian atmosphere is relatively dry, with concentrations of water vapour being measured in precipitable microns (i.e., if all the water contained in a vertical column of atmosphere were condensed to a liquid state, it would form a layer typically only a few microns thick), as compared with Earth where water vapour constitutes around 1% of the atmospheric mass. Thus, the Martian atmosphere is not breathable, and has extremely low relative humidity. In fact over much of Mars the pressure of the atmosphere is below that of the triple point of water. This means that, even if the temperature could get up to the normal freezing point of water (around 273 K), ice would not melt into liquid water but would sublime directly into water vapour. The resulting environment on Mars is thus extremely dry, and it comes as no surprise to find that the typical landscape (outside the polar regions) closely resembles that of a desert region on Earth. This is clearly apparent, for example, in the striking images obtained by the Viking Lander (VL) spacecraft (see Figure 1.1, colour section).

Mars therefore offers the prospect of a variety of landscapes, ranging from extensive sand-covered deserts in the northern tropics and mid-latitudes to arctic ice-fields, especially near the North Pole. Mars also has some spectacular examples of mountainous terrain, including the largest volcanic mountain in the Solar System (Olympus Mons, rising up to 27 km above the surrounding plains). Each of these kinds of landscape may be expected to exhibit their own form of local weather, with patterns of wind, temperature and even occasional clouds. Water clouds are much rarer on Mars than on Earth, but do occur at times in the form of suspended ice crystals, typically at altitudes around 10 km or above.

1.2 PAST AND PRESENT CLIMATES

The thin, dry, CO_2-rich atmosphere on present-day Mars provides a modest degree of greenhouse warming, allowing the atmosphere to retain a little of the heat provided by the Sun, but raises the temperature by no more than around 5 K above the basic radiative equilibrium temperature which would be obtained in the absence of an atmosphere. Coupled with the much lower solar constant at the Sun–Mars distance, surface temperatures on Mars are typically colder than on Earth. The absence of oceans or lakes on Mars means that the surface has a relatively low thermal capacity and so responds rapidly to changes of insolation (e.g., as night falls). As a result, the range of temperature variations on Mars is large. At their warmest, the surface temperatures in low-lying regions can reach those of a warm summer day in the UK. At their coldest, however, temperatures can drop to as low as 145 K (−130°C). This is remarkable not only because it is much colder than anywhere on the Earth's surface, but also because CO_2 itself freezes to form dry ice at a temperature of around 150 K at the pressures found on Mars. Thus, during the depths of polar winter the ground temperature on Mars can reach the freezing point of the major constituent of the atmosphere. The implication of this observation is that a substantial fraction of the mass of the atmosphere (around 30%) typically condenses onto the surface at the winter pole, forming a seasonal ice cap up to 2 m deep.

From the viewpoint of a person on Earth, many aspects of Mars will thus look somewhat familiar at least in relation to desert landscapes on Earth, though with other features which appear alien. The less familiar aspects of Mars arise partly from the weaker heating from the Sun (allowing much lower surface temperatures) and also from the lack of water and very low humidity of the atmosphere and surface. An

Table 1.2. Major geological epochs on Mars.

Epoch	Time before present (Gyr)
Noachian	4.6–3.5
Hesperian	3.5–1.8
Amazonian	1.8–present

intriguing additional aspect of Mars, however, discovered quite early in the exploration of the planet by spacecraft, is the occurrence of various landforms which indicate the possibility of erosion and modification of the surface by the presence of large quantities of a liquid – believed by many scientists to have been water. This contrasts dramatically with the present environment of Mars, raising the possibility that the climate of Mars may have been very different in the distant past, perhaps even going through a relatively warm and wet phase which would have resembled conditions on Earth much more closely than at present.

Further investigation has indicated, however, that this warm wet phase was probably relatively short-lived in geological terms, and occurred very early in Mars' history. Geologists generally refer to various epochs of Mars' past history in relation to the times at which major land units were laid down. Table 1.2 lists the main epochs and the generally accepted times to which they refer. The Noachian epoch encompasses the period immediately after the formation of the planet, during which the planet cooled following the gradual decay of the impact rate from infalling meteorites and planetesimals as the planet accreted to its final size. The oldest surviving landscapes in the southern highlands date from the Noachian epoch, which are densely cratered with the scars of the early accretion when impacts were much more common than now. The warm, wet climate mentioned above most probably occurred during the late Noachian and Hesperian epochs, after which episodes of warm climates became much less frequent and long-lived. The final epoch (Amazonian) has lasted for the past 1.8 Gyrs or so (where 1 Gyr $= 10^9$ yr). Its long duration reflects the general impression that erosion rates on Mars are currently extremely slow (though not negligible). The Martian landscape thus represents a relatively old, well-preserved (and hence precious) environment from the early years of the Solar System itself.

It would be incorrect, however, to assume that the Martian climate has been completely static since the late Hesperian and Amazonian epochs. Various features on the ground, especially near the two polar caps on Mars, appear to represent much more recent climate variability, possibly on a cyclic timescale. The subsequent discovery that various orbital and rotational parameters of Mars, such as its orbital eccentricity and obliquity, seem to undergo cyclic variations on timescales of 10^4–10^6 yr raises the possibility that the climate of Mars may undergo cyclic variations which are direct analogues of the Milankovitch cycles on Earth, which are at least partially responsible for alternating glacial and interglacial episodes at mid-latitudes.

We begin to see, therefore, that the climate and meteorology of Mars have in fact many features in common with the Earth, at least at a qualitative level. In this book, we will explore the various components of the atmospheric circulation, meteorology and climate of Mars in some detail in the following chapters, both from a fundamental point of view, seeking to understand through observations and a range of models the distinctive and dynamic environment which Mars presents, and also wherever appropriate to look for similarities and contrasts with the Earth. By looking closely at such comparisons, much can be learnt about either planet, in the genuine spirit of comparative planetology.

1.3 OBSERVATIONS OF THE MARTIAN CLIMATE AND METEOROLOGY

Before embarking on a detailed survey of Mars' meteorology and climate, however, it is important to gain some appreciation of the sources of information from which our knowledge of this planet is derived. As with all of the other planets, the earliest information came from ground-based astronomical observations, from which it was possible to determine most of the gross characteristics of Mars such as its size, orbit and rotation, and to obtain coarse-resolution maps of the main surface features. As Figure 1.2 shows, however, such images are quite fuzzy and only show the largest and most obvious variations in surface albedo. Even so, it is possible to pick out features such as the polar ice caps and dark albedo features such as Syrtis Major, and hence to observe changes associated with the passage of the seasons such as the growth and decline of the polar ice. As a result of the relatively low resolution of these images, early studies relied on the visual acuity of individual observers to pick out the main features which could be seen. This led to problems at times, since some observers were more imaginative than others with what they thought they could see. Both Giovanni Schiaparelli and Percival Lowell claimed to see various linear features connecting some of the larger dark patches on the surface, which

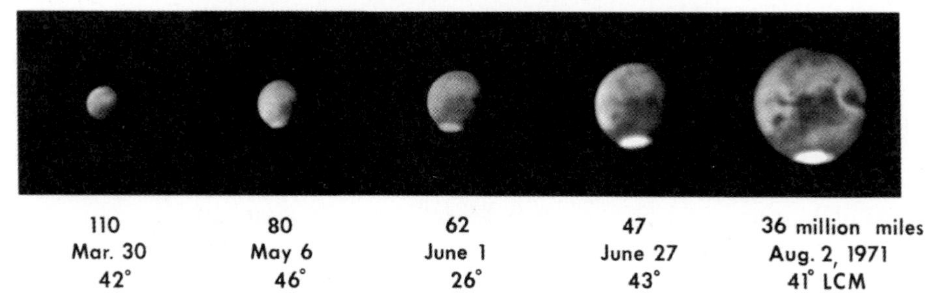

Figure 1.2. Sequence of ground-based optical images of Mars, showing the varying appearance of the planet with distance from the Earth and the North Polar cap (images are inverted).

Images courtesy of the International Planetary Patrol Program, Lowell Observatory, Flagstaff, AZ.

Schiaparelli called 'canali' and Lowell subsequently called canals. The latter suggested artificial constructions, at least to Lowell and his followers (Lowell, 1895, 1906, 1908), which led to a prevailing belief around the end of the 19th century that Mars was inhabited by intelligent beings. This belief persisted well into the 20th century before more detailed observations became available which made such a belief untenable. Later ground-based studies made extensive use of photography. Earl Slipher was perhaps the most prolific of Mars astronomers who obtained more than 100,000 photographs of Mars from various sites on Earth (Slipher, 1962) before he died in 1964.

The presence of an atmosphere on Mars was also deduced from ground-based telescopic observations, with historic records of various clouds appearing, which are now recognized to have been large-scale dust storms (e.g., see McKim (1999)). The presence of CO_2 in the Martian atmosphere was deduced from infrared spectroscopic measurements made as early as 1947 by Gerard Kuiper (Kuiper, 1952). Kuiper even attempted to determine the mean surface pressure of CO_2, though this was difficult to do with any accuracy. Based on the erroneous assumption that CO_2 was a minor constituent, as on Earth ($\sim$0.03% rather than 95%), early estimates of total atmospheric pressure for Mars were around twice that for Earth, with the bulk of the atmosphere assumed to be N_2. The presence of other gases on Mars, including water, were also deduced from spectroscopic measurements made from the ground, though there were considerable uncertainties as to amounts of these species, which made it difficult to determine what conditions were like at the Martian surface.

1.3.1 Beginnings of the space age

The direct exploration of Mars by spacecraft began (following some earlier failures by the USSR and USA) in 1964 with the launch by NASA of Mariner 4. This was a relatively simple fly-by mission, but gave the first close-up pictures of the Martian surface, revealing a barren, cratered, Moon-like landscape. This was a critical watershed in finally disproving Lowell's concept of an Earth-like, inhabited world covered in vegetation. Mariner 4 was also equipped with infrared sensors able to measure the temperature of the surface. It was thus able to show that the surface temperature at the ice caps was close to 150 K, hence confirming that the ice was most likely frozen CO_2 (Leighton and Murray, 1966), and not water ice as had been suspected before.

Several other missions followed Mariner 4 in the next 4 years from both the USA and USSR, including a disconcerting number of failures, culminating in the success of Mariner 9 which arrived at Mars in late 1971. Mariner 9 was the first spacecraft to get successfully into orbit around Mars, enabling for the first time a systematic survey of the Martian surface and atmosphere from close-up. Unfortunately when Mariner 9 first arrived at Mars, the planet was completely shrouded in a dense dust cloud produced by a planet-encircling dust storm. As a result, hardly any of the surface was actually visible. During the following few months, however, the atmosphere gradually cleared, enabling a substantial collection of clear images of the

Martian surface to be obtained. But Mariner 9's observational legacy for the atmosphere of Mars comprised not only images but also high-resolution infrared spectra from the InfraRed Interferometer Spectrometer (IRIS) instrument which enabled measurements of atmospheric composition and temperature to be obtained. In addition, measurements of the refraction of radio telemetry signals from the spacecraft as it passed behind the planet relative to the Earth (a form of radio-occultation) enabled vertical profiles of density and temperature to be measured between the surface and altitudes of 30–40 km. Thus, for the first time scientists could measure directly the vertical and horizontal structure of the atmosphere.

This early phase of Martian exploration continued until the hugely ambitious Viking mission, which placed two spacecraft simultaneously into orbit around Mars and deposited two massive and sophisticated landers onto the Martian surface in 1977. The Viking Landers were not, in fact, the first craft to land on the Martian surface – the USSR had managed this in 1969 with their Mars 2 spacecraft, though it did not survive long enough on the surface to return data to Earth. They were easily the most sophisticated, however, each being equipped with 14 instruments designed to measure various aspects of the Martian environment. Although both Landers were equipped with a range of meteorological sensors to measure atmospheric pressure, temperature and wind speed and direction, their main priority was in fact to measure aspects of the Martian soil in an attempt to detect evidence of living organisms. This turned out to be only partially successful, since the geochemical tests they were able to carry out were not sufficiently refined to distinguish between organic and inorganic causes. The long duration of the Landers on the Martian surface actually enabled a remarkable meteorological dataset to be acquired, providing a unique record of seasonal, daily, and even hourly variations of the Martian meteorology and climate over several Mars years. The pressure record, in particular, from these missions remains to this day an invaluable dataset for studies of atmospheric waves and tides on Mars, and for the calibration of models.

1.3.2 Martian exploration renewed: the present campaign

Together with the orbiters, which were also equipped with cameras, infrared sensors and other instruments for viewing the surface and measuring the space environment in orbit, the Viking mission is now seen as representing the end of a 'golden age' of Martian exploration. The vast dataset it produced over several years provided a major resource which dominated research on Mars for more than a decade after it had come to an end (see the book by Kieffer *et al.* (1992)). The 1980s were a relatively lean time for Martian space science, as the USA turned its attention elsewhere.

The early 1990s, however, saw a renewal of interest in mounting a campaign of exploration of Mars. Viking had enabled dramatic advances in the knowledge of Mars, but had also raised many questions which could only be resolved by a new generation of instruments which needed to be placed either in close orbit around Mars or deposited on the surface itself. This culminated initially in the ill-fated Mars

Observer spacecraft, designed by NASA to place a comprehensive package of very powerful remote sensing instruments into low orbit around Mars, including (for the atmosphere) very high-resolution imagery, infrared spectroscopy and atmospheric limb sounding, and radio-occultation measurements of unprecedented vertical resolution (Malin *et al.*, 1992; Christensen *et al.*, 1992; McCleese *et al.*, 1992; Tyler *et al.*, 1992). Despite a successful launch in 1992 and subsequent cruise to Mars, the spacecraft unfortunately failed to reach Mars orbit because of a malfunction in its propulsion system and was lost in 1993.

With the loss of a billion dollar mission, NASA might have been forgiven if it had given up on Mars for a while. But it responded by resolving to mount an extended campaign of smaller missions, spreading the risk of losing instruments and science objectives across a range of smaller spacecraft which were to be launched at every available launch window for the following decade. Thus was begun the present highly productive campaign of Martian exploration, in which most of the complement of instruments planned for Mars Observer were rebuilt and flown on a series of new spacecraft, beginning with Mars Global Surveyor (MGS). This was launched in November 1996, successfully reaching Mars orbit in September 1997. It carried roughly half of the original Mars Observer payload, as illustrated in Figure 1.3, including the Mars Orbiter Camera (MOC – see Malin *et al.*, 1992), the Mars Orbiter Laser Altimeter (MOLA – see Zuber *et al.*, 1992) capable of measuring the absolute surface topography relative to the areoid with metre precision, and the high-resolution Thermal Emission Spectrometer (TES – cf

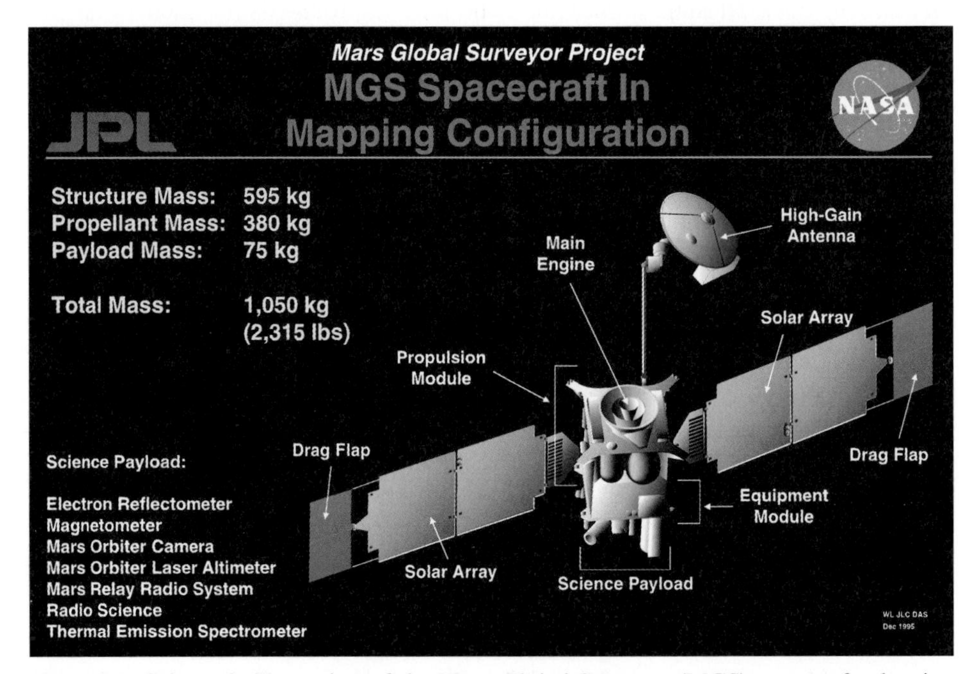

Figure 1.3. Schematic illustration of the Mars Global Surveyor (MGS) spacecraft, showing the main complement of remote-sensing instruments and their location on the orbiter.

Christensen *et al.*, 1992). The spacecraft was intended eventually to reach a circular, pole-to-pole, Sun-synchronous orbit around Mars with an altitude of approximately 300 km above the surface and an orbital period of just under 2 h. To achieve this orbit at relatively low cost in fuel, MGS used the novel technique of aerobraking, in which the spacecraft was deliberately flown through the upper atmosphere of Mars during periapse to use the aerodynamic drag forces to modify its orbital parameters. In practice, this turned out to be less than straightforward, and the early manoeuvres led to excessive decelerations. This meant that MGS could not achieve its originally planned orbit, but had to wait in a parking orbit while Mars rotated underneath it until the spacecraft could place itself into the mirror image of this orbit with the ascending and descending nodes interchanged. This was finally achieved in early 1999, inaugurating a new 'golden age' of Martian exploration (though instrumental observations were able to begin operations much earlier than this, functioning during much of the aerobraking hiatus in a 'science phasing' period).

MGS was not the only spacecraft to be launched by NASA in 1996, but was accompanied by the Mars Pathfinder (MPF), a new concept in lander technology which included both a base station and a small rover ('Sojourner'; see Figure 1.4(a)) to explore the surface around the landing site under remote control from Earth. This mission also tried out some new technology, using inflatable bags to cushion the impact of the lander onto the Martian surface (visible in the image of the base station from the Sojourner camera in Figure 1.4(b)). This turned out to be extremely successful, and MPF landed in Ares Vallis on the 4 July, 1997. MPF was never intended to be a flagship science mission, but was primarily intended to prove some novel lander technology. In the event, however, it was highly successful as a science mission, enabling some important *in situ* studies of Martian geology and geochemistry. Pathfinder was also equipped with some meteorological instrumentation, much as the Viking landers were. It lacked a reliable means of measuring wind speed, though it could measure wind direction. It was able, therefore, to yield both a detailed *in situ* vertical profile of atmospheric variables during its descent phase and a sequence of surface measurements from a 1.2-m boom on the lander. The MPF surface operations lasted for just 83 days, after which the lander's battery failed. Despite this, MPF obtained a wealth of high-quality data which is still being analysed at the time of writing this book.

MGS and MPF were succeeded in 1998 with the launch of two further ill-fated missions, Mars Climate Orbiter (MCO) and Mars Polar Lander (MPL). MCO was intended to enable the reflight of the Pressure Modulator Infrared Radiometer (PMIRR), an atmosphere-sounding infrared instrument originally on Mars Observer which was to obtain detailed daily maps of atmospheric temperature, dust and water vapour via limb-sounding techniques. In the meantime, MPL was planned to land in the southern polar layered terrains, with the intention of exploring the potentially ice-rich subsoil. In the event, both missions were lost because of various technical problems on arrival at Mars.

The reflight of Mars Observer instrumentation has since continued with the launch of Mars Odyssey in April 2001. Mars Odyssey was primarily intended to carry the Gamma Ray Spectrometer (GRS) from Mars Observer, capable of

(a)

(b)

Figure 1.4. Two views of the Mars Pathfinder spacecraft, showing (a) the Sojourner rover making measurements at a boulder, and (b) a view of the Pathfinder base station from the Sojourner rover itself.
From NASA/JPL.

detecting emissions from the Martian surface and providing information on the composition of the surface material. It is also capable of detecting epithermal neutrons released mainly from hydrogen-rich material within a metre or so below the surface by interactions with galactic cosmic rays. Since the most common material containing large amounts of hydrogen on Mars is likely to be water, the

GRS is in fact capable of detecting subsurface water ice stored in the uppermost metre of the ground. Together with its own camera and infrared sensors (Blasius *et al.*, 2002; Titus *et al.*, 2002), Mars Odyssey provides a complementary platform for systematic mapping observations of Mars simultaneously with MGS.

Additional Mars exploration was continuing at the time of writing this book with the launch of ESA's Mars Express orbiter and Beagle 2 lander in June 2003, and the two NASA Mars Exploration Rovers. Missions planned for the future include a reflight of an atmospheric limb sounder (Mars Climate Sounder) in 2005 and a range of other orbiter and lander spacecraft for the rest of the decade.

We cannot conclude this section without reference to one other important source of observational information, namely the Hubble Space Telescope (HST). Equipped with a 1-m aperture telescope and a variety of sensors, the HST is capable of acquiring relatively high-resolution ($\lesssim 10$ km), diffraction-limited images of Mars when it is in opposition to Earth. Such images are easily able to resolve major surface features, the polar ice distribution and even clouds of dust and water ice in the atmosphere. The timing and position of such images may actually complement those obtainable from Mars orbit, usefully extending the coverage of such orbital missions.

1.4 PLAN OF THE BOOK

The discussion above indicates that the observational information now available to Martian scientists is fast approaching the extent and sophistication of information available to Earth scientists on our home planet. It is at last becoming possible, therefore, to begin the serious task of unravelling the mysteries of the Martian environment in almost as much detail as has been done for the Earth, and to start to make some relatively sophisticated comparative studies between both planets.

This book is mainly concerned with efforts to understand the present climate and meteorology of Mars, especially in light of the new wave of information being acquired from the most recent generation of spacecraft operational in the Martian environment. In writing this book, however, we have not set out simply to review the latest findings 'hot off the press' from NASA, but to provide a systematic and fundamental introduction to Mars atmospheric and climate science, almost from first principles. Our hope, therefore, is that, even if some of the topical references will eventually become outdated, much of the introductory material will continue to be useful for students and others interested in Mars science for some time to come.

Apart from the next chapter, which introduces the other major tool in the Martian atmospheric scientist's armoury (the general circulation model, or GCM), the remainder of the book is structured to introduce each major component of the Martian atmospheric circulation in turn. As well as discussing results from observations and highly detailed numerical simulation models of Mars, we also make extensive use of relatively simple conceptual models to provide valuable insights and knowledge of the basic processes underlying the physics of each component of the circulation and climate. Finally, in the latter chapters we briefly examine factors which may cause the Martian climate to vary over various time-scales, and conclude with a discussion of how, and whether, actions of man might be able to influence the future state of the Martian climate.

2

Mars climate models

2.1 MODELLING THE MARTIAN ATMOSPHERE

As intimated in the previous chapter, much of our present-day knowledge and understanding of the Martian atmosphere and its circulation has been obtained through numerical modelling and simulation. Although real observations are, of course, always crucial, spacecraft coverage of the planet has been sparse in both time and space and tends to be restricted in terms of the range of atmospheric variables which can be measured. The precision and effective resolution of remotely-sensed observations must also be considered. A numerical model, which is adequately capable of simulating observations, provides a means of making intelligent extrapolations and predictions for regions of the atmosphere or to times of day or year which have not yet been observed. The complete, physically consistent set of atmospheric fields which may be generated by a model can then be subjected to detailed diagnoses and to further experiment in order to test hypotheses, in ways which are impossible in a real atmosphere.

Much understanding of terrestrial weather and climate processes has been built up in a similar way, through the development of a hierarchy of models, from analytic solutions of simplified equations to large-scale numerical simulations. Among the most powerful tools for weather and climate prediction for the Earth are the general circulation models (GCMs). These attempt to model the whole of the lower (and often the middle and even upper) atmosphere, from global scales down to the smallest horizontal scales which can be represented (typically 10 s to 100 s of km) consistent with the computational resources available and the timescale on which a prediction is required. For example, a weather forecast of a few days may be conducted with a model which is able to resolve much finer scales than that which would be practical for a climate prediction ensemble for the next few hundred years. The physical principles on which each prediction is based are, however, generally identical. In neither case is a GCM able to resolve down to the smallest scales of motion, which may still have a large net impact upon the large-scale synoptic weather systems. There is, therefore, the need to develop a complex set of parameterizations which attempt to predict the large-scale impact of small-scale processes, such as convective overturning and mixing. A very significant part of the computational expense of a GCM lies in the representation and parameterization of physical processes (e.g., radiative heating and cooling, cloud microphysics, and interaction and coupling of the atmosphere with the solid surface, the oceans, and with small-scale, unresolved atmospheric waves). These parameterizations are often more time consuming than the explicit representation of the large-scale dynamical and thermodynamic equations in the GCM.

The detailed physical modelling which is required for a complete GCM is perhaps hard to justify at present for many planets other than the Earth, since there are relatively few observations of extraterrestrial planets with which such a model could be constrained, or against which it could be validated. Mars is the most notable exception, because of both its proximity to Earth, resulting in a relatively large set of observations being accumulated with time, and the fact that it is in many ways the most Earth-like planet. It thus forms an almost ideal test bed for extending terrestrial GCMs to a new setting and for validating the techniques used under another set of conditions.

Some of the astronomical and atmospheric parameters which can be used to compare the Earth and Mars are summarized in Table 1.1. As discussed in the previous chapter, Mars rotates with almost the same period as the Earth (one Martian sol is about 40 min longer than one Earth sol), and has an atmosphere which is generally transparent to visible radiation, which can fall upon and heat the solid surface of the planet. The atmosphere is stably stratified with a similar vertical scale height H_p (e.g., Andrews, 2001) to that of the Earth. The similarities in rotation rate and stratification means that the typical horizontal scale of mid-latitude weather systems on both planets, the Rossby radius of deformation L (e.g., Andrews, 2001), is comparable, at around 1000 km. In other words, Martian weather systems are expected to have a similar scale to those we see on Earth and a similar set of physical balances (e.g., geostrophic balance between the horizontal pressure gradient and Coriolis force) should hold to around the same level of approximation. The axis of rotation of Mars is currently inclined at a similar angle to the plane of the ecliptic as that of the Earth (25.2° versus 23.5°), and so a familiar pattern of seasons results. Despite the rather obvious differences in mean surface pressure (that on Mars being less than 1% of that on Earth), the differences in atmospheric composition, the smaller planetary radius and the longer year on Mars, the essential dynamical processes which determine atmospheric circulation, and which transport heat and momentum between the equator and poles, exhibit some striking similarities.

There are, of course, also significant differences between the planets which must be taken into account when developing a Mars GCM. Some, such as the primarily CO_2-based atmosphere, can make a GCM more straightforward in principle, since the infrared spectrum is dominated by absorption due to one species, although a new radiative transfer algorithm must be calibrated for Martian conditions. Another example of a significant difference is the lack of oceans on the surface of present-day Mars, removing the need for complex ocean–atmosphere coupling. Many parameterization schemes should be similar for both planets, but may require retuning for the Martian environment (e.g., the gravity wave drag scheme described below). Finally, there are also unique and more exotic parameterizations that must be developed for Mars. For instance, large amounts of CO_2 ice form in the polar winters, with CO_2 ice clouds and snow. Dust lifting, transport, and deposition schemes, which are able to produce realistic seasonal variations in atmospheric dust loading, also form a major focus of current Mars GCM development. More detail on the individual components of a successful GCM will be discussed in following sections.

Despite these challenges, and the smaller resources available without such pressing operational needs as weather forecasting and climate change prediction, there has been a history of GCM development for the atmosphere of Mars which has largely paralleled developments in terrestrial modelling and advances in computational capacity. The handful of leading Mars GCMs in the world are now comparable in complexity and resolution with mainstream terrestrial climate models.

2.2 MARS GCM BACKGROUND

Comprehensive global atmospheric modelling of Mars began at almost the same time as successful global atmospheric models of Earth with the work of Conway Leovy and Yale Mintz (1969), who successfully adapted the then recently developed terrestrial GCM of the University of California, Los Angeles (UCLA) to Martian conditions. The model predicted atmospheric condensation of CO_2 and the presence of transient baroclinic waves in the winter mid-latitudes. Developments of this model continue at the NASA Ames Research Center up to the present day. The Ames GCM has provided many insights into the Martian climate and has been used to help interpret much of the spacecraft observational record and to conduct climate experiments for the planet (e.g., Pollack *et al.*, 1981, 1990; Zurek *et al.*, 1992; Haberle *et al.*, 1993; Barnes *et al.*, 1993, 1996; Murphy *et al.*, 1995; Hollingsworth and Barnes, 1996; Haberle *et al.*, 2003).

By 1990, other efforts were under way to model Mars. Work had begun on adapting the French Laboratoire de Météorologie Dynamique (LMD) terrestrial climate model to Mars, with the development of a new radiation scheme (Hourdin, 1992). This model was the first from which a simulation of a full Martian year, without any forcing other than insolation, was published (Hourdin *et al.*, 1993, 1995). Prior to this date, most Mars GCM experiments consisted of limited 'spin-ups' from rest over about 50 sols at a fixed time of year. The LMD GCM was able to reproduce in a realistic way the seasonal and transient pressure variations as observed by the Viking landers (Hourdin *et al.*, 1995; Collins *et al.*, 1996).

A Mars GCM was also developed at Oxford University in the early 1990s, originally in the form of a simple dynamical model in collaboration with Reading University (Collins and James, 1995). A spectral dynamical solver was used, as explained below, whose heritage was from a terrestrial model (Hoskins and Simmons, 1975), in conjunction with a simplified set of physical parameterizations. Original results were obtained on the dynamical regime of baroclinic waves (Collins *et al.*, 1996) and on the boundary–current nature of the low-level cross-equatorial branch of the Hadley Circulation (Joshi *et al.*, 1994b, 1995).

By the mid-1990s, the LMD and Oxford Mars GCMs were being developed in close collaboration and, although both groups continue to retain their different dynamical cores (which has lead to some productive intercomparisons), a complete physical parameterization package appropriate to Mars has been accumulated which is shared between the models, and from which both groups can select routines appropriate to the modelling task at hand (Forget *et al.*, 1999). Some results from these models have been combined to form a publicly available Mars Climate database (Lewis *et al.*, 1999), as described in more detail in Appendix A. Since Forget *et al.* (1999), the GCMs have been further improved to include full, interactive dust lifting, transport, and deposition schemes (Newman *et al.*, 2002a, b), to model water ice clouds (Montmessin and Forget 2003) and water transport (Böttger *et al.*, 2003), and to extend the top of the model above the thermopause (at about 120 km altitude) into the upper atmosphere (Angelats i Coll *et al.*, 2003).

A fourth Mars GCM was constructed by John Wilson at the Geophysical Fluid Dynamics Laboratory (GFDL) in Princeton with inclusion of a physical package similar to the Ames model at the time and a radiation scheme derived largely from the LMD model. The model was first used to study the role of thermal tides (Wilson and Hamilton, 1996), and to indicate the importance of extending the vertical coverage of GCMs in order to create a polar warming (Wilson, 1997). Recently, this model has been used in particular to investigate the Martian water cycle (Richardson and Wilson, 2002) including the formation of water ice clouds (Richardson *et al.*, 2002). In recent years, Allison *et al.* (1999) have adapted a version of the NASA Goddard Institute for Space Studies (GISS) terrestrial GCM for Mars. Also, Takahashi *et al.* (2003a, b) have recently begun work with a sixth MGCM in Japan, although this model is not yet as fully developed as the four main models described above.

A range of other Mars models of intermediate complexity have also been used. Some make use of more highly idealized dynamical equations or do not yet contain the full range of detailed physical schemes of the comprehensive GCMs described above (e.g., Mass and Sagan, 1976; Moriyama and Iwashima, 1980; Nayvelt *et al.*, 1997; Segschneider *et al.*, 2003; Takahashi *et al.*, 2003a, b). Although attention is focussed upon full GCMs in this chapter, simpler models such as these can and do still play an extremely useful role, as a stepping-stone toward a full GCM, as a model which is able to isolate individual processes in a less complex setting than a full GCM, and as a more computationally efficient tool in certain circumstances where detailed physical processes are not necessarily required (e.g., for some data assimilation experiments, as explained later in this chapter, see also Banfield *et al.*, 1995; Lewis and Read, 1995; Lewis *et al.*, 1996; Houben, 1999).

2.3 MESOSCALE MODELLING

While attention in this chapter is focused upon global atmospheric models of Mars, it is also worth noting that progress has been made, and continues to be made, towards more local, or mesoscale models (Savijärvi and Siili, 1993; Rafkin *et al.*, 2001; Toigo and Richardson, 2002; Tyler and Barnes, 2002). These can be of particular interest in the study of horizontally small-scale (about 100 km and smaller) phenomena, such as dust lifting processes, slope winds or volatile transport in the planetary boundary layer, or when considering spacecraft landing sites. Such regional or limited area models allow the possibility of finer horizontal and vertical resolution than a global GCM is able to achieve with similar computational resources, partly at the cost of requiring atmospheric boundary conditions around the edges of their limited domain to be specified. Such boundary conditions can either be prescribed in an arbitrary way, or taken from a global GCM. Again, the development of mesoscale models follows a practice common in terrestrial meteorology, where a more detailed local weather prediction model may be 'embedded' within an Earth GCM. The mesoscale model is able to provide more detailed local, short-range forecasts; the GCM is required for the boundary conditions and predicts the longer timescale changes (e.g., due to weather systems which may travel across ocean basins and continents over a period of a few days).

An important feature of mesoscale models, in addition to their finer resolution, is that they do not always have to make some of the simplifying assumptions which full GCMs often do. For example, mesoscale models need not impose hydrostatic balance (Rafkin *et al.*, 2001; Toigo and Richardson, 2002), as briefly discussed in Section 2.5.1, and may impose less strong constraints on vertical convective mixing (Michaels and Rafkin, 2003). The question of how a GCM should parameterize small-scale processes, from the larger-scale variations in fields which it is able to represent, is always a vexed one. Mesoscale models also require subgridscale parameterizations; they can resolve scales down to, say $\sim$100 m, but this is still far greater than true viscous scales. The opportunity to resolve to several orders of magnitude better than a GCM, however, opens the possibility of a more explicit representation of some processes which operate within a GCM 'grid box'. The use of mesoscale models for increasing resolution may assist in calibrating, or at least highlighting possible weaknesses of, some GCM parameterization schemes.

2.4 UPPER ATMOSPHERE MODELLING

A further class of 3-D atmospheric model has been developed with the specific aim to model the Martian upper atmosphere, above the thermopause (Bougher *et al.*, 1990, 1999, 2000). Upper atmosphere models require significantly different physical schemes to the lower and middle atmosphere GCMs, which mainly seek to model the neutral, homogeneous atmosphere. A thermosphere model must keep track of different species and ion densities separately; typically, prognostic equations are included for at least CO_2, CO, N_2, O, Ar, He, O_2, O_2^+, CO_2^+, O^+, and NO^+. In addition to the usual seasonal cycle in insolation, it is necessary also to consider variability over the solar cycle when modelling the thermosphere. Such models are computationally intensive and tend to be run for limited periods of time (often just a few sols, at fixed times of year and solar cycle), rather than for full seasonal cycle integrations as is now standard for GCMs.

One major problem with upper-atmosphere modelling is the question of how to couple such a model to the lower atmosphere. This usually means that either very simple, prescribed boundary conditions for wind and temperature (fixed at, say, 60–80 km altitude) have been applied, or, more recently, that middle-atmosphere GCM output is taken offline at regular intervals and introduced to the bottom of a thermosphere model (Bougher *et al.*, 2002). Note that this procedure still means that there is no possibility of the upper atmosphere affecting the lower, and so the coupling is incomplete, although the degree of downward influence from the thermosphere on the middle atmosphere, if any, is questionable. The upward influence of the lower atmosphere on the thermosphere is, however, very clear both through inflation and contraction of the lower atmosphere, as it warms and cools, and through upward-propagating tides, gravity waves, and planetary waves (Keating *et al.*, 1998; Forbes and Hagan, 2000; Wilson, 2000; Bougher *et al.*, 2001). The offline coupling procedure is perhaps still less than ideal, although it is a practical necessity at present. Another possibility is that, as the upper boundary of the neutral atmosphere GCMs is raised, a single model might be able to represent the upper atmosphere as well as the lower. This has obvious attractions, and is at present being pursued with the LMD-Oxford Mars GCM (Angelats i Coll *et al.*, 2003), but there are many problems in combining both upper and lower atmosphere in one model and the computational cost of the upper atmosphere restricts the length of model time over which experiments may be conducted. Preliminary results from the extended model are shown in Figure 2.1.

The validation of upper-atmosphere models remains an outstanding issue, since spacecraft missions to Mars have not yet sampled all combinations of season and solar cycle, and the Martian lower thermosphere is known to be a highly variable region (Bougher *et al.*, 1999, 2000).

2.5 GCMs

We now return to the subject of Mars GCMs (MGCMs) for the lower and middle atmosphere. This section outlines the main components required for a successful MGCM, and indicates some of the areas in which current research progress is being made, illustrated with some examples from the LMD and Oxford MGCMs described above.

2.5.1 Fluid dynamics

Solution of the so-called 'primitive equations' of meteorology forms the core of a numerical atmospheric GCM. Derivation of the primitive equations begins with the Navier–Stokes equation for fluid flow, essentially expressing Newton's Second Law in a rotating frame of reference (e.g., Andrews, 2001; Holton, 1992):

$$\frac{D\mathbf{u}}{Dt} = -2\boldsymbol{\Omega} \times \mathbf{u} - \boldsymbol{\Omega} \times (\boldsymbol{\Omega} \times \mathbf{r}) - g^{*}\mathbf{k} - \frac{1}{\rho}\nabla p + \frac{\eta_d}{\rho}\nabla^2\mathbf{u} \tag{2.1}$$

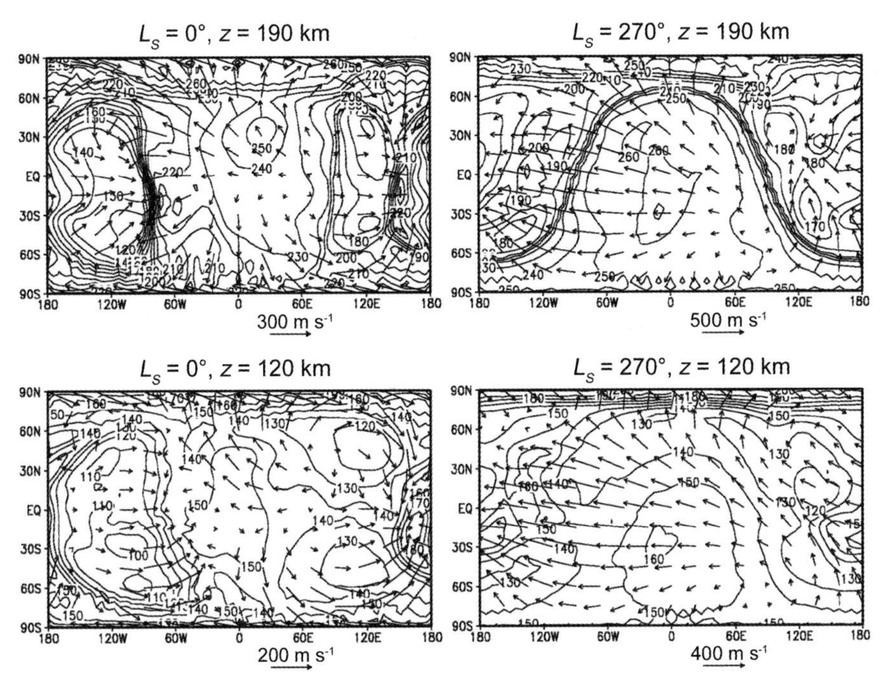

Figure 2.1. Temperature (contours) and wind (vectors, with a scale arrow in $\mathrm{m\,s^{-1}}$ shown below each plot) for two geometric height levels in the LMD MGCM with the top raised well into the thermosphere. The lower two panels are at a height of 120 km, close to the current top of the LMD and Oxford GCMs, and the upper panels at 190 km. The left pair of plots are for northern hemisphere spring equinox, the right pair for northern hemisphere winter solstice. All plots have been made for noon at 0° longitude and show the strong day–night temperature contrast and winds seen in the thermosphere.

where the substantive derivative operator, $D/Dt \equiv \partial/\partial t + \mathbf{u} \cdot \nabla$; $\mathbf{u}$ is the 3-D velocity vector; $\mathbf{r}$ is the 3-D position vector; the first term on the right is the Coriolis acceleration in the frame rotating with angular velocity $\boldsymbol{\Omega}$; the second term on the right is the centripetal acceleration, which is normally included with the true Newtonian gravity, g^*, to form an effective gravity experienced in the rotating frame, $\mathbf{g} = -g^*\mathbf{k} - \boldsymbol{\Omega} \times (\boldsymbol{\Omega} \times \mathbf{r})$; ρ is the density; p is the pressure; and η_d is the dynamic viscosity. To the Navier–Stokes equation can be added equations for mass conservation or continuity:

$$\frac{D\rho}{Dt} + \rho \nabla \cdot \mathbf{u} = 0 \tag{2.2}$$

the equation of state for a perfect gas:

$$p = R\rho T \tag{2.3}$$

where T is the temperature and R the gas constant per unit mass; and a thermodynamic energy equation:

$$\frac{D\theta}{Dt} = Q \tag{2.4}$$

where $\theta = T(p_0/p)^{(R/c_p)}$ is the potential temperature (Andrews, 2001; see also Section 3.1.3); c_p the specific heat at constant pressure per unit mass; and Q represents diabatic heating (with $Q = 0$ for adiabatic flow).

The form of the primitive equations commonly used in GCMs involve several simplifications to the full forms of Eqs (2.1–2.4), justified through scaling arguments (e.g., Gill, 1982; Andrews *et al.*, 1987; Holton, 1992; Andrews, 2001). In particular, the atmosphere is generally taken to be perfectly spherical and thin compared to the radius of the planet, a, the Coriolis force from the locally-horizontal component of the planet's rotation vector is neglected (along with some of the terms resulting from spherical curvature to preserve energy, angular momentum, and potential vorticity conservation properties of the equations), and the vertical component of the momentum equations is replaced by hydrostatic balance. This means that the three components of Eq. (2.1) can be rewritten as:

$$\frac{Du}{Dt} - \left(2\Omega \sin\phi + \frac{u \tan\phi}{a}\right) v + \frac{1}{\rho a \cos\phi} \frac{\partial p}{\partial \lambda} = F_\lambda \tag{2.5}$$

$$\frac{Dv}{Dt} + \left(2\Omega \sin\phi + \frac{u \tan\phi}{a}\right) u + \frac{1}{\rho} \frac{\partial p}{\partial \phi} = F_\phi \tag{2.6}$$

$$g + \frac{1}{\rho} \frac{\partial p}{\partial z} = 0 \tag{2.7}$$

with spherical coordinates $(\lambda, \phi, z) =$ (longitude, latitude, height) and with F_λ and F_ϕ as the horizontal components of the frictional force:

$$(u, v, w) \equiv \left(a \cos\phi \frac{D\lambda}{Dt}, a \frac{D\phi}{Dt}, \frac{Dz}{Dt}\right) \tag{2.8}$$

and

$$\frac{D}{Dt} \equiv \frac{\partial}{\partial t} + \frac{u}{a \cos\phi} \frac{\partial}{\partial \lambda} + \frac{v}{a} \frac{\partial}{\partial \phi} + w \frac{\partial}{\partial z} \tag{2.9}$$

It is worth noting that there are now terrestrial GCMs which do not make the standard hydrostatic primitive equation simplifications, and which include the full Coriolis terms and a full vertical momentum equation. White (2003) covers the derivation of these terms in more detail.* For large-scale flows on both Earth and

*For comparison with Eqs (2.5–2.7), the unapproximated forms of the momentum equations on a sphere may be written:

$$\frac{Du}{Dt} - \left(2\Omega \sin\phi + \frac{u \tan\phi}{r}\right) v + \left(2\Omega \cos\phi + \frac{u}{r}\right) w + \frac{1}{\rho r \cos\phi} \frac{\partial p}{\partial \lambda} = F_\lambda \tag{2.10}$$

$$\frac{Dv}{Dt} + \left(2\Omega \sin\phi + \frac{u \tan\phi}{r}\right) u + \frac{vw}{r} + \frac{1}{\rho r} \frac{\partial p}{\partial \phi} = F_\phi \tag{2.11}$$

$$\frac{Dw}{Dt} - 2\Omega u \cos\phi - \frac{u^2}{r} - \frac{v^2}{r} + g + \frac{1}{\rho} \frac{\partial p}{\partial r} = F_r \tag{2.12}$$

with the variable radial distance, r, replacing the fixed radius, a, in these equations and the definitions given in Eqs (2.8) and (2.9).

Mars, however, the hydrostatic forms given in Eqs (2.5–2.7) have consistent conservation properties, are reasonably accurate, at least compared to other uncertainties in the models, and are employed by current versions of the MGCMs described in this chapter. Hydrostatic balance, Eq. (2.7), acts to remove w as a prognostic variable (w can still be diagnosed from the equations via the continuity relation, Eq. 2.2) and also rapidly filters acoustic waves from the system. Hydrostatic balance is an extremely accurate approximation on large scales (Andrews, 2001; Holton, 1992), but is more questionable at scales smaller than Mars GCMs currently often represent (Rafkin *et al.*, 2001; Toigo and Richardson, 2002).

2.5.1.1 *Boundary conditions and vertical coordinate*

It is often most convenient, and natural, to re-write the primitive equations (Eqs 2.2–2.7) in a vertical coordinate system using pressure, rather than geometric height, z. Details are provided by Holton (1992) and Andrews *et al.* (1987). A major simplification is the replacement of Eq. (2.2) with an equation with no time derivatives:

$$\left(\frac{\partial u}{\partial x}+\frac{\partial v}{\partial y}\right)+\frac{\partial \omega}{\partial p}=0 \tag{2.13}$$

with $\omega = Dp/Dt$ replacing w (note that positive ω implies downward motion, or negative w).

The pressure coordinate system is in many ways a natural choice, being directly related to mass, but suffers from the problem that pressure surfaces can intercept the ground at the bottom of the atmosphere. This problem is particularly acute in the case of Mars, which has large amplitude topography. For this reason many GCMs, including most of the MGCMs described in this chapter, make use of a terrain-following σ-coordinate system in the vertical (Phillips, 1956), instead of pressure, where $\sigma = p(x, y, z, t)/p_s(x, y, t)$, where p_s is the surface pressure. Thus the surface is always at $\sigma = 1$, and the 'top' of the atmosphere at $\sigma = 0$, or zero pressure. The vertical velocity in pressure coordinates, ω, is replaced by $D\sigma/Dt$. The use of the σ-coordinate allows the (impermeable) boundary conditions at the top and bottom of the atmosphere to be written in a particularly simple form:

$$\frac{D\sigma}{Dt}=0 \quad \text{at} \quad \sigma = 0, 1 \tag{2.14}$$

Note that the lower boundary of the model, $\sigma = 1$, is neither constant in pressure nor a uniform geometric height surface. A topographic data set is required to specify the geometric height of the surface, compared to a constant geopotential surface. A very accurate topographic data set for Mars is now available from the Mars Orbiter Laser Altimeter (MOLA) aboard MGS (Smith *et al.*, 1999), and is further described in Chapter 4. For a rotating planet, a surface of constant effective geopotential is not spherical but an ellipsoid; the difference between the polar and equatorial radii of which on Mars is about 20 km. Once referenced to a constant geopotential surface, the curvature terms for flow relative to this surface are well approximated by those for flow on a sphere in the primitive equations.

A potential disadvantage of the σ-coordinate occurs when a GCM extends above the lower atmosphere, where the terrain-following nature of σ is less desirable and a pressure coordinate would be more suitable. For example, a field which was constant on isobaric surfaces might have steep gradients on a σ-surface, especially if it varies strongly with height (or p), and it is necessary to be careful when applying quasihorizontal diffusion. In this case it is possible to move to a more complex so-called 'η-coordinate', which transforms smoothly from σ-surfaces in the lower atmosphere to p-surfaces in the upper atmosphere. This introduces new complications in the formulation of the equations of motion, but is now employed by many terrestrial models. The LMD MGCM has conducted preliminary experiments with an η-coordinate, and the differences found have not been a serious concern when compared to the usual σ-coordinate system model, at least below about 100 km altitude.

The upper boundary of a GCM can cause problems and it is common to increase friction artificially in the upper two or three levels, sometimes through linear friction on eddy terms only and sometimes through a strong viscous damping term. This 'sponge' layer is intended to reduce spurious downward reflections of upward-propagating waves back from the top of the model.

2.5.1.2 Numerical representations

A numerical solution of the partial differential equations described above (Eqs 2.2–2.7), demands that they are discretized in some way to form a large set of ordinary differential equations. There are many ways of doing this, but two approaches dominate terrestrial GCMs and both are in use with the different Mars GCMs.

The first approach, often known as a 'finite difference' or 'grid point' model, simply employs a large 3-D grid of points in longitude–latitude–sigma at which values are held for each of the prognostic atmospheric fields (typically u, v, and either T or θ in 3-D and the 2-D field surface pressure, p_s). Horizontal and vertical gradients in fields can be calculated by differencing between grid points. All the fields need not be held at the same points in space, and a grid which is staggered in the horizontal between the wind and temperature fields can have advantages for calculating advective terms, although it is a little less convenient for parameterization of vertical column processes. For example, the LMD MGCM (Forget *et al.*, 1999) uses a horizontal discretization scheme which is designed to conserve quantities such as potential entropy for barotropic non-divergent flows and angular momentum for axisymmetric flow (Sadourny, 1975). The model resolution, or number of horizontal and vertical points which are specified for the grid, is flexible and determined by constraints such as computational resources and the aims of any particular experiment. A typical resolution for a long MGCM integration is currently around 64×48 points in the horizontal (or 5.625° longitude by 3.75° latitude) with 32 levels in the vertical. The vertical grid is normally highly stretched with many levels close to the surface (up to ten within 1 km, with the lowest at 4 m above the surface), to resolve the boundary layer, and a resolution closer to 5 km in the middle and upper atmosphere. It is worth remembering that, since the radius of

Mars is about half that of the Earth, but typical weather scales are similar, such a model effectively resolves synoptic scale processes about as well as one with twice the number of horizontal grid points in each direction in a terrestrial context. The Ames and GFDL MGCMs also employ finite difference techniques, with variable resolutions, again broadly comparable with current terrestrial climate models (but coarser than typical weather forecast models).

The second approach, used by the Oxford MGCM and known as a 'spectral' model, is to store the atmospheric fields in the the form of coefficients of a truncated series of spherical harmonic basis functions (sines and cosines in longitude and Legendre polynomials in latitude). The fields are stored on discretized vertical levels in the same way as the finite difference models, and, in fact, the same 32-level vertical grid is used for both the LMD and Oxford MGCMs for many experiments. The spectral approach has the advantage that fields can be smoothly interpolated and differenced, since the analytic form of the basis functions is known, and would be highly efficient for a purely linear model. The advantage of this approach is less clear cut for non-linear differential equations, which involve products of fields and/or their horizontal gradients (e.g., the advective ($\mathbf{u} \cdot \nabla(u, v)$) terms in the Lagrangian differentials in Eqs (2.5) and (2.6), and cause coupling between different wavenumbers. This is solved by transforming atmospheric fields (and their gradients) onto a grid, calculating non-linear products in real space, and transforming the resultant tendencies back into spectral space contributions. Such a procedure is feasible for large models thanks to the development of Fast Fourier Transforms and related techniques (Cooley and Tukey, 1965). The spectral model also requires a transformation of the primitive equation prognostic fields to real space for the physical parameterization step.

The Oxford MGCM core is based on the terrestrial solver of Hoskins and Simmons (1975), with a vertical differencing scheme designed to conserve energy and angular momentum (Simmons and Burridge, 1980). The prognostic fields held in spectral space are actually vorticity, $\nabla \times \mathbf{u}$, divergence, $\nabla \cdot \mathbf{u}$, temperature, and log-surface pressure. Re-writing the momentum equations, Eqs (2.5) and (2.6), in terms of vorticity and divergence allows the two fields to be treated differently. This is convenient since most of the large-scale, slowly-varying flow is non-divergent, so the faster, small-scale divergence field can be integrated implicitly (Hoskins and Simmons, 1975), allowing a longer time step. The typical climate integration resolution at which the Oxford model is currently run is a triangular truncation at a total (longitudinal plus latitudinal) wavenumber 31 (T31). The triangular truncation is intended to provide an approximately isotropic resolution, in contrast to a rectangular truncation to a maximum longitudinal and latitudinal wavenumber. A rectangular truncation would include modes with high wavenumber in both directions which had a greater total wavenumber than those primarily oriented in either longitudinal or latitudinal directions. The T31 spectral fields are transformed onto a 96×48 real space grid ($3.75^\circ \times 3.75^\circ$), which is over-sampled to reduce aliasing of high wavenumber modes when transforming the non-linear products back to spectral space; those modes which are retained after the reverse transform should be accurate.

All GCMs, whether finite difference or spectral, tend to require some scale-selective dissipation to remove energy at small scales or high wavenumbers. Almost paradoxically, this is because of the models' ability to conserve energy. As different scales of motion interact, some energy cascades to small scales, and this process would, in reality, continue to smaller scales until eventually motions are dissipated by friction. In a GCM, this energy simply accumulates near the model resolution limit, which is still very much larger than true viscous scales, and causes an unrealistic 'bump' in the energy spectrum. In the case of the Oxford MGCM, a $\nu\nabla^6$ or $\nu\nabla^8$ term is added to the right-hand side of the vorticity, divergence, and temperature equations. The constant ν is chosen to dissipate the smallest scales retained in less than one day, and the order of the operator is chosen to leave larger scales almost unaffected (MacVean, 1983). Similar procedures are applied to the finite difference models.

2.5.2 Physical processes and environmental factors

The dynamical part of a GCM, as described above, is largely the same whether a model represents Earth or Mars, except for the choice of basic parameters such as the planetary rotation rate and gas constant. However, the detailed representation of other physical processes requires careful consideration and represents a large part of the work required to build a realistic Mars model. Physical parameterization schemes also take up a significant proportion of the computational resources needed to run a full GCM, both in terms of processing and memory requirements, by introducing additional prognostic variables for the differential equations used to describe each scheme (dust distribution, sub-gridscale kinetic energy, soil temperatures, ice cover, etc.). Some processes may be based on code used for terrestrial problems, which has been retuned for Martian conditions (e.g., convection and gravity wave drag). Other processes are particular to Mars and require schemes not present in a terrestrial model (e.g., CO_2 condensation).

2.5.2.1 Radiative heating and cooling

A radiative transfer scheme for a GCM is a significant challenge, since it must calculate heating and cooling rates sufficiently quickly to be run on several thousand vertical profiles (one for each horizontal grid point in a model) many times per model day. A detailed spectral line-by-line calculation is too slow for this purpose, so simpler approximations must be developed but which are still accurate to within a few % under most circumstances encountered.

A radiative transfer scheme for the Martian atmosphere must primarily consider the presence of CO_2 gas, mineral dust, water vapour, water ice particles, and CO_2 ice particles. The radiative effects of other species, such as ozone, are probably negligible in the lower and middle atmosphere (Zurek *et al.*, 1992). The main effects which are included in most Mars GCMs are those of gaseous CO_2 and suspended dust at both solar visible and infrared wavelengths. The main sources of uncertainty in a Martian radiative transfer scheme are almost certainly related to the dust radiative properties and spatial distribution. The radiative effects of water vapor are commonly neglected, since the cold Martian atmosphere cannot sustain more than a few precipitable microns of water (see Chapter 8). Similarly, the effect of water ice clouds has often been neglected, although it is currently an area of active development. CO_2 ice clouds have been parameterized indirectly via modifications to polar ice emissivity (Forget *et al.*, 1998, 1999), although this is an oversimplification and further work remains to be done to improve this.

In the thermal infrared, absorption and emission in the lower atmosphere is dominated by the strong CO_2 15-μm band (Hourdin, 1992). Around 70 km, and above, departures from local thermodynamic equilibrium (LTE) become significant (López-Valverde and López-Puertas, 1994). These are now incorporated in the LMD and Oxford GCMs through a simplified model, reduced to two emission bands (one weak and one strong), which is accurate to around 120 km altitude and somewhat higher. The non-LTE scheme uses a cool-to-space approximation with tabulated escape functions, which have been precomputed as a function of pressure using a more detailed non-LTE model (Lopez-Valverde *et al.*, 1998). Work is ongoing to extend and improve the representation of non-LTE effects further throughout the thermosphere.

Direct atmospheric heating through near-infrared absorption of solar radiation by CO_2 is small in the lower atmosphere but becomes significant above 50 km altitude. A simple parameterization for calculating these heating rates as a function of pressure is described by Forget *et al.* (1999) based on radiative transfer calculations which include non-LTE effects (López-Puertas and López-Valverde, 1995).

2.5.2.2 Dust

Dust is the major absorber of solar radiation in the lower Martian atmosphere and, through feedbacks, is probably the main source of interannual variability on Mars. Even in the absence of major dust storms, there always appears to be a significant background level of dust on Mars and its representation is vital to a successful Martian GCM.

The scattering and absorption of solar radiation by dust in a Martian GCM (e.g., that described by Hourdin *et al.*, 1993, 1995) is closely based on the methods used for the treatment of dust in a terrestrial GCM (e.g., Fouqart and Bonnel, 1980). The code calculates upward and downward fluxes for each atmospheric layer from the reflectances and transmittances, using the Delta–Eddington approximation. As with the infrared CO_2 scheme, a detailed multispectral calculation would be too expensive for a GCM, so two broadband regions are used, split at 0.5 μm, near the peak of the incident flux. As described by Forget *et al.* (1999), GCM heating rates are unfortunately very sensitive to the dust properties, such as single-scattering albedo, and these properties are not known with any certainty. Hence, there is ambiguity in the combination of total dust optical depth and dust properties for a given heating rate.

Dust scattering must also be considered in the thermal infrared, at least outside the main CO_2 15-μm band, as described by Forget *et al.* (1998, 1999). Infrared dust scattering is commonly ignored in terrestrial GCMs and is a peculiarly Martian issue. The present LMD and Oxford GCMs include a radiative transfer scheme which accounts for multiple scattering by dust particles and uses a set of synthetic single scattering spectral properties derived in order to match the Mariner 9 observations of the 1971 global dust storm in the infrared, without making any particular assumption about the actual dust composition (Forget *et al.*, 1999).

The ratio of dust opacity in the infrared compared to the visible is another important, and poorly known, dust parameter, which is important in determining the local radiative balance of the atmosphere and surface. The ratio is sensitive to particle size distribution, which is difficult to determine (Zurek, 1982) and probably variable. Martin (1986) and Clancy *et al.* (1995) measure ratios of 2.0–2.5 between 0.67 μm and 9.0 μm, so a ratio of 2 is commonly used in GCMs, but this is another feature of current dust uncertainty.

2.5.2.3 Dust distribution

One eventual objective of GCM modelling on Mars is to have a MGCM which can self-consistently lift, transport, and deposit dust, spontaneously forming a realistic dust distribution as a function of location, height, and time. Despite current progress with schemes to represent all these processes, described in more detail in Chapter 7, there are still many uncertainties and sensitivities (e.g., to model resolution and to thresholds in the lifting schemes) aside from the additional computational burden. Historically, MGCMs have used simple, prescribed dust distributions, and these models are still of great value for performing straightforward, controlled experiments which remain close to the present Martian climate.

The simplest assumption to make is that dust can be represented as a function of pressure alone, being well mixed in the lower atmosphere and decaying above a certain height to a clear upper atmosphere. Typically dust mass mixing ratio has been represented analytically as:

$$q = q_0 \exp\{\nu[1 - p_0/p]\} \tag{2.15}$$

for pressure $p \leq p_0$, with $q = q_0$ for $p > p_0$, following Conrath (1975) who estimated $\nu = 0.007$ after the global dust storm observed by Mariner 9. Most recent GCMs (e.g., Pollack *et al.*, 1990) have taken lower values of $\nu = 0.03$ for non-dust storm conditions, and can produce plausible results using this vertical distribution and an optical depth which is horizontally uniform on a fixed pressure surface.

Forget *et al.* (1999) and Lewis *et al.* (1999) made a refinement to the vertical distribution in Eq. (2.15) to improve the shape in the case of thinner dust layers, trapped close to the ground, but still to match Conrath (1975) in the $\nu = 0.007$ case when the dust is mixed high in the atmosphere, with $z_{max} = 70$ km:

$$q = q_0 \exp\{0.007[1 - (p_0/p)^{(70\,\mathrm{km}/z_{max})}]\} \tag{2.16}$$

The dust 'top' parameter, z_{max}, is the altitude where the dust falls to one thousandth of its well mixed, lower atmosphere value, and was allowed to vary as a function of latitude and season. Lewis *et al.* (1999) described four dust scenarios, a cold year, a dustier year without storms and two levels of global dust storm, in which the optical depth could also vary as a function of time of year.

Since 1999, many new observations of Martian total dust optical depth (Smith *et al.*, 2000) and temperatures (Hinson *et al.*, 1999; Conrath *et al.*, 2000) have been obtained from the Mars Global Surveyor (MGS) spacecraft instruments. It is now possible to run a Mars GCM with a dust distribution based solely on a smoothed version of the MGS observations, or to use data assimilation, as described in Section 2.6, to produce an evolving 3-D dust distribution for GCM use. Despite this it still seems desirable to have a simpler, prescribed dust scenario for many purposes, not least because it will remove individual dust events and small storms that are unlikely to occur at exactly the same time and place every Mars year and provide a more controlled way to run a Mars GCM. One initial possibility, derived for the LMD and Oxford GCMs, is illustrated in Figure 2.2. This shows the total dust optical depth at a fixed pressure (700 Pa), from which q_0 is derived, and the value of $z_{\max}$ for Eq. 2.6, both of which are allowed to vary with season and latitude, but not longitude for this simplified scenario.

The total dust optical depth shown in Figure 2.2 has the same shape as seen in Thermal Emission Spectrometer (TES) dust observations, with a dusty southern summer and clear northern pole, although all the individual details seen in the real data have been removed in order to simplify the prescribed state. There are certainly further improvements which can be made to this prescription, but it forms a useful background state for model integrations of the Martian climate in the absence of major dust storms. Because of the uncertainties in dust optical properties, the magnitude of the total dust optical depth has been adjusted to provide a good match between the GCM and many of the MGS radio science profiles, taken at the same time. A few examples are illustrated in Figure 2.3. While many of the radio occultations are fitted remarkably well by the GCM using this simple dust prescription there are certainly also regions where the GCM is less successful, as illustrated in Figure 2.4. It is possible that the regions of poor agreement are where the model is misrepresenting strong tidal features, or that the inversions seen are the result of clouds, which were not included in the GCM run for this comparison.

In addition to prescribed dust experiments, most Mars GCMs can now also perform experiments with a full dust cycle incorporated explicitly, sometimes including transport of multiple dust particle sizes (Murphy *et al.*, 1995; Wilson and Hamilton, 1996; Newman *et al.*, 2002a, 2002b). Problems in obtaining a realistic multiannual cycle from first principles persist (e.g., Newman *et al.*, 2002b), largely because of the model sensitivities and uncertainties. Earlier studies (Murphy *et al.*, 1995; Wilson and Hamilton, 1996, 1997), therefore, have tried an approach in which the GCM is free to model dust transport and deposition, but in which the injection of dust into the atmosphere is prescribed. This approach is intermediate compared to a simple prescribed dust distribution and a fully interactive dust experiment in which dust lifting is also modelled. Some success has now been achieved in reproducing a full dust cycle, generally lifting dust through

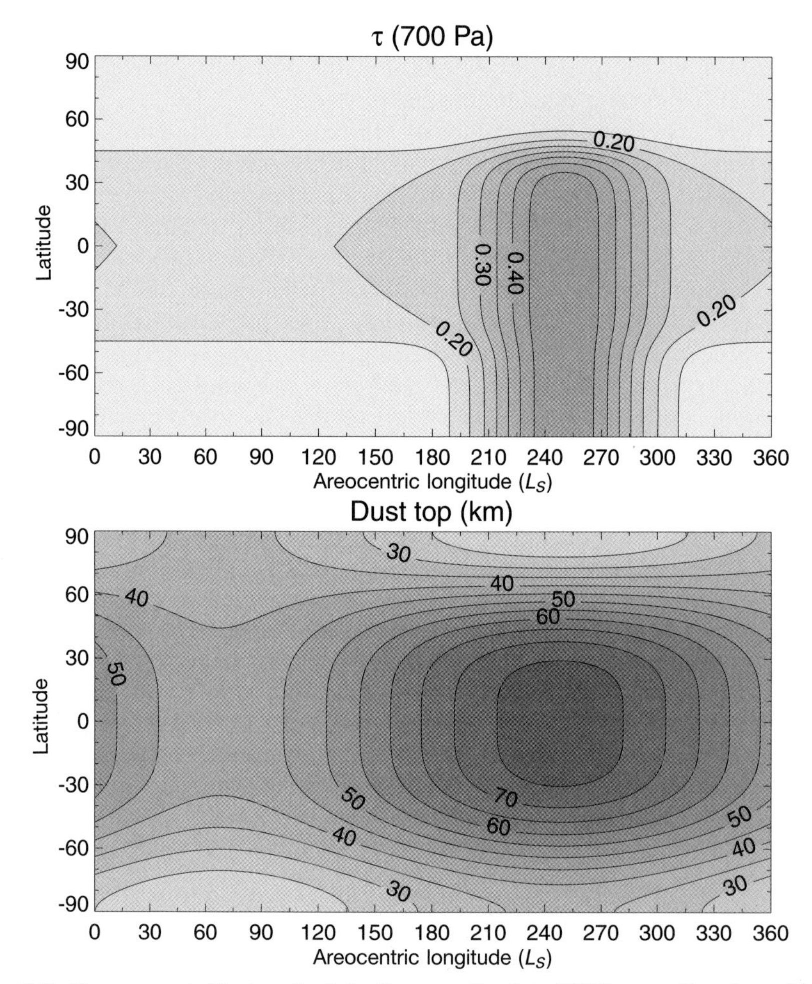

Figure 2.2. Upper panel: Dust optical depth normalized to 700 Pa as a function of latitude and time of year (L_S) for an idealized, smooth prescribed dust scenario which simulates many Mars Global Surveyor (MGS) radio occultation temperatures well. Lower panel: The height of the dust 'top' in km, where the mixing ratio falls to 1/1,000th of its lower atmosphere value, as explained in the text, for the same dust scenario. The dust field is axisymmetric in longitude.

near-surface wind stresses and saltation and/or dust devils, the effects of both of which can be parameterized from the large-scale GCM atmospheric fields (e.g., Newman *et al.*, 2002a, b).

2.5.2.4 Water and CO_2 ice clouds

Although the Martian atmosphere is generally clear, except for dust, clouds of both water and CO_2 ice can form at certain regions and times of day and year, and reach significant opacities. Particular examples are water ice clouds in the equatorial belt,

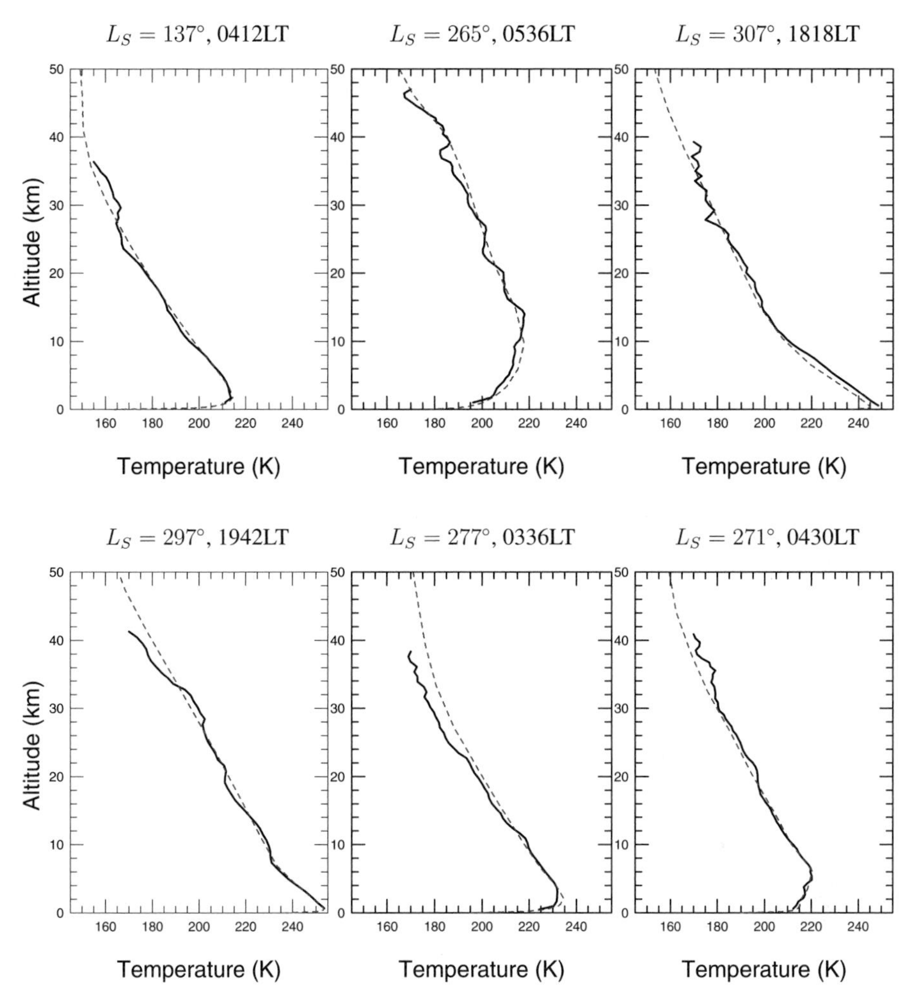

Figure 2.3. Six typical examples of MGS radio science temperature profiles for the Martian lower atmosphere, compared with profiles extracted at the same time and place from the MGCM. The upper panels are from the tropics, showing various times of year, and the lower from the southern hemisphere in summer, showing different times of day. Observations are solid lines and model results are dashed lines; plots are labelled with areocentric longitude and local time of day.

which are prominent around the time of aphelion in the northern hemisphere summer, and CO_2 ice clouds over the polar winter ice caps. The Martian water cycle will be discussed in more detail in Chapter 8.

It is only recently that clouds and microphysical schemes have been introduced to Mars GCMs. Although there is now very active work under way incorporating CO_2 and water ice clouds into all the major Mars models (Colaprete and Toon, 2002; Richardson *et al.*, 2002; Böttger *et al.*, 2003; Montmessin and Forget, 2003), results at the time of writing are still preliminary but indicate some successes in reproducing

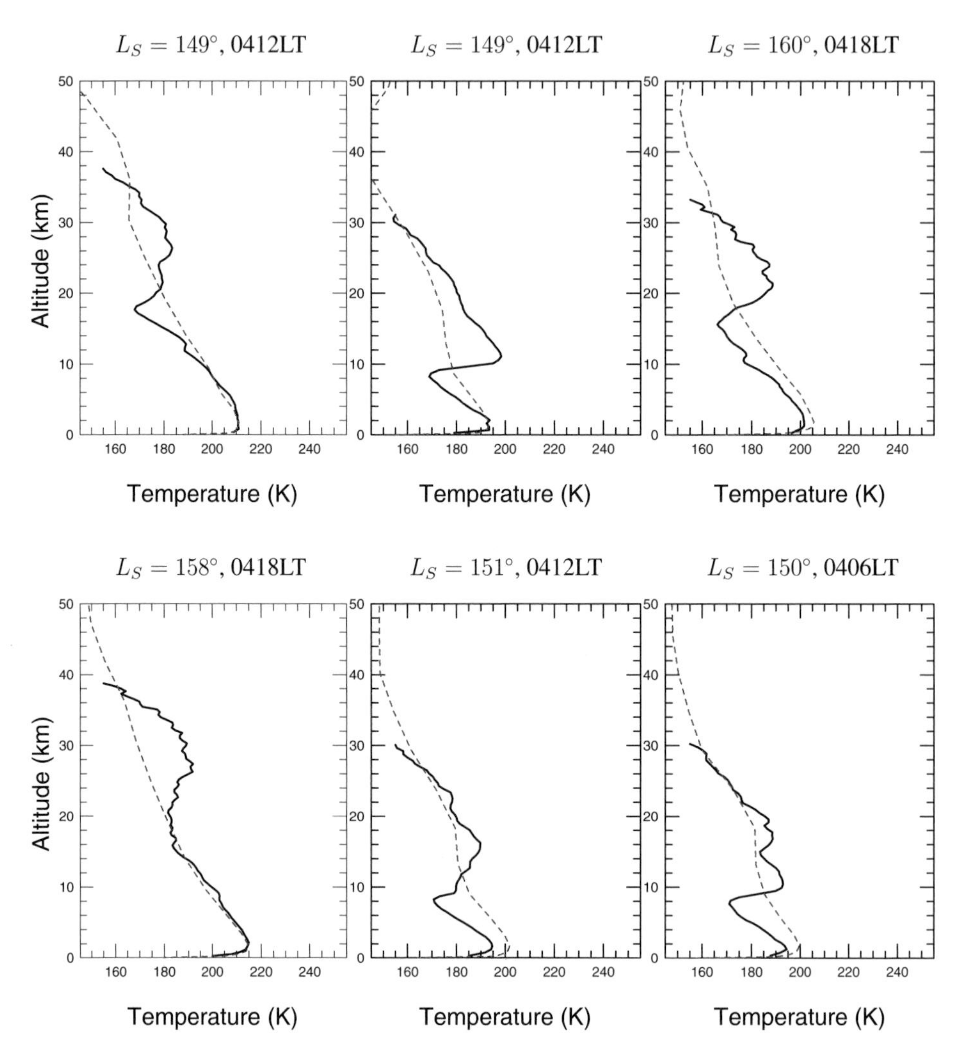

Figure 2.4. Six MGS radio science temperature profiles and model comparisons, as in Figure 2.3, chosen where the agreement is poor. Strong temperature inversions in these observations are not well reproduced by the model. The observations in these cases all come from the early morning in tropical regions in northern hemisphere summer.

the observed cloud bands. A very simple treatment of the net radiative effect of CO_2 ice clouds is included in the discussion of CO_2 ice physics in Section 2.5.2.8.

Schemes in the LMD and Oxford GCMs currently transport water vapour and ice in the same way as dust particles and other tracers are transported. If water vapour becomes saturated, it condenses to ice and falls out of the the atmosphere onto the surface, with regular tests being made to see whether each atmospheric layer is saturated on the way down, allowing for the possibility of resublimation. Microphysical schemes can represent water ice clouds in a similar way to that in which they are treated in terrestrial GCMs. Water in three forms, as vapour, ice, and as an adsorbate on regolith grains, is also tracked within each level of the Martian regolith, or 'soil model' as described in the following section, where it is modelled using a diffusion equation similar to that employed for soil temperature. Water can diffuse into the upper part of the Martian regolith from the atmosphere, or water at depth can diffuse upward and be released into the atmosphere. Some results from a GCM employing the full water model will be illustrated in Chapter 8.

2.5.2.5 *Surface processes and the soil model*

The surface temperature is an extremely important driver for the Martian atmosphere. Its evolution is determined by the balance between incoming solar radiation, incoming infrared radiation from the atmosphere, outgoing infrared radiation from the surface, turbulent atmospheric fluxes, and thermal conduction in the soil below the surface. Although the term 'soil' is used here for convenience, the Martian surface is unlikely, of course, to be soil in the sense used on Earth (containing biological material), but rather is made up of mainly fine-grained particles with some coarser rock fragments. In order to model the Martian soil, the LMD and Oxford GCMs employ an 11-layer diffusive soil model (Hourdin *et al.*, 1993), with the number of levels chosen to allow the soil to reproduce temperature variations on timescales from less than one sol to several Martian years. The soil is assumed to be homogeneous in the vertical, with surface thermal inertia and albedo derived from spacecraft measurements, as shown in Chapter 4. Within each vertical column of soil, the temperature evolves according to a simple conduction equation:

$$\frac{\partial T}{\partial t} = -\frac{1}{C}\frac{\partial}{\partial z}F_{\mathrm{C}} = -\frac{1}{C}\frac{\partial}{\partial z}\left(-\lambda\frac{\partial T}{\partial z}\right) \tag{2.17}$$

where F_{C} is the conductive heat flux and C and λ are the soil specific heat and thermal conductivity per unit volume, respectively. The soil thermal inertia, $I = \sqrt{\lambda C}$.

2.5.2.6 *Convection, turbulence, and the planetary boundary layer*

With current GCM horizontal resolutions generally no finer than about 100–200 km, near the equator, there are many scales of atmospheric motion which cannot be resolved explicitly. Whilst the subgrid-scale diffusion, described in Section 2.5.1.2, could be seen as one attempt to represent at least horizontal diffusive processes, vertical convection, small-scale 3-D turbulence, and the interaction of the atmosphere with the planetary surface in the boundary layer all require more detailed and careful treatment. Similar schemes for these processes are present in both terrestrial and Martian GCMs.

Convective adjustment schemes rapidly mix heat and momentum in unstable atmospheric layers, where potential temperature (Eq. 2.4) decreases with height, until the atmosphere is brought to neutral stability, or constant potential temperature, within the previously unstable layer. Emanuel (1994) reviews atmospheric convection in detail. Momentum mixing occurs because of mass fluxes in the convection and its intensity is a function of the strength of the initial vertical instability (Hourdin *et al.*, 1993). GCM parameterization schemes generally impose a strong requirement that all vertical profiles be brought immediately to neutral stability by convection, but there is some evidence that this approach may be over-aggressive and the real Martian atmosphere may not be able to convect so efficiently (Michaels and Rafkin, 2003).

The simplest parameterization of the atmospheric boundary layer is to apply a linear 'Rayleigh friction' drag term, $-ru, -rv$, to the right-hand sides of Eqs (2.5) and (2.6) in the lowest atmospheric layer only, with the value of r (typically of order $1\,\text{sol}^{-1}$) chosen according to the thickness of the bottom layer and which determines the strength of the viscous dissipation which tends to bring the atmosphere in contact with the ground to rest. Such a scheme can produce useful results in simplified circulation models (e.g., Joshi *et al.*, 1994, 1995a).

A better description of atmospheric turbulence and the boundary layer in a full GCM requires considerably more sophistication. Modelling the boundary layer for Mars is particularly important since there is a strong day–night contrast near the desert-like surface, with a very shallow (10 s of m), strongly-stratified nocturnal boundary layer and a deep (often more than 10 km) convective layer which forms above the surface during the day.

In most GCMs, both terrestrial and Martian, boundary layer dynamics are modelled by a turbulence closure scheme (Mellor and Yamada 1974), with convective adjustment performed on any unstable profiles. Turbulent mixing of a field, s, is governed by:

$$\frac{\partial s}{\partial t} = \frac{1}{\rho}\frac{\partial}{\partial z}\left(K\rho\frac{\partial s}{\partial z}\right) \tag{2.18}$$

where K can take different values for different variables (e.g., K_U for wind components and K_θ for potential temperature). The turbulent surface flux is also needed, taken as $\rho C_d U_1(s_1 - s_s)$, where C_d is the drag coefficient, U_1 is the wind speed in the first atmospheric layer, and s_1 and s_s are the field values in the first layer and at the surface. For a shallow first layer, it is reasonable to assume that the wind profile varies logarithmically with height immediately above the surface, $U \propto \log(z/z_0)$, with z_0 the roughness length, which can be approximately taken as 0.01 m on Mars, based on the Viking Lander (VL) sites (Sutton *et al.*, 1978). This gives $C_d = (k/\log(z/z_0))^2$, with $k = 0.4$, the von Kárman constant.

Forget *et al.* (1999) describe how a boundary layer scheme can be derived for Mars based on the non-stationary 2.5-level turbulence closure scheme of Mellor and Yamada, with modifications to solve numerical instability problems. The scheme essentially provides a parameterization of the values of K_U and K_θ to use in Eq. (2.18). Computation of these mixing coefficients is based on a differential equation for the time evolution of the turbulent kinetic energy, E:

$$\frac{\partial E}{\partial t} = \frac{q^3}{l}(S_U G_U + S_\theta G_\theta - 1/b_1) + \frac{\partial}{\partial z}\left(qlS_E\rho\frac{\partial E}{\partial z}\right) \tag{2.19}$$

where $q^2 = 2E$; $G_U = l^2((\partial u/\partial z)^2 + (\partial v/\partial z)^2)/q^2$; $G_\theta = -l^2 g(\partial\theta/\partial z)/q^2\theta_0$; S_U and

S_θ are functions of G_θ; and S_E is a constant, all specified by the details of the scheme. The turbulent diffusion coefficients are then $K_U = qlS_U$, $K_\theta = qlS_\theta$, and $K_E = qlS_E$. The mixing length, $l = kz/(1 + kz/l_0)$, with $l_0 = 160$ m. The implementation of such a scheme involves introducing a new 3-D prognostic variable, the subgrid-scale turbulent kinetic energy E. Forget *et al.* (1999) give more details on the practical implementation of a turbulence closure scheme for Mars. An example of the diurnal evolution of the planetary boundary layer using this scheme is shown in more detail in Chapter 5.

2.5.2.7 *Gravity waves and topographic drag*

Gravity waves are atmospheric disturbances whose restoring force is buoyancy and which are able to propagate in a stably-stratified atmosphere. Gravity waves can be excited by various means including weather fronts, and, of most relevance to the Martian atmosphere, by air flow over a rough topographic surface. Many of these waves would be of too small a horizontal and vertical scale to be resolved in a typical GCM, and would be of little consequence for the large-scale flow if it were not for the fact that they grow in amplitude exponentially as they propagate upwards, eventually leading to instability and wave breaking (e.g., Andrews 2001; Holton 1992). When gravity waves break, they cause turbulent mixing and deposit their momentum locally. In the case of topographically-generated waves, which are locked to surface features, this momentum mixing from many wave-breaking events tends to decelerate the large-scale wind towards zero, hence the common use of the term 'gravity wave drag'.

It has been known for more than 40 years that gravity wave mixing is important in the terrestrial mesosphere (Hines, 1960), and the effect of breaking gravity waves is considered a major factor in explaining the mean flow structure, and hence mesospheric horizontal temperature gradients (Fels and Lindzen, 1974; Andrews *et al.*, 1987). Like the Earth, Mars also has a stably-stratified atmosphere and has large surface topographic variations, which would be expected to generate strong gravity wave activity, given typical Martian wind speeds. There are, indeed, many direct observations of Martian lee waves associated with topography, made from spacecraft since the Mariner 9 mission (Briggs and Leovy, 1974). Several studies have attempted to model gravity wave effects in idealized Mars atmospheric models, ranging from 2-D studies to GCMs (Barnes, 1990; Théodore *et al.*, 1993; Joshi *et al.*, 1995b, 1996; Collins *et al.*, 1997). All conclude that gravity wave drag might be a large, and potentially even a dominant, effect on Mars above 45–50-km altitude. These altitudes coincide with the apparent requirement in 1-D chemical studies for a sudden increase in eddy diffusion coefficient in this region to reproduce observed ozone profiles (Blamont and Chassefière, 1993); one plausible source of extra mixing is gravity wave breaking.

One of the simplest ways to incorporate gravity wave drag is as Rayleigh friction in the upper levels of a model, as described in the section above for the planetary boundary layer. Such a linear drag term is simple and imposes no significant additional computational cost on a GCM. The problems with such a scheme are that it is almost impossible to calibrate the arbitrary drag coefficient, r, to apply at each model level and it is difficult to relate this simple drag to a diffusive process, such as expected to be the result of breaking and mixing.

A parameterization which is based more strongly on physical principles is desirable, since the drag should be related to the roughness of the surface, to the strength of the surface winds, and to the changing atmospheric vertical structure, which governs how gravity waves grow and break as they propagate. Several parameterizations of gravity wave drag have been developed for terrestrial models and are in common use in most current GCMs which include the middle atmosphere (Andrews *et al.*, 1987). It is difficult to tune a gravity wave parameterization for Mars, however, given the relative dearth of observations of the atmosphere above 50 km; this situation may improve as more spacecraft limb-sounding retrievals become available to greater altitudes (e.g., Collins *et al.* (1997) made a preliminary attempt to tune a Martian gravity wave drag scheme using Mariner 9 temperature retrievals (Santee and Crisp, 1993)). Tuning is necessary not only because of potential inaccuracies in a gravity wave parameterization, but also because of the fact that some larger gravity waves will be represented explicitly by the GCM and should not be accounted for twice. For these reasons, most Mars GCMs still use simple Rayleigh friction in the upper layers near the model top, essentially for model stability and justified as an *ad hoc* representation of gravity wave drag.

The LMD and Oxford MGCMs are currently exceptions, and have a more detailed gravity wave and low-level topographic drag formulation (Collins *et al.*, 1997; Forget *et al.*, 1999), based on the terrestrial scheme of Lott and Miller (1997) and Palmer *et al.* (1986). This scheme treats drag arising both from gravity wave breaking, mainly in the middle and upper atmosphere, and from small-scale, unresolved topography, in the lower atmosphere. In order to calculate these, it is necessary to form fields which describe statistical properties of the subgrid-scale topography, as well as the mean topography used by the GCM, at the model resolution. These are compiled from high-resolution topographic data sets which have many points per model grid box (e.g., topography resolved to more than 64 points/degree is currently available from the MOLA instrument aboard MGS (Smith *et al.*, 1999), which provides excellent subgrid-scale information compared to a typical GCM grid box of order $5^\circ \times 5^\circ$). If n of the data points, with topographic height, h, are included in a box surrounding a model grid point, the topographic parameters are: the mean topographic height; $\bar{h} = \sum h/n$, the standard deviation of the topographic height; $\mu = \sqrt{\sum (h - \bar{h})^2/n}$; the mean ridge direction of the subgrid-scale mountains:

$$\theta = \frac{1}{2}\arctan\left[2\overline{\frac{\partial h}{\partial x}\frac{\partial h}{\partial y}}\left\{\overline{\left(\frac{\partial h}{\partial x}\right)^2} + \overline{\left(\frac{\partial h}{\partial y}\right)^2}\right\}\right] \tag{2.20}$$

the subgrid-scale slope:

$$\sigma = \sqrt{\overline{\left(\frac{\partial h}{\partial x'}\right)^2}} \tag{2.21}$$

where $x' = x\cos\theta + y\sin\theta$ and $y' = y\cos\theta - x\sin\theta$; and the subgrid-scale anisotropy:

$$\gamma = \sqrt{\overline{\left(\frac{\partial h}{\partial y'}\right)^2} \bigg/ \overline{\left(\frac{\partial h}{\partial x'}\right)^2}} \tag{2.22}$$

The stress per unit area induced by gravity waves generated by the subgrid-scale mountains can be estimated by summing the result of Phillips (1984), for a single elliptical mountain, over all the mountains in the grid box and taking the height of the envelope passing through the mountain tops to be 2μ:

$$\tau_{\text{GW}} = \rho_s N_s U_s \mu \sigma G (B\cos^2\psi + C\sin^2\psi, (B - C)\sin\psi\cos\psi) \tag{2.23}$$

where the subscript s is used to indicate near-surface atmospheric values; ρ_s is the density; N_s is the buoyancy or Brunt–Väisälä frequency; U_s the wind speed; ψ the angle between the wind and mean ridge directions; G a number of order unity; and $B = 1 - 0.18\gamma - 0.04\gamma^2$, $C = 0.48\gamma + 0.3\gamma^2$. The stress at any vertical level in the model is then taken to be parallel to the surface stress, and its magnitude, τ_{GW}, can be written in the form (Palmer *et al.*, 1986):

$$\tau_{\text{GW}} = \kappa\rho N U \delta h^2 \tag{2.24}$$

where $\kappa = \sigma G(B\cos^2\psi + C\sin^2\psi, (B - C)\sin\psi\cos\psi)/\mu$ plays the role of a horizontal wavenumber for the gravity wave; ρ, U, and N are the values applicable at the current height; and δh is the vertical displacement of material surfaces in the fluid associated with the wave. δh will tend to grow exponentially with height as ρ decreases, $\rho = \rho_s \exp(-z/H_p)$ from Eqs. (2.3) and (2.7), with $H_p = RT_s/g$ the scale height based on a reference temperature T_s. If all other factors are equal, $\delta h \propto \exp(-z/2H_p)$, and gravity waves grow in amplitude with a vertical e-folding scale of $2H_p \approx 20$ km on Mars, until they reach an amplitude at which they become convectively unstable and break.

The Richardson number of a fluid is a non-dimensional parameter, which measures the stability of a stratified fluid in the presence of vertical shear, and which can be defined as $Ri = N^2/(\partial U/\partial z)^2$. Local convective instability is implied if $Ri < 0$ and Kelvin–Helmholtz instability if $0 < Ri < 0.25$. A gravity wave with isentropic displacement δh 'feels' a minimum Richardson number based on Ri calculated at that point in the atmospheric column:

$$Ri_{\text{min}} = Ri \frac{1 - N\delta h/U}{(1 + \sqrt{Ri} N\delta h/U)^2} \tag{2.25}$$

To calculate the gravity wave drag in the model, the stress in the atmosphere is taken to be equal to that near the surface and δh is calculated from Eq. (2.24) as the wave propagates upwards. At each model level, Ri_{min} is calculated and tested against a critical value, generally of order unity. If this critical value is reached, the wave is assumed to break, and a new isentropic displacement, δh, and thereby local stress, are calculated to maintain Ri_{min} at the critical value. The growth and saturation of a gravity wave when it breaks are illustrated later in Appendix A (Figure A.5). Once

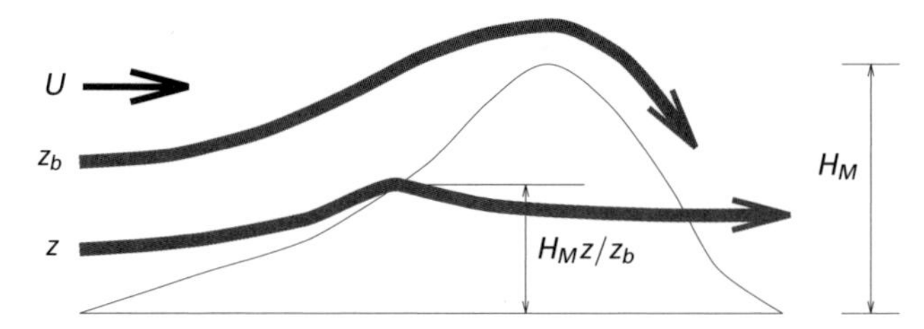

Figure 2.5. Schematic representation of low-level flow in the topographic drag parameterization scheme.

the entire stress profile has been calculated, tendencies for the momentum equations can be calculated from the vertical derivative of the stress profile.

Should the component of the wind in the direction of the surface stress become small, the wave amplitude will become very small, with the wave absorbed at the critical level and the stress set to zero above that point.

The Lott and Miller (1997) scheme also provides for low-level drag, as illustrated by Figure 2.5. In this scheme, mountain heights, H_M, are non-dimensionalized as $H_N = NH_M/U$. For $H_N < 1$, the flow all goes over the mountain and may generate gravity waves as above. For larger mountains, the lowest flow which will go over the mountain begins at upstream height, $Z_b = H_M(H_N - 1)/H_N$. Below Z_b the low-level-form flow will go around the mountain and experience low-level form drag. This drag can be written as:

$$D(z) = -\rho C_l l(z) U^2/2 \tag{2.26}$$

and acts in the opposite direction to the low-level flow, with C_l a drag coefficient of order unity. The horizontal length of the intersection of the mountains with the flow $l(z)$ is calculated using the statistical parameters for subgrid-scale topographic standard deviation, direction, and anisotropy given above (details are in Collins *et al.*, 1997 and Forget *et al.*, 1999).

2.5.2.8 CO_2 ice physics

A unique feature of the Martian atmosphere is that the major constituent may condense and, in fact, over a third of the total Martian atmospheric mass may be exchanged with ice on the planet's surface. This process can be included in a Martian GCM in a relatively straightforward way. If the atmospheric temperature falls below the CO_2 condensation temperature:

$$T_{CO_2} = 149.2 + 6.48 \ln(0.00135p) \tag{2.27}$$

where T_{CO_2} is given in Kelvin and pressure, p, is specified in Pascals, then atmospheric CO_2 condenses into surface ice until the latent heat release brings the atmospheric temperature back to T_{CO_2} (Pollack *et al.*, 1981; Hourdin *et al.*, 1993). T_{CO_2} is typically around 145 K for Martian surface pressures and atmospheric temperatures above the polar ice caps are held at this level by the condensation process. Once the surface begins to be heated again, surface temperatures are held at T_{CO_2} by subliming CO_2 ice back into the atmosphere, until the surface is once again clear and ice-free. The budget of CO_2 ice on the surface is held within the model, with the total CO_2 mass and the total energy in the system conserved, and the surface albedo is modified while covered with ice.

Forget *et al.* (1999) extended the scheme above to include the possibility that CO_2 ice particles which formed higher in the atmosphere would resublime on their way down, if they encountered warmer air, rather than appearing on the surface instantaneously. Further small, but not negligible, energy terms were also included, such as release of potential energy by the ice particles as they fall and corrections to the heat required to warm CO_2 ice as the frost temperature increases with changes to local pressure (Eq. 2.27). A significant correction to the radiative effects of CO_2 ice particles can be applied following Forget *et al.* (1998) by decreasing the surface emissivity, ϵ, when atmospheric condensation is predicted by the model to mimic the effects of ice clouds and fresh snow. With time, after the snowfall, the emissivity relaxes towards $\epsilon = 1$ for old surface ice:

$$\frac{\partial \epsilon}{\partial t} = -0.00015 \frac{\epsilon^4}{r} \frac{\partial m}{\partial t} + \frac{1}{\tau_{CO_2}}(1 - \epsilon) \tag{2.28}$$

where r is an effective ice particle radius ($\sim$100 μm); $\partial m / \partial t$ is the total atmospheric mass condensation rate per unit area; and τ_{CO_2} ($\sim$0.5 sol) is the characteristic timescale taken for snow evolution.

Various aspects of the output from the Oxford MGCM are illustrated throughout this book. The Martian global circulation is described in more detail in Chapter 3, which shows an annual-mean GCM atmospheric state and longitudinal means for three of the cardinal times of year: northern hemisphere spring equinox, and the summer and winter solstices.

2.6 DATA ASSIMILATION FOR MARS

Data assimilation is a powerful technique by which information from both present and past observations and knowledge of physics, embodied in a numerical model such as described in this chapter, may be combined to produce a 'best guess' analysis of the present state of an atmosphere. Data assimilation for the Earth has largely developed from the operational need to initialize a terrestrial model with the best possible estimate of the correct initial atmospheric state for a weather forecast, based on a variety of recent observations at different times and places. Progress has been made with data assimilation techniques in the terrestrial community since the 1960s (e.g., Daley, 1991), based on improvements to earlier objective analyses made by simple function fitting to data (e.g., via least-squares techniques) and still earlier subjective analyses of synoptic charts by forecasters, before the advent of numerical models.

Assimilation is also attractive in meteorological contexts other than weather forecasting, however, especially where a time-evolving map of synoptic-scale features is required from data acquired asynchronously. A straightforward gridding analysis of observations from a single satellite, for example, can lead to ambiguities in the interpretation of observations when studying synoptic-scale phenomena which change significantly over the course of a day or two. Assimilation is valuable when it is desirable to produce a full, physically-consistent analysis of all atmospheric fields, including those not directly observed. The examination of misfits between model forecasts and observations may also be useful in helping to identify potential deficiencies in the model. For these reasons, data assimilation is also used in the terrestrial context for subsequent systematic reanalyses of observational data for both the atmosphere (Lorenc, 1986; Daley, 1991) and the oceans (Ghil, 1989), where the aim is often to extract the most benefit from a relatively sparse observational record.

In order for data assimilation to be effective we require both a sufficient quantity of data to reasonably constrain the problem and a sufficiently realistic model to represent the behaviour of the atmosphere. Until recently, such a procedure was not feasible for other planets, but with the advent of new, larger atmospheric data sets from orbital missions such as MGS and associated developments in Mars GCMs, data assimilation has been proposed for Mars (Banfield *et al.*, 1995; Lewis and Read, 1995; Lewis *et al.*, 1996, 1997) and has begun to be applied to the MGS TES (Conrath *et al.*, 2000; Smith *et al.*, 2000) temperature retrievals (Houben, 1999; Zhang *et al.*, 2001) and to simultaneous TES temperature and dust measurements (Lewis *et al.*, 2003). An assimilation of the full data set from the TES instrument is one way of summarizing a vast number of individual retrieved profiles, as a continuous model history record, as well as obtaining the other advantages mentioned above. A further advantage of data assimilation is that it is an efficient way to combine information from more than one data source. Although initial studies have concentrated upon the TES instrument alone, the future possibility of combining information from more than one source, including a satellite and lander network (Lewis, *et al.*, 1996), can be beneficial. Other data sets, such as radio science occultations (Hinson *et al.*, 1999), are also valuable as a means of validating the assimilations.

The essence of any data assimilation procedure is to obtain a model state which minimizes the misfit between the observations and the GCM predictions of the observations. This can often be formalized (e.g., Lorenc, 1986) as a least-squares minimization problem in terms of the penalty functional J:

$$J = \tfrac{1}{2}(\mathbf{y} - H(\mathbf{x}))^T \mathbf{R}^{-1}(\mathbf{y} - H(\mathbf{x})) + \tfrac{1}{2}(\mathbf{b} - \mathbf{x})^T \mathbf{B}^{-1}(\mathbf{b} - \mathbf{x}) \tag{2.31}$$

where $\mathbf{x}$ is the model state vector, which completely describes the current state of the GCM (e.g., every parameter and every variable, saved at each grid point); $\mathbf{b}$ is the background or a priori model state vector; $\mathbf{y}$ the current observation vector; $H(\mathbf{x})$ is the forward model, which predicts the observations from the model; and $\mathbf{R}$ and $\mathbf{B}$ are covariance matrices for experimental error (observational and forward model) and a priori error respectively. In other words, there is a penalty for the difference between the model predictions and the current observations and for the difference between the model and its forecast made before the new observations were introduced, each weighted by their relative error. The reason for the second term on the right-hand side is that the background state effectively contains the information from all previous observations, forecast forward in time to the present by the model, and may therefore contain much more useful information than a few current observations, especially if these are subject to large errors. The minimum of J occurs when:

$$\left(\frac{\partial J}{\partial \mathbf{x}}\right)^T = -\mathbf{H}^T\mathbf{R}^{-1}(\mathbf{y} - F(\mathbf{x})) - \mathbf{B}^{-1}(\mathbf{b} - \mathbf{x}) = 0 \qquad (2.32)$$

if $\mathbf{Hx}$ is the linearization of $H(\mathbf{x})$.

Assimilation schemes can be divided into sequential and variational methods, although both have similar aims. A sequential optimal solution which minimizes Eq. (2.31) is the Kalman filter (Kalman, 1960) from the engineering literature, or the extended Kalman filter in meteorology or oceanography (Gelb, 1974; Ghil and Malanotte-Rizzoli, 1991). The Kalman filter consists of a forecast step, in which the model and its forecast-error covariance matrix are integrated forward in time, and an analysis step, in which both model and forecast-error covariance matrix are updated with the new information available from current observations. This procedure is then repeated throughout the model integration period. A problem with the explicit calculations of the Kalman filter for meteorological applications is that the model status vector is already (almost by definition) made as large as can be treated comfortably in the computer time available. As computers get faster, models are run at higher resolution and with more detailed physical schemes. Typically $\mathbf{x}$ would be of order $n = 10^6$–10^7 for a modern Mars GCM, even run at moderate resolution. Hence the requirement to operate repeatedly on covariance matrices with n^2 elements can become prohibitive, unless the scheme is simplified. This has lead, in the terrestrial forecast community, to a variety of suboptimal sequential schemes, most notably those known as 'successive corrections' and 'optimal interpolation' (Daley, 1991) which use approximate forms of the forecast-error covariance.

Variational assimilation methods operate slightly differently in that J (Eq. 2.31) is minimized over a fixed model time interval. An efficient means of doing so is known as '4D-Var', which requires the adjoint of the model and observation operator (e.g., Talagrand and Courtier, 1987); generally, a linearized dynamical adjoint is used in practice. Variational methods can either be formulated with the model as a weak constraint (Sasaki, 1970), analogous to sequential methods, or, if the model is believed to be accurate, more often as a strong constraint. In the strong constraint case, the problem reduces to 3-D, in which only an initial state for the assimilation interval must be found; the rest of the time evolution is determined by the model.

Assimilation schemes in current use for Mars are derived from these terrestrial techniques. Banfield *et al.* (1995) proposed a variant on the sequential Kalman filter in which computational economies are achieved by only calculating the gain matrix once, and then holding it steady in time. This type of scheme was applied to the Ames GCM, but initial success with TES data was limited (Zhang *et al.*, 2001), perhaps owing to temporal changes in the Martian atmosphere. Lewis and Read (1995) and Lewis *et al.* (1996) retuned for Martian conditions a form of successive corrections, known as the 'analysis correction scheme' (Lorenc *et al.*, 1991), which was in operational use with the U.K. Meteorological Office forecast model throughout the 1990s, replacing the previous optimal interpolation scheme. This is an iterative, sequential scheme which interleaves an analysis step between every model time step and which approximates covariance matrices by continuous covariance functions. Balance relationships are applied, if one meteorological variable is incremented, to generate corresponding increments in other fields. This scheme was tuned empirically with artificial data from twin model experiments (both identical and non-identical Mars GCMs). The scheme is suboptimal from a formal point of view, but appears to be able to be tuned to produce useful and robust assimilations with a variety of data (Lewis *et al.*, 1997, 2003), using the Oxford Mars GCM. In contrast, Houben (1999) has employed a variational scheme with a GCM which has a full dynamical representation, but highly simplified physical processes. In this approach, the model radiative heating rates are assimilated with the other meteorological variables, rather than being calculated by the GCM. These different methods should prove complimentary in the analysis of current and future Mars data sets.

Assimilation is illustrated here with a few results from the Oxford GCM, with assimilations conducted as in Lewis *et al.* (2003), but using TES nadir temperature retrievals from the first MGS mapping year. Further assimilation results appear in later chapters. Figure 2.6 illustrates the zonal mean temperature from two initial assimilations using the Oxford GCM. Both use TES thermal data, one from the first MGS mapping year and the second from the same time in the previous Mars year, when some TES data was available during the MGS aerobraking hiatus. During this period in the aerobraking year, $L_S = 225°–233°$, a regional dust storm occurred in the Noachis area at around 30°S. The effects of the storm on the lower and middle atmosphere are immediately apparent. The mapping phase year is more than 6-K cooler over most latitudes between 10-km and 40-km altitude and up to 30-K cooler over the South Pole. The only region of the atmosphere which remains warmer is the North Polar warming above 40-km, but there are no direct temperature observations to confirm this, since the TES nadir profiles do not allow temperatures above 40-km to be retrieved.

A major advantage of assimilation techniques is in deriving information about the transient wave behaviour. Ensemble techniques are one method which can be used in order to test whether the transient wave information is self-consistent, or simply represents waves of random phase which might be produced by the model anyway without observational information. In the case shown here, a small ensemble, consisting of nine members drawn from model fields at $L_S = 210°$, was used to provide a set of initial conditions for nine independent control experiments and for nine assimilations.

Figures 2.7 and 2.8 show the surface pressure at one mid-latitude site (0°E, 47.5°N) from both ensembles. The difficulty an assimilation might have in controlling surface pressure from atmospheric temperature data alone, especially when the temperature profiles have poor vertical resolution (>10 km) and suffer from particular problems near the surface, should not be underestimated. Despite this stern test, the assimilations appear to produce a consistent picture.

Figure 2.7 first shows the surface pressure from the control ensemble. The seasonal surface pressure cycle is apparent as a trend. The evidence for transient waves of other periods is weak and poorly correlated. Some experiments have a weak

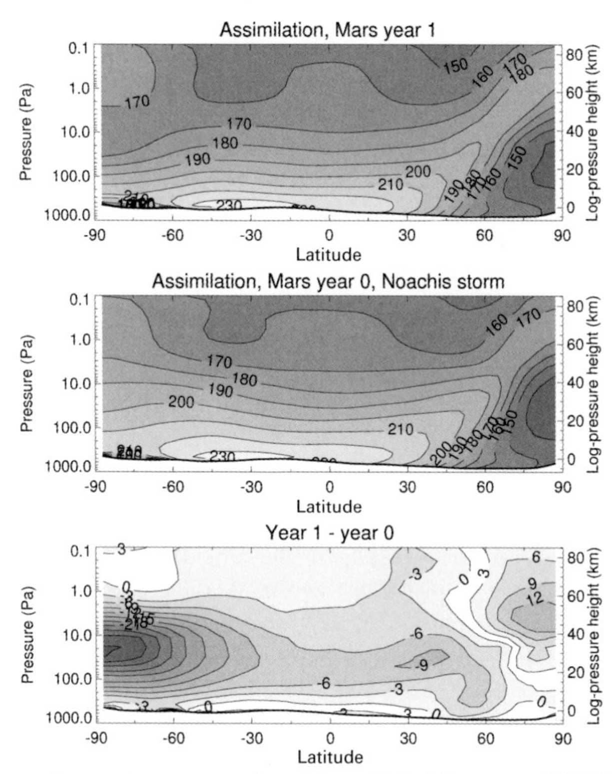

Figure 2.6. The zonal-mean temperature in a Mars Global Surveyor (MGS) mapping phase assimilation (upper panel, Mars year 1) compared with that for the same period of time in the aerobraking year assimilation (middle panel, Mars year 0) with the difference shown (lower panel). A time mean has been taken over the period $L_S = 225°–233°$. In the aerobraking year this was the time of the Noachis dust storm.

zonal wavenumber 2 perturbation with a period of around 2 sols, others have some evidence of longer periods around 5 sols associated with zonal wavenumber 1, while many have little sign of strong transient waves at all.

The assimilation ensemble, Figure 2.8, shows a different picture, with a broadly coherent 2–3-sol wave at the beginning, in transition to a very strong 5–7-sol wave around $L_S = 216°$. It is worth noting that this is almost exactly the same time of year that a transition was seen from a zonal wavenumber 2, 2.3-sol period wave to a zonal wavenumber 1, 5-sol wave in the aerobraking year assimilation. In the aerobraking year, the longer period wave was not quite so large in amplitude, but the situation was complicated by a period of missing data around this time and by the subsequent Noachis dust storm event. Although the assimilation ensemble here is not in perfect agreement, there is clearly a strong correlation between all the members with one of the control simulations showing the development of such a strong wave at this time of year.

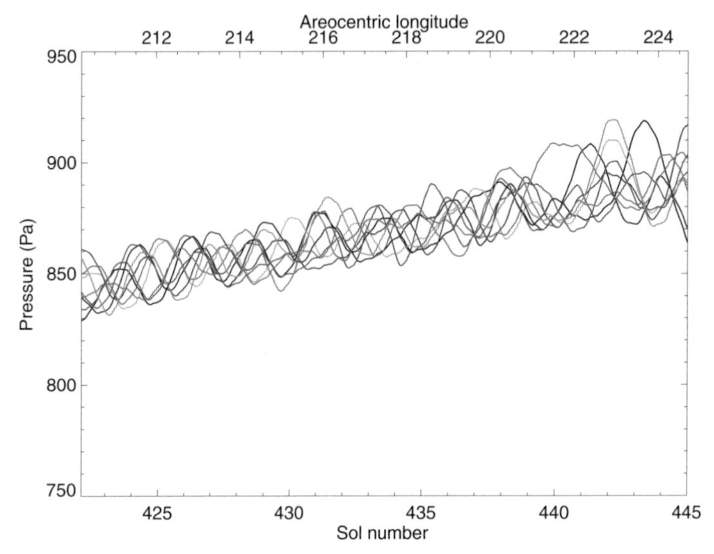

Figure 2.7. Surface pressure at a northern mid-latitude point (0°E, 47.5°N) for an ensemble of nine control integrations over the period $L_S = 210°–225°$. The data has been low-pass filtered, to remove diurnal and semidiurnal tides for clarity.

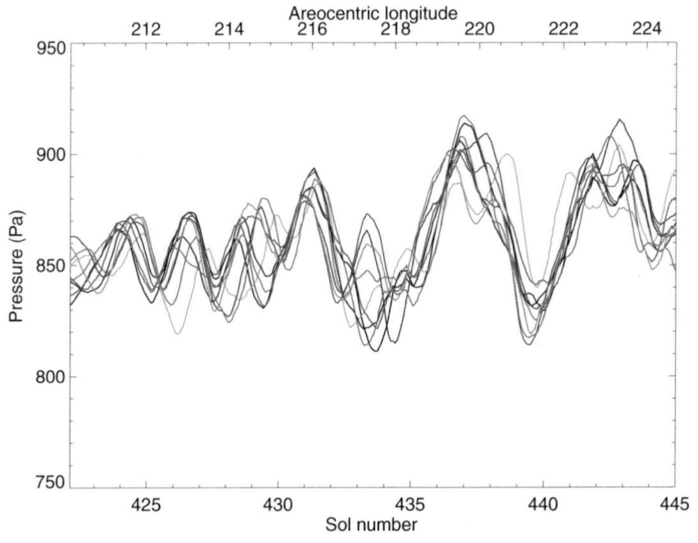

Figure 2.8. As Figure 2.7 for an ensemble of nine assimilations started with the same set of initial conditions as the control integrations at $L_S = 210°$.

3

Mars' global-scale atmospheric structure

This is the first in a series of chapters in which we will dismantle the circulation of Mars' atmosphere and examine the building blocks which go together to determine the strength and direction of the winds, the temperature structure, and other aspects of the weather throughout the atmosphere, and how they vary with time and from place to place. We begin by considering the largest possible scales on Mars, and ask questions such as what determines the overall structure of the atmosphere, and what causes the atmosphere to move around?

3.1 SOME BASICS: THE ROLE OF BUOYANCY FORCES

The simplest and most direct answer which a physicist would give to this last question is – because of the effects of buoyancy forces – which are essentially the same forces which generate circulating convection currents in the air near a domestic radiator in a room, cause the vigorous motions seen in a saucepan of water being heated over a stove, or force bubbles to rise in a liquid such as carbonated water. The main factor which leads to these motions is the variation in the density of a fluid, such as air or water, when it is heated. Almost every substance will tend to expand if heated, so that its density reduces. If a body of fluid is not heated uniformly throughout its volume, such as via conduction from the hot surface of a radiator or saucepan, then the fluid closest to the hot surface will heat up more intensely than elsewhere, and hence become less dense than its surroundings.

3.1.1 Stable and unstable stratification

It is perhaps most helpful to think of performing a 'thought experiment', in which, in our mind's eye, we take two equal volumes of fluid, initially in two different places, and swap them over, whilst preserving their properties (temperature, density, etc.).

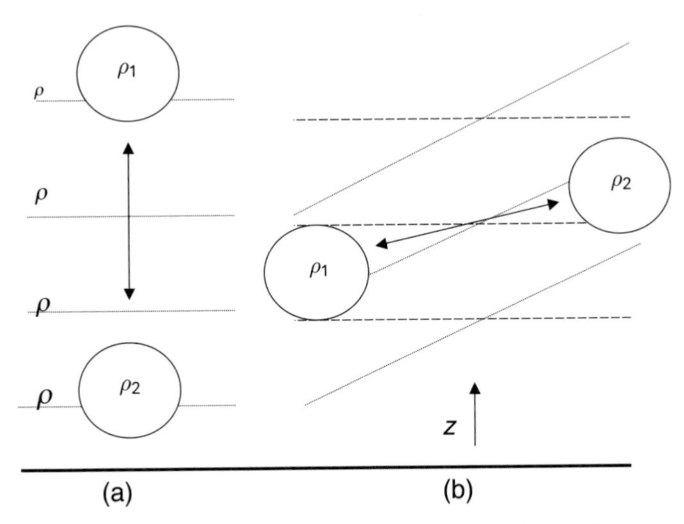

Figure 3.1. Schematic illustration of the energetics of convection, when parcels of air of different density are exchanged by fluid motions: (a) stable stratification without horizontal gradients; (b) stable stratification with horizontal gradient.

The easiest case to think about is if the two volumes lie one immediately above the other, as in Figure 3.1(a). In this case, if blob 1 lies above blob 2, and the densities are such that ρ_1 is greater than ρ_2 (e.g., because blob 2 has been heated), then the interchange of the two blobs results in a net movement of mass downwards – the centre of mass (and hence the centre of gravity) of the two blobs moves downwards as a result of the exchange. In this situation, the potential energy of the configuration has been reduced as a result of the exchange, and this excess energy can be released (at least in principle) in the form of the kinetic energy of motion of the two blobs – so energy is available to make the interchange happen. In this case, we have demonstrated that if differential heating produces a top-heavy distribution of density, then there is potential energy available to allow the fluid to move and overturn. The resulting motion is what we know as *convection*, and will typically arise if a body of fluid is heated from below or cooled from above. The force which leads to this motion is due to the action of gravity on the density contrast in the fluid, and is what is understood as the *buoyancy force*.

The converse of this case, however, in which ρ_1 is less than ρ_2, will not, in general, lead to spontaneous motion. This is because when we exchange blob 1 and blob 2 in our mind's eye this time, the centre of mass (and gravity) of the two blobs has to rise as a result of the exchange. This means that the exchanged state has *more* potential energy than it started with, and this energy has to come from somewhere. Thus, unless we apply another force externally, the exchange cannot take place spontaneously, and the fluid will remain stationary. The initial configuration, with the lighter blob 1 lying above the denser blob 2, is thus in a stable equilibrium. For this reason, if we find a body of fluid in which the density stratification is such that vertical exchanges of elements of fluid leads to a lifting of the centre of gravity of pairs of elements, such as when cold, dense water underlies water which is warmer and lighter, then the stratification is said to be *statically stable*. In this case, if the fluid starts out at rest, there is no tendency for the fluid to start moving.

3.1.2 Horizontal *and* vertical density contrasts

The use of these kinds of 'thought experiment' can also be applied to the more general situation where the density of a fluid varies in both the vertical *and* horizontal directions. In this case, simple vertical exchanges of pairs of fluid elements may be insufficient to determine whether potential energy is available to drive the motions needed to bring about such an exchange.

Consider the situation, for example, in Figure 3.1(b). In this case, density decreases with height everywhere, and also decreases towards the left. Thus, if we exchange any pair of fluid elements which lie one above the other, we will always encounter the *statically stable* situation discussed above, in which the centre of gravity has to rise as a result of the exchange. This would suggest that the density distribution is stable, and would therefore inhibit motion. If we consider exchanging pairs of fluid elements along a sloping path, however, such as between blob 1 and blob 2 in Figure 3.1(b), then we again effectively lower the centre of gravity of the pair of blobs and can release potential energy. Thus, any flow pattern which can bring about the exchange of 1 and 2 will release potential energy, and can therefore develop spontaneously without an external injection of energy.

Thus, we see here a counter-example to the statically-stable case discussed above. Even though the vertical density distribution is statically stable to vertical exchanges, an initial state in which the fluid starts out at rest and has the density distribution shown in Figure 3.1(b) will not, in general, remain stationary, but will tend to move around to bring about the exchange of blobs such as 1 and 2. The corollary of this result is that, if we want to set up a stable state in which the whole fluid is permanently stationary, we need to eliminate horizontal density contrasts and ensure the vertical density distribution is statically stable. This is a result which was first realized independently by the British geophysicist Harold Jeffreys and the Norwegian meteorologist Jacob Bjerknes in the 1920s, and is sometimes referred to as the *Jeffreys–Bjerknes theorem*.

3.1.3 Stratification in a compressible atmosphere

The ideas discussed in the foregoing sections apply most obviously for the case of an incompressible liquid, such as water, in which the density depends solely on the temperature. For a gas, such as air or (in the case of Mars) CO_2, however, the density depends not only on temperature, but also on the ambient pressure. If we squeeze a sample of air whilst insulating it from any heat input, for example, the same mass of air can be made to occupy a smaller volume (though it will also increase its temperature, due to adiabatic compression), and hence increase its density.

A stationary layer of compressible gas under gravity will adjust its density distribution towards a hydrostatic equilibrium in which the local pressure at a given height above the ground represents the weight of the column of gas lying above that height. When we consider exchanging blobs of a compressible gas in our 'thought experiment', therefore, by analogy with the sections above, we must make sure each blob encloses a volume which is equivalent at the two heights under comparison. On making the exchange, we must take into account the differing hydrostatic pressure at the two points, 1 and 2, and adiabatically adjust the density, temperature and volume of each blob of gas accordingly via the well known relationships:

$$p_1 V_1^\gamma = p_2 V_2^\gamma \tag{3.1}$$

$$p_1^{1-\gamma} T_1^\gamma = p_2^{1-\gamma} T_2^\gamma \tag{3.2}$$

$$p_1 \rho_1^{-\gamma} = p_2 \rho_2^{-\gamma} \tag{3.3}$$

for an ideal gas, where γ is the ratio of specific heat capacities c_p/c_v.

Thus, if blob 1 has a density ρ_{11} and temperature T_{11} at pressure p_1, and occupies a volume V_{11}, it will have a density ρ_{12} (and temperature T_{12}) at p_2 and occupy volume V_{12}. We then take blob 2 to have an initial volume $V_{22} = V_{12}$ and density ρ_{22} (but at temperature $T_{22} \neq T_{12}$), and take it adiabatically to p_1, where it will have density ρ_{21} and occupy volume V_{21}. If the exchange results in a movement of the centre of mass of the pair, then the same considerations apply as before – if the centre of mass loses altitude as a result of the exchange, the stratification is unstable and will result in spontaneous convective motion.

Since $\rho = m/V = RT/p$ for an ideal gas, where m is the molar mass of the gas and R the gas constant, we can see that the centre of mass will drop if $M_{11} = V_{11}\rho_{11}$ is less than $M_{21} = V_{21}\rho_{21}$. Provided m does not vary from place to place (which would imply the added complication of variable composition of the air), this condition reduces to $T_{11} < T_{21}$, which we can write in terms of T_{11} and T_{22} using Eq. (3.2) as:

$$T_{11} < T_{22} \left(\frac{p_2}{p_1} \right)^{(\gamma-1)/\gamma} \tag{3.4}$$

The quantity $T(p/p_0)^{(\gamma-1)/\gamma} = T(p/p_0)^{R/c_p}$ (where c_p is the specific heat at constant pressure) is often referred to as the *potential temperature* θ with respect to the 'standard pressure' p_0, and is also related to the specific thermodynamic entropy of the gas at pressure p and temperature T (e.g., see Andrews, 2001). In terms of θ, Eq. (3.4) can be written as $\theta_1 < \theta_2$. Thus, the distribution of potential temperature (or an equivalently defined potential density via Eq. (3.3)) fulfills the same role for a compressible gas in determining the stability of the distribution to parcel exchange as simple density in an incompressible fluid. In this case, therefore, we will only observe a planet with no winds if potential temperature increases monotonically with height, and there is no variation of potential temperature in the horizontal direction. In practice, this is rather unlikely.

Since pressure varies with height according to the hydrostatic relation, it is straightforward to show that the vertical condition for stability on θ implies that the normal temperature may typically decrease with height (as generally observed on the Earth, for example, in the troposphere). However, the convective stability criterion limits the vertical temperature gradient to the condition:

$$\frac{\partial T}{\partial z} \geq \frac{-g}{c_p} \tag{3.5}$$

where g is the acceleration due to gravity. The limit g/c_p is often referred to as the *dry adiabatic lapse rate*.

3.2 MARS' ZONAL MEAN ATMOSPHERIC STRUCTURE

In Mars' case, both observations (e.g., from spacecraft entry profiles such as from Mars Pathfinder (MPF), the Thermal Emission Spectrometer (TES) instrument onboard the Mars Global Surveyor (MGS) spacecraft and from radio-occultation profiles of temperature such as those obtained by the Mariner, Viking, and MGS missions – see Figure 3.2) and numerical models show that the typical distribution of temperature satisfies the condition for vertical stability almost everywhere (except very near the ground under daytime conditions in the summer). However, temperature (both actual and potential) does vary in the horizontal direction.

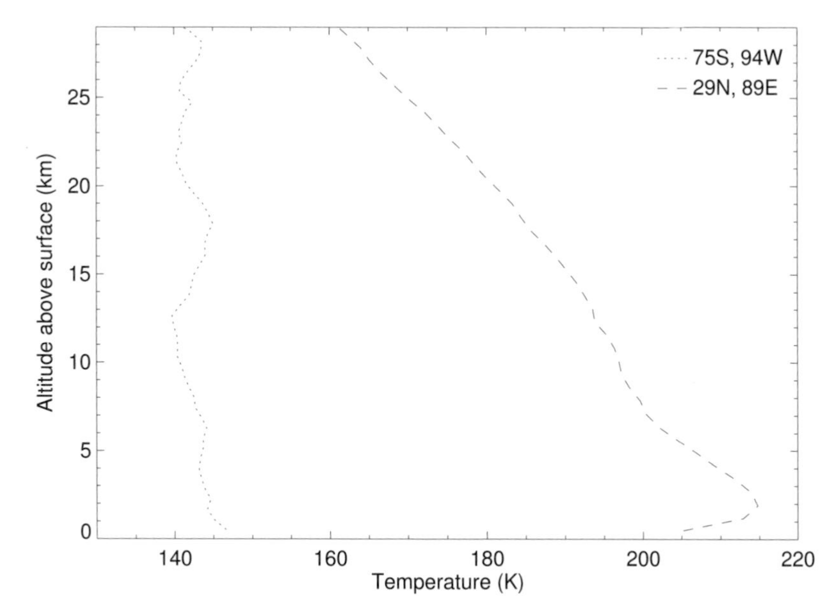

Figure 3.2. Typical radio-occultation profiles of temperature from the Martian surface to around 40 km altitude, taken from the Mars Global Surveyor (MGS) orbiter spacecraft.

Figure 3.3 (see colour section) illustrates these variations in the observations from the TES instrument on MGS. It shows the annual mean variation in latitude and height of the observed temperature structure of the Martian atmosphere, from near the surface up to around 45-km altitude and averaged around latitude circles. The data are taken from the first Martian year of the mapping phase of MGS, so represent a typical year without a major global dust storm. In this view, the temperature decreases with height at all latitudes, and generally decreases away from the equator towards both poles – much as found on the Earth (e.g., see Gill, 1982). The reason for this variation with latitude is quite similar to the reason behind the same qualitative variation on the Earth, namely, the variation in the net heating/cooling with latitude.

3.2.1 Annual mean thermal structure and net heating

Heating takes place largely as a result of absorption of visible sunlight, both within the atmosphere (e.g., due to suspended dust) and at the ground. Cooling occurs through a mix of direct infrared cooling from the surface and indirect infrared cooling from the atmosphere itself – mainly by emission in the 15 μm band of CO_2. The latter is a function of the temperature of the surface or atmosphere, and tends to be reasonably uniform with latitude. The intensity of solar heating at the surface, however, is a strong function of latitude, owing to the varying projection $\int \mathbf{S} \cdot \mathbf{ds}$ of the incident solar flux $\mathbf{S}$ onto the ground surface. Since the direction of $\mathbf{S}$ is fixed in space (parallel to the plane of the ecliptic), $\mathbf{S} \cdot \mathbf{ds}$ will peak at the latitude at which the Sun appears to be directly overhead at local noon (the *subsolar latitude*). This will vary with season, owing to the obliquity of Mars' rotation relative to the normal to the ecliptic plane but, when averaged over the year, will tend to peak on the equator itself and reduce sharply towards either pole. The infrared cooling, however, depends on the temperature of the surface and lower atmosphere. Since this does not (in practice) vary greatly from pole to equator, to a first approximation the infrared cooling is roughly uniform with latitude. The *net heating* is then the difference between the solar heating and infrared cooling, and shows a net heating in the tropics and net cooling towards both poles.

This is clearly reflected in the annual mean temperature structure, which shows a typical horizontal difference of some 30–50 K between the equator and poles at most heights. This difference is much smaller, however, than would be expected if the atmosphere were stationary and came into direct, local equilibrium with the distribution of radiative heating and cooling. Theoretical calculations (Pollack *et al.*, 1979) suggest that a difference of more like 200 K would then be expected, much as is also found for the Earth. The observed difference is much smaller because the atmosphere actually moves in response to this difference in net heating, due to the buoyancy forces discussed above, and the resulting motions transport heat both horizontally and vertically in the sense required to even out large temperature contrasts. We have seen above in Section 3.1.2 that the Jeffreys–Bjerknes theorem shows that a static equilibrium is not possible if temperature varies both horizontally and vertically, even if the vertical stratification is stable. So an active atmospheric circulation, carrying heat and momentum from one latitude to another, is seen to be an inevitable consequence of the inhomogeneous distribution of solar heating.

3.2.2 Vertical structure and the atmospheric energy budget

Of course the actual situation on Mars is more complicated than the simple conceptual description given above. This is because of a variety of factors, ranging from the imperfect transparency and scattering/absorbing properties of the atmosphere itself to the variable properties of the underlying surface, and includes other thermodynamic effects such as the condensation/sublimation of CO_2 and H_2O at the polar ice caps and in transient clouds. Some indication of the complexity of the atmospheric energy budget is apparent in Figure 3.4.

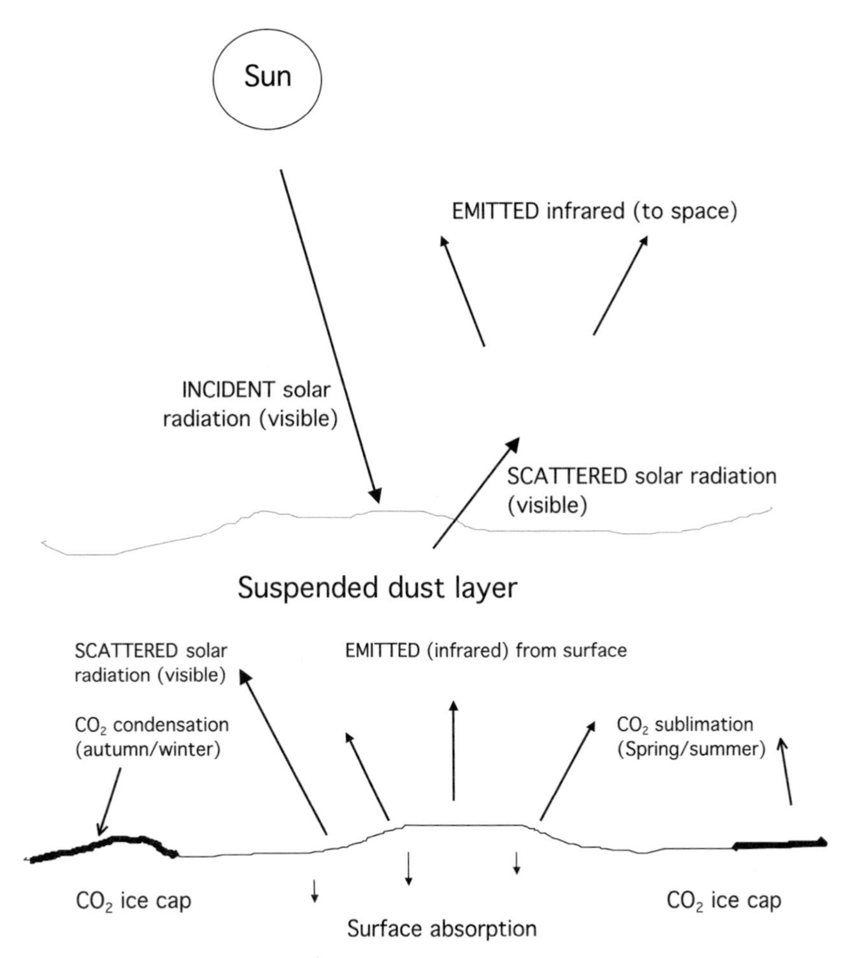

Figure 3.4. A schematic representation of the thermodynamic energy budget of the Martian atmosphere.

Table 3.1. Typical optical properties in the visible (0.1–0.5 μm) and near-infrared (0.5–5 μm) for atmospheric dust on Mars, as used in the EuroMars GCM described by Forget *et al.* (1999).

	Band	Q_{ext}/Q_{ext} (0.67 μm)	ω	$\bar{g}$
'Normal' dust	0.1–0.5 μm	0.878	0.665	0.819
(Ockert-Bell *et al.*, 1997)	0.5–5 μm	1.024	0.927	0.648
'Bright' dust	0.1–0.5 μm	1.0	0.920	0.55
(Clancy and Lee, 1991)	0.5–5 μm	1.0	0.920	0.55

Absorption of solar radiation takes place substantially (though not exclusively) at the surface, at a rate which depends on the locally variable albedo, which in turn is affected by the structure and composition of the surface, ice cover, and so on. The albedo is observed to vary between lows of around 0.1 in the darkest regions of the surface to around 0.9 in the ice-covered polar regions. Thus, the fraction of incident sunlight absorbed directly at the ground varies between around 85% and 5% or so, depending on location.

Some incident radiation is scattered and/or absorbed within the atmosphere itself, either by air molecules, or by dust particles and/or aerosol particles in CO_2 or water ice clouds. The fraction of solar energy absorbed can be estimated using a delta–Eddington code (such as those used in atmospheric models described in Chapter 2, see Forget *et al.*, 1999) for a typical vertical distribution of dust and surface albedo. If we assume a surface albedo of around 0.24, and use values for the dust optical properties for the 'bright dust' of Forget *et al.* (1999) as given in Table 3.1, these calculations suggest that around 45% of incoming solar energy is absorbed within the atmosphere under relatively clear conditions (visible optical depth $\tau \sim 0.6$), but this rises to ~60% at optical depth 1 and to more than 85% during major dust storms (when $\tau \sim 5$ or even greater) – see Chapter 7 for further discussion.

Infrared emission, in turn, emerges from the surface at a rate determined by the wavelength, surface temperature, and emissivity, and may be affected by factors such as stored heat (due to the finite thermal inertia of the ground), local geology, aerosol or dust clouds, and ice cover. Some typical infrared spectra observed at the top of Mars' atmosphere are shown in Figure 3.5.

The shape of the emission spectra show an overall resemblance to a black-body spectrum for a temperature of around 200 K, superimposed upon which are absorption or emission bands due to atmospheric constituents, such as CO_2 and dust particles, but also spectral features associated with the underlying surface. The atmospheric features can appear either in emission or absorption, depending upon whether the atmosphere is warmer or cooler than the underlying surface. Some infrared emission escapes directly from the surface to space (especially at wavelengths outside the 15 μm CO_2 molecular absorption band and the 9 μm silicate dust feature), but some is also absorbed and/or scattered in the atmosphere and

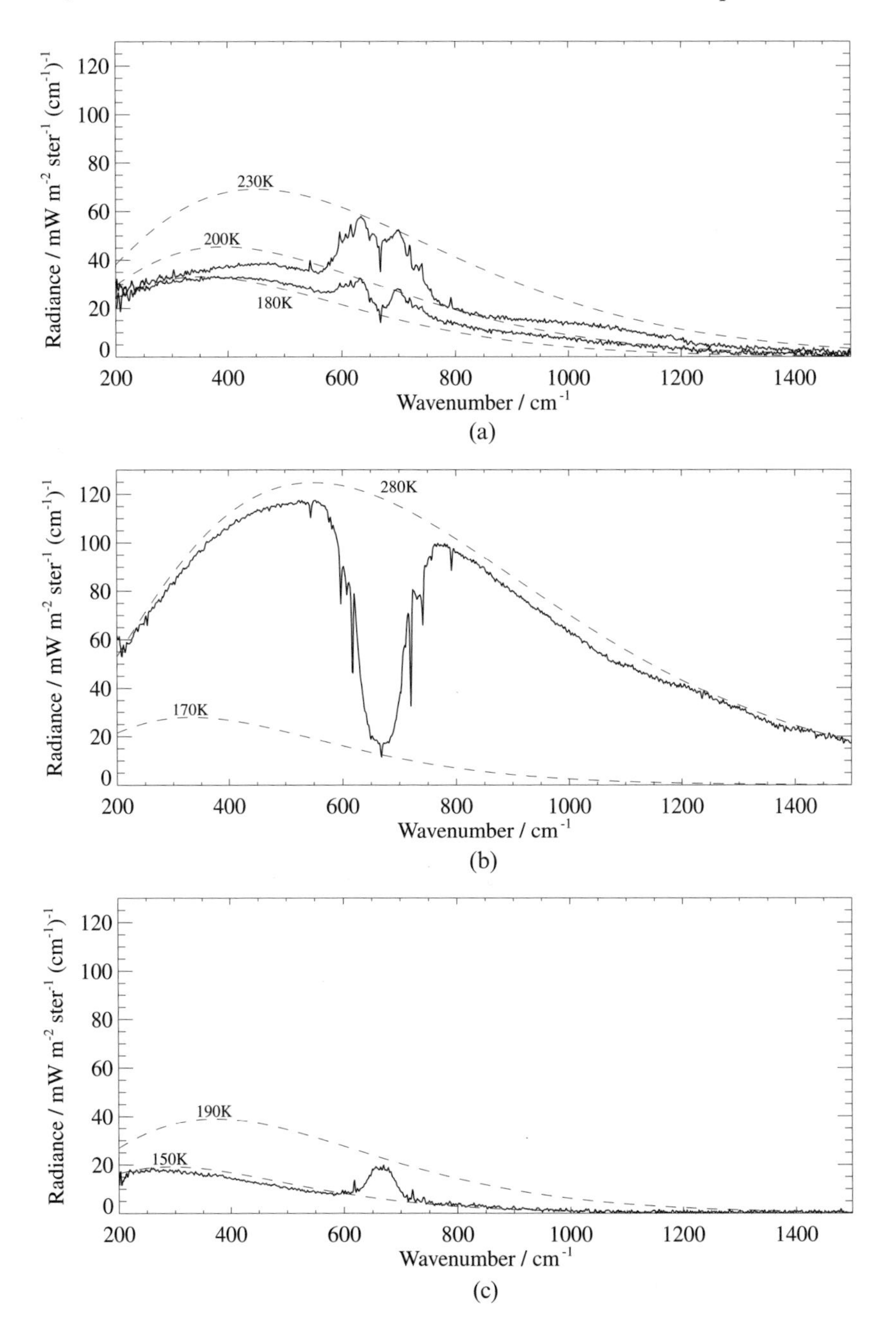

Figure 3.5. Typical infrared emission spectra from Mars, obtained by the Infrared Interferometer Spectrometer (IRIS) instrument on the Mariner 9 orbiter (adapted from Hanel *et al.* 1992): (a) over the south polar region (the upper curve includes a smaller fraction of the polar ice cap than the lower curve); (b) near 21°S, over a relatively warm surface; and (c) near 61°N over polar ice.

subsequently re-emitted. The atmosphere itself is thus only partly transparent, and this finite transparency at infrared wavelengths leads to a modest greenhouse warming of the surface, typically by around 5–10 K (cf the value of around 30 K for the Earth (Andrews, 2001), owing to the additional opacity due mainly to water vapour, and other greenhouse gases).

Other factors which influence the thermal structure close to the surface include the seasonal condensation and sublimation of CO_2 to form ice deposits in the polar regions, which can store and release energy in the form of the latent heat of condensation (around 597 $kJ\,kg^{-1}\,K^{-1}$ at 150 K) as well as affecting the surface albedo and emissivity, and the radiative effects of minor constituents such as water vapour or ozone. Whilst the effects of water and ozone are relatively small, CO_2 condensation/sublimation has a major influence on the thermodynamic energy budget, and, because (as we shall see later) it leads to substantial changes in the atmospheric mass over seasonal timescales, is also an important modifying driver of the atmospheric circulation itself. This will be discussed in greater detail in Section 3.4. Finally, the thermal structure in the vertical is also modified by the dynamical motions of the circulation itself. In particular, vertical motion will result in significant adiabatic cooling (upward motion) or warming (downwards motion) which can change the temperature locally by 10–20 K. Figure 3.6 shows an example of a calculation using a simple 1-D radiative–convective model with solar heating and infrared cooling, for the location of the Viking lander 1 site. The circles indicate the observed state (during Lander entry) and the dashed lines show calculations assuming no dynamical motion, which clearly indicate that the model temperatures are $\sim$10–20 K too warm above 5–10 km or so. When the effect of adiabatic cooling in weak upwelling is included, however, as shown by the solid lines, the model is more closely in agreement with the observations, suggesting the importance of dynamical uplift in determining the thermal structure.

3.3 ANNUAL AND ZONAL MEAN ATMOSPHERIC CIRCULATION

We have seen above that variations in the net heating/cooling across the planet are expected to lead to the generation of a pattern of motions which will transfer heat, momentum, and other tracers both vertically and horizontally. Although some features of this pattern of circulation can be discerned in observations, the clearest way of determining this pattern of winds is probably via a comprehensive numerical circulation model, such as that described in Chapter 2. Figure 3.7 (see colour section) shows some results from such a simulation representing the annual and zonal mean circulation and temperature structure for a 'typical' Mars year without a major global dust storm.

Figure 3.7(a) (see colour section) shows the zonal mean temperature structure, for comparison with Figure 3.3 (see colour section), and shows a similar pattern of cooling towards both poles with warming aloft at high latitudes. The observations are reproduced to within 10 K or so over most of the planet, except perhaps over the

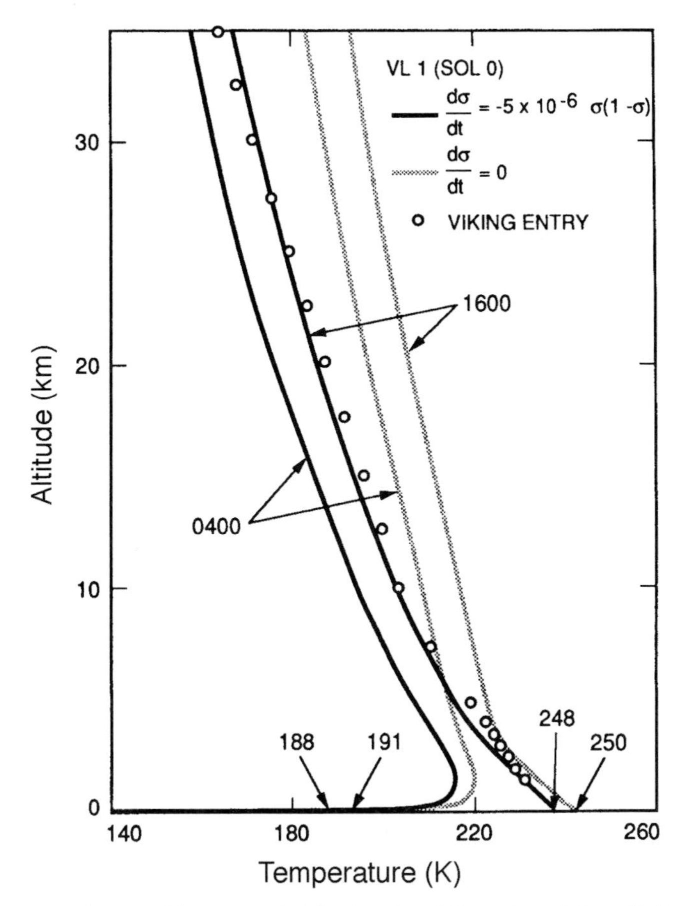

Figure 3.6. Temperature profiles computed for two local times by a 1-D radiative–convective model (Pollack *et al.*, 1979) during early summer at the Viking Lander 1 site without (dashed lines) and with (solid lines) inclusion of a simple profile of adiabatic cooling due to a regional upwelling.

poles where the actual dust distribution probably departs significantly from that assumed in the model simulations.

Because Mars is a rotating planet, north–south motion is expected to result in a pattern of east–west winds, owing to the tendency of moving air to conserve its absolute angular momentum about the axis of rotation. Thus, equatorward flow tends to be associated with westward zonal winds (retrograde relative to the rotation of Mars itself) and poleward flow with eastward (prograde) winds. This is approximately consistent with Figure 3.7(b) (see colour section), though the situation at equilibrium is somewhat more complicated (e.g., see Andrews, 2001; Holton, 1992; etc.). In fact the pattern of zonal wind u is reasonably consistent with a geostrophic equilibrium on large-scales, so that u approximately obeys the *thermal wind* equation (e.g., Holton, 1992; Andrews, 2001)

$$\frac{\partial u}{\partial z} \simeq \frac{1}{H_p}\frac{\partial u}{\partial \ln p} = \frac{g}{af\theta}\frac{\partial \theta}{\partial \phi} \tag{3.6}$$

where H_p is the pressure scale height ($= RT/g$); θ is potential temperature; a is the planetary radius; f the Coriolis parameter ($=2\Omega \sin \phi$); and ϕ is latitude. Thus, u becomes increasingly eastward with height at middle latitudes in both hemispheres, where the (actual and potential) temperature decreases with increasing latitude, forming circumpolar jetstreams at high altitudes. These jetstreams are most intense at the latitudes corresponding to the annual mean boundaries of the CO_2 polar ice caps in each hemisphere, where the horizontal thermal gradient is also strongest. Over the equator, however, the situation is more complicated, because the geostrophic thermal wind equation is no longer valid (since $f \rightarrow 0$). In practice, the zonal winds at low latitudes are the result of a number of complex factors, including tides and planetary and gravity waves, some of which will be discussed in more detail later in this book.

Figure 3.7(c) (see colour section) shows the mass streamfunction Ψ for the mean meridional circulation corresponding to this flow, where the meridional and vertical winds blow parallel to the contours of Ψ. Thus, rising motion occurs over much of the tropics, where there is net heating, and descending motion occurs at high latitudes in both hemispheres, where there is net cooling. Continuity and conservation of mass requires there to be horizontal flow connecting these regions of upwelling and downwelling, resulting in the overturning cells seen in the map of Ψ. Some more complicated perturbations are also seen in the pattern of Ψ at high latitudes, which are associated with the effect of large-scale eddies (to be discussed in Chapters 4–6), in much the same way as Ferrell cells are found in the meridional circulation of the Earth's atmosphere (Gill, 1982; Holton, 1992; Andrews, 2001).

The main thermally-direct overturning circulation is often referred to as the *Hadley circulation* or *Hadley cell* (after the 18th century English pioneer James Hadley, who first explained the pattern of trade winds in the Earth's tropics), and represents a notionally axisymmetric circulation pattern driven by the imbalance of net heating between the equator and higher latitudes, and approximately conserving angular momentum at high altitudes (though not near the ground). Some basic properties of the Hadley circulation can be understood with reference to a simple analytical model, due to Isaac Held and Arthur Hou of Princeton University (Held and Hou, 1980), which we present in outline here (further details can be found in the original paper by Held and Hou, and also in Chapter 4 of James (1994)).

3.3.1 The Held–Hou model

The atmospheric circulation is envisaged schematically (see Figure 3.8) as comprising two layers of air, one near the ground (where surface friction keeps the zonal velocity $u \sim 0$) and the other at high altitude. The angular momentum per unit mass in the upper layer is assumed to be conserved by the meridional Hadley circulation, which

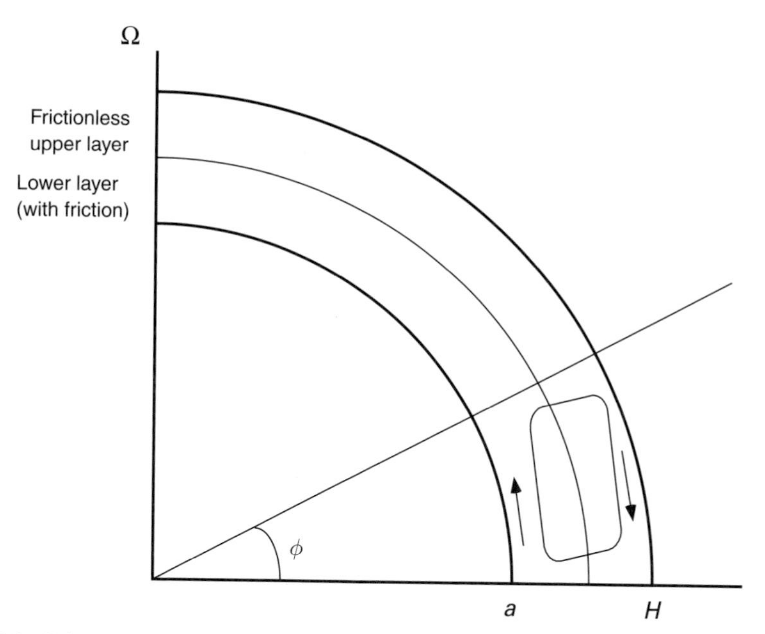

Figure 3.8. Schematic illustration of the Held–Hou model of the angular-momentum conserving Hadley circulation.

is poleward at upper levels and equatorward in the lower layer. If we make the small angle approximation for latitude ϕ, such that $\sin\phi \sim y/a$, where y is the northward coordinate, z the upwards coordinate, and a the planetary radius, then this will imply that the angular momentum conserving zonal flow in the upper layer takes the form:

$$u = U_M = \frac{\Omega y^2}{a} \tag{3.7}$$

where Ω is the planetary rotation rate. The zonal flow in the upper layer is further assumed to be in thermal wind balance with the latitudinal gradient of potential temperature (at the height of the interface between the two atmospheric layers), so that:

$$\frac{\partial u}{\partial z} = -\frac{ga}{2\Omega\theta_0 y}\frac{\partial\theta}{\partial y} \tag{3.8}$$

From Eq. (3.7) and taking $\partial u/\partial z \simeq U_M/H = \Omega y^2/(aH)$, where H is the vertical scale of our two-layer atmosphere, we have:

$$\frac{\partial\theta}{\partial y} \simeq -\frac{2\Omega^2\theta_0}{a^2 gH}y^3 \tag{3.9}$$

This can now be integrated in y to obtain the variation of $\theta = \theta_M$ with y which is consistent with the thermal wind balance for the angular momentum conserving

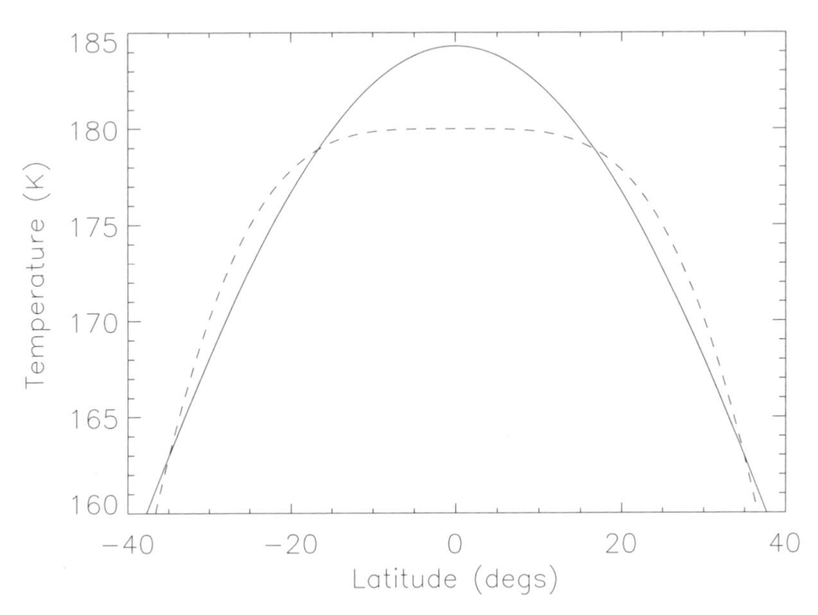

Figure 3.9. Respective variations of θ_E and θ_M close to the equator in the Held–Hou model (in the small latitude angle approximation used in the text).

form of U_M in the upper layer:

$$\theta_M = \theta_{M_0} - \frac{\Omega^2 \theta_0}{2a^2 gH} y^4 \tag{3.10}$$

where θ_{M_0} is a constant of integration which remains to be determined.

The problem is closed by supposing that the Hadley circulation is driven by a meridional heating function proportional to $\theta_E - \theta_M$, where $\theta_E(y)$ is the radiative equilibrium temperature distribution. For simplicity, this is assumed to take the form:

$$\theta_E = \theta_{E_0} - \frac{\Delta\theta}{a^2} y^2 \tag{3.11}$$

where θ_{E_0} is a constant (corresponding to the global mean potential temperature at this level) and $\Delta\theta$ is the equilibrium equator–pole temperature difference. By choosing a suitable value for θ_{M_0}, the curves of θ_M and θ_E can be made to cross twice in each hemisphere (e.g., see Figure 3.9), with the second crossing at $y = \pm Y$. This value of y effectively defines the poleward edge of the Hadley circulation, since air is effectively heated close to the equator ($|y| \leq Y_1$, see Figure 3.9) where $\theta_E > \theta_M$, and cooled for $Y_1 \leq |y| < Y$. By requiring that there is no net heating or cooling of air parcels within the Hadley circulation, so that:

$$\int_0^Y \theta_E \, dy - \int_0^Y \theta_M \, dy = 0 \tag{3.12}$$

Table 3.2. Parameters appropriate to Earth and Mars for the Held–Hou model of axisymmetric Hadley circulations.

Parameter	Earth	Mars
$\Delta\theta$ (K)	40	65
H (km)	8	11
$\Omega(s^{-1})$	7.3×10^{-5}	7.3×10^{-5}
θ_0 (K)	255	210
Y (km)	1960	2060
$\Delta\phi$	18°	35°
$\theta_{E_0} - \theta_{M_0}$ (K)	0.8	4.3
Max $(u(\lvert y \rvert = Y)$ (m s^{-1})	55	85

we can solve for the extent Y of the Hadley circulation in latitude of the form:

$$Y = \left(\frac{5\Delta\theta g H}{3\Omega^2\theta_0}\right)^{1/2} \tag{3.13}$$

enabling a prediction for Y in terms of various planetary parameters.

Taking values appropriate for the Earth, for example (see Table 3.2), leads to $Y \simeq 2000$ km, corresponding to around 18° in latitude as the nominal width of each Hadley cell. This corresponds quite well to the observed Hadley circulation on Earth (if perhaps a little on the small side), lending some confidence to this simple model.

For Mars, appropriate values (see Table 3.2) lead to $Y \simeq 2060$ km or around 35° latitude. This also compares reasonably well with the GCM simulation illustrated in Figure 3.7 (see colour section). This deceptively simple model therefore accounts reasonably well for several of the main features of the annual and zonal mean circulation pattern on Mars. In addition to the width of the Hadley circulation in each hemisphere, it also predicts the maximum strength U_M of the (angular momentum-conserving) zonal winds at the top of the Hadley circulation as:

$$U_M = \frac{\Omega Y^2}{a} \tag{3.14}$$

where a is the planetary radius. For Mars, this gives $U_M \sim 85\,\mathrm{m\,s^{-1}}$, which compares quite well with the winds predicted by the GCM. The model also, however, predicts that $u \to 0$ on the equator, which clearly is not the case in Figure 3.7(b) (see colour section), and makes no clear prediction of what the circulation will do at higher latitudes than $y = Y$. A more complete understanding of these features requires a consideration of the non-axisymmetric components of the circulation, which will be the subject of later chapters.

3.4 CO_2 SUBLIMATION AND THE SEASONAL CYCLE

The very cold temperatures reached at high latitudes during winter lead to some unusual features of Mars' atmospheric circulation and seasonal cycle. Temperatures at the winter pole are regularly observed as low as 140 K, which has particular significance since this is close to the frost point of CO_2 at the typical surface pressure of around 600 Pa. Once this temperature is reached, CO_2 ice will start to condense onto the ground (either directly, or by formation of icy precipitation or 'snow' from CO_2 ice clouds). Since CO_2 is the principal constituent of the atmosphere, relatively large amounts of CO_2 are available for condensation, and this can lead over time to a substantial fraction of the atmospheric mass being deposited on the surface as ice.

Figure 3.10 (see colour section), for example, shows the variation in density of highly reflective clouds obtained by the Mars Orbiter Laser Altimeter (MOLA) instrument on MGS. These returns are dominated by the polar hood clouds in winter time, which almost certainly represent thick clouds of CO_2 ice which form over the poles, leading to dry ice snow which falls to the surface to build up the seasonal polar ice caps.

The rate of ice deposition is limited primarily by the energetics and thermodynamics. As CO_2 ice is formed directly from the gas, latent heat of sublimation is released (at a rate L of around 597 kJ kg^{-1}). Further deposition cannot then take place until that energy is removed, which happens most effectively via infrared radiation to space. Thus, neglecting other sinks of heat energy (such as thermal conduction into the ground or the surrounding air), the increase in condensate mass per unit area M_c can be written as:

$$L\frac{\partial M_c}{\partial t} = \epsilon\sigma T_c^4 \tag{3.15}$$

where σ is the Stefan–Boltzmann constant; ϵ is the emissivity of the ice; and T_c its temperature. Over the winter season in either hemisphere, this can evidently lead to around a third of the entire atmospheric mass being deposited at the surface. The process is only arrested when the polar night ends in local spring, when solar heating allows the polar temperatures to start to rise again. During spring and into summer, therefore, the polar ice deposits can sublime back into the atmosphere to replenish the atmospheric reservoir of CO_2 – whilst in the opposite hemisphere, the onset of autumn leads to the start of condensation on the other pole (see Figure 3.11).

Thus, over the annual cycle, around a third of the total atmospheric CO_2 mass is exchanged between the polar ice caps. This leads to a number of observable effects at the surface of Mars. Pressure variations at the surface were measured over long periods by the two Viking Lander (VL) spacecraft, and Figure 3.12 shows a time series from VL 1 over nearly three Mars years. The effects of the seasonal condensation of CO_2 are clearly visible as a slow cyclic variation of the surface pressure with an amplitude of nearly 100 Pa. This implies a peak-peak variation of around 30% of the mean value (cf the typical variation at the Earth's surface of perhaps 5% peak-peak, due solely to thermally-driven local meteorology).

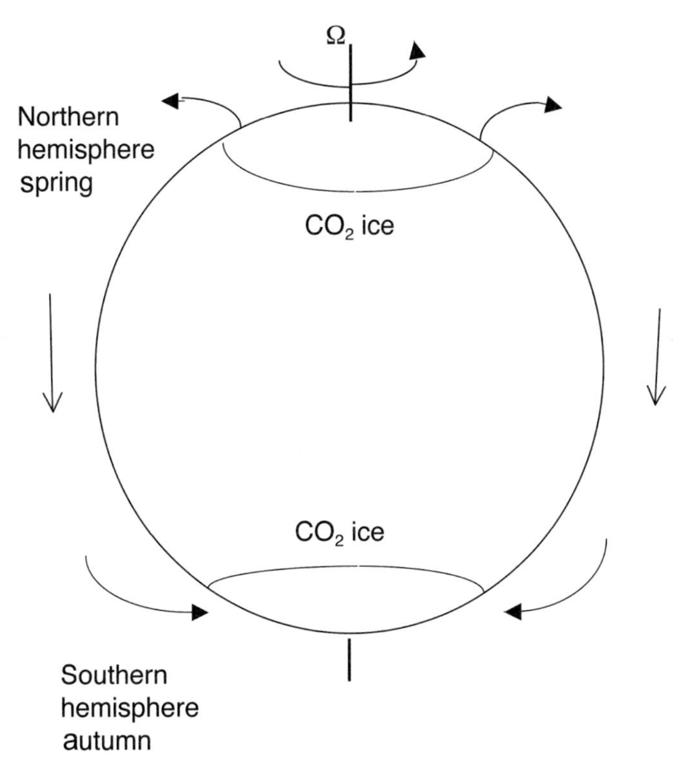

Figure 3.11. Schematic circulation produced by the seasonal condensation/sublimation cycle of CO_2, showing the exchange of mass between the northern and southern polar caps during northern hemisphere spring.

The periodic growth and decline of the polar ice caps has been observed for more than 150 years from Earth in ground-based telescopic observations, and can be seen particularly clearly in Hubble Space Telescope (HST) images from Earth orbit, taken at recent oppositions of Mars. Figure 3.13 (see colour section), for example, shows a composite of HST images, projected to show the North Polar region of Mars at differing seasons. These show the polar ice cap extending to as far south as 40°–50° N during early spring, yet almost disappearing altogether during late summer.

Similar behaviour can also be reproduced in some detail in the current generation of Mars GCMs. Figure 3.14 (see colour section) shows a comparable set of images derived from such simulations, using a simple quantitative representation of CO_2 condensation based on the discussion below (see Chapter 2). Many features of the observed variations in the polar caps are clearly seen in the simulations.

The quantitative accuracy of these simulations is further evident in measurements of the size of the polar cap during the Martian year. Figure 3.15 shows the variation of the *mean equivalent latitude* of the polar cap, derived by measuring the area of ice cover in both Viking observations (from Pollack *et al.*, 1993) and GCM

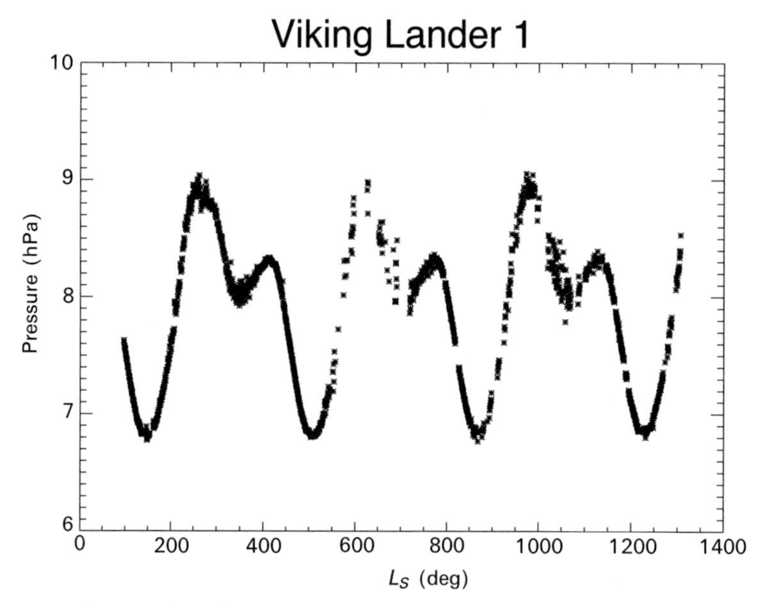

Figure 3.12. Variation of surface pressure at the Viking Lander 1 site at 22.5°N, 48.0°W measured intermittently over more than three Mars years and shown plotted as a function of L_S.

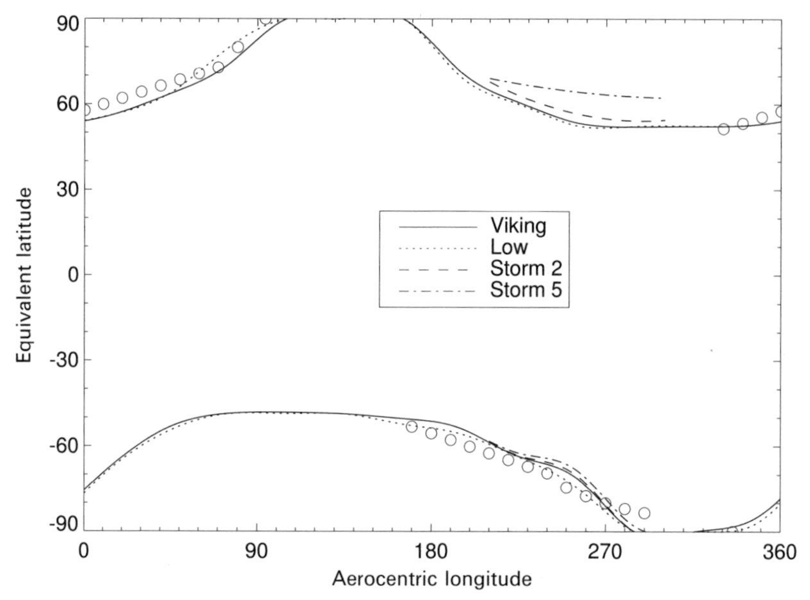

Figure 3.15. Mean equivalent latitudes for the edges of the CO_2 polar ice caps on Mars for the various climate scenarios simulated using the Oxford and LMD Mars GCM by Lewis *et al.*, (1999). The circles indicate observations of the equatorward boundaries of the Martian polar caps (after Pollack *et al.*, 1993).

simulations, and determining the latitude to which a perfectly circular ice cap of the same area, centred on the pole, would extend. The observations were only feasible when each polar cap was illuminated by the Sun, hence are only available during part of the year. From these measurements, it is clear that the GCM captures the variation reasonably well, indicating that both CO_2 caps disappear altogether in their respective summer seasons. Interestingly, the advance and retreat of the ice edge does not seem to depend strongly on how much dust is included in the simulation. This is partly because the assumed dust distribution in each simulation places relatively little dust over the poles themselves, but also because the rate of sublimation is mainly determined by the infrared radiative energy flux, which is only weakly affected by atmospheric dust.

3.4.1 CO_2 sublimation and the atmospheric circulation

Such massive exchanges of CO_2 between the two polar ice caps are bound to lead to a major perturbation to the meteorology and circulation of the atmosphere (cf Figure 3.11), and it is of interest to examine these implications in more quantitative detail (albeit in a highly idealized model). To begin with, we will estimate the average north–south wind associated with the mass flow forced by this condensation cycle. We consider the growing polar ice cap over the winter pole to consist of a circular cap of radius r_0, centred on the pole, and regard the cap as growing in thickness (so M_c is increasing) by condensation from the air above it. We will further suppose that the mass condensing onto the ice cap is continually replenished by a flow into the cylindrical region above the cap at radial speed v_0 (see Figure 3.16). Mass is conserved across the vertical cylindrical surface bounding the air above the polar cap, so that:

$$\int_0^{r_0} \frac{\partial M_c}{\partial t} 2\pi r \, dr = \int_0^{\infty} v_0 \rho 2\pi r_0 \, dz \tag{3.16}$$

where ρ is the atmospheric mass density at height z. For simplicity we take the density to be in approximate hydrostatic balance, so that $\rho = \rho_0 \exp(-z/H_p)$, where ρ_0 is the mass density of the atmosphere at $z = 0$. The integrals in Eq. (3.16)

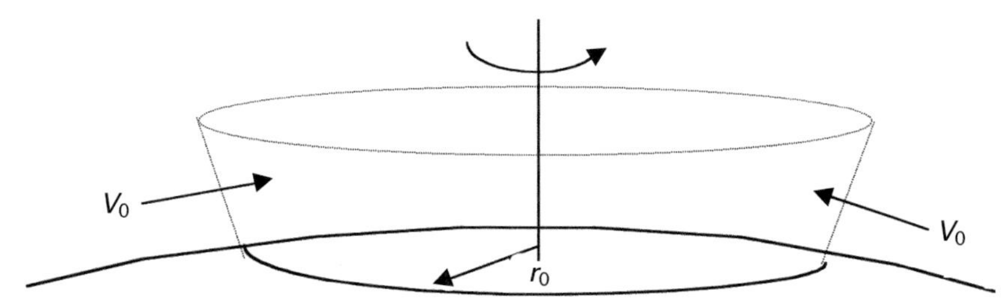

Figure 3.16. Schematic model for the condensation or sublimation of CO_2 ice at Mars' polar caps.

can then be evaluated (also using Eq. 3.15) to give:

$$\frac{\pi r_0^2 \epsilon \sigma T_c^4}{L} = 2\pi r_0 H_p \overline{v_0} \rho_0 \tag{3.17}$$

where the overbar on v_0 represents a (mass-weighted) vertical average. Rearranging Eq. (3.17), therefore, leads to our estimate for $\overline{v_0}$

$$\overline{v_0} = \frac{\epsilon \sigma T_c^4 r_0}{2\rho_0 H_p L} \tag{3.18}$$

For Mars, we take $\epsilon = 1$, $T_c = 140\,\mathrm{K}$, $\rho_0 \simeq 2.2 \times 10^{-2}\,\mathrm{kg\,m^{-3}}$, $H_p \simeq 10\,\mathrm{km}$, and $r_0 = 1500$–$3000\,\mathrm{km}$. This leads to an estimate for $\overline{v_0}$ of around 0.2–0.5 m s^{-1}.

How significant is this flow for the circulation of Mars' atmosphere? A mean meridional velocity of ~0.5 m s^{-1} is not negligible compared with the thermally-driven flow, though is somewhat smaller than the peak meridional flow velocities in the main Hadley circulation (e.g., as evident in Figure 3.7(c), see colour section). This equatorward or poleward mass flow will also lead to a perturbation to the zonal winds, through the tendency of the air to conserve angular momentum. In order to estimate the magnitude of this perturbation, however, it is not sufficient to apply angular momentum conservation too naively. Starting an air parcel at rest on Mars from 60° latitude and taking it to the equator, for example, whilst conserving angular momentum would lead to westward zonal winds ~350 m s^{-1}, which is clearly unrealistic. This is partly because as the air moves towards the equator and spins up, it will interact (via friction in the boundary layer) with the underlying surface, reducing the flow relative to the planet. A better estimate, therefore, comes from simplifying the zonal equation of motion, neglecting zonal (and vertical) variations and representing the effects of surface friction by a simple 'Rayleigh friction' drag law.

$$\frac{\partial u}{\partial t} - fv = -u/\tau_{\mathrm{drag}} \tag{3.19}$$

where τ_{drag} represents the timescale for frictional drag. In a steady state, therefore, $\partial/\partial t \sim 0$ and so:

$$u \sim f v_0 \tau_{\mathrm{drag}} \tag{3.20}$$

Given suitable values for Mars ($f \sim 10^{-4}\,\mathrm{s^{-1}}$, $\tau_{\mathrm{drag}} \sim 2$ days) and taking $v_0 \sim 0.5\,\mathrm{m\,s^{-1}}$ leads to $u \sim 10\,\mathrm{m\,s^{-1}}$. This may be compared with typical zonal winds in Mars' middle atmosphere of up to 150 m s^{-1}, though near the ground velocities are much less than this. Thus, we would expect the condensation flow to have a relatively minor influence on the main pattern of thermally-driven zonal winds, though this influence may be substantially more important near the surface, especially over and near the edge of the polar cap itself. In fact the action of the Coriolis term in Eq. (3.19) will lead to the air following spiral paths towards or away from the pole. It is interesting to note that both polar caps typically show spiral structures in the ice deposition (as can be clearly seen in both visible images of the polar ice caps and in MOLA maps of the surface topography; see Figure 3.17),

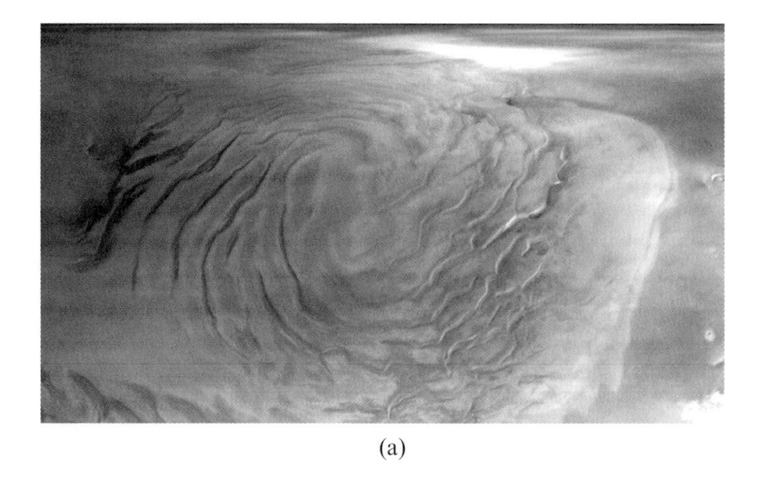

(a)

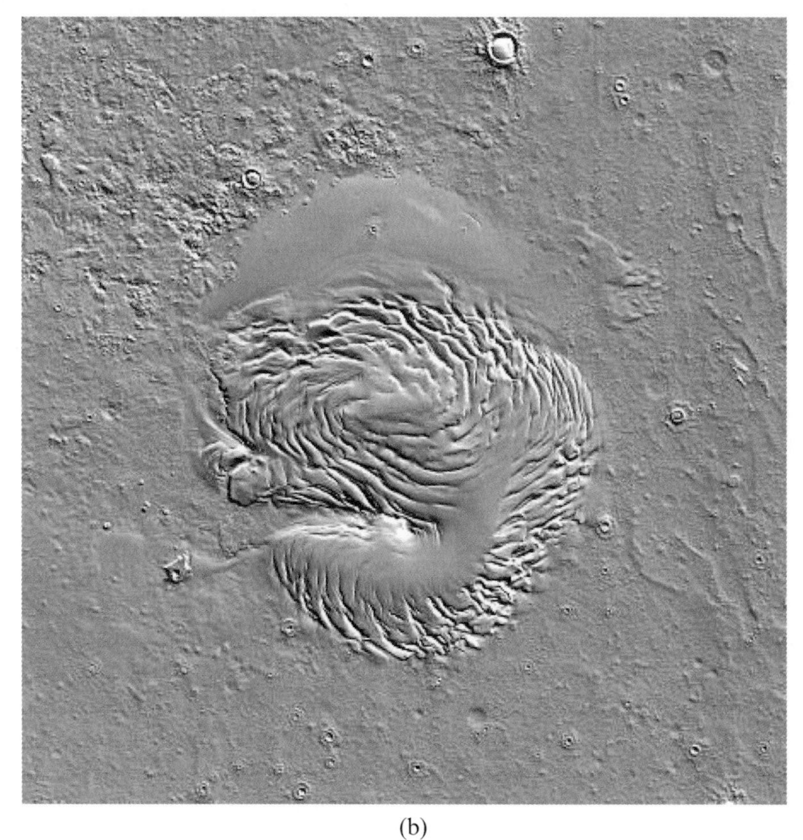

(b)

Figure 3.17. (a) Image of Mars' North Polar ice cap, from Mars Orbiter Camera (MOC) wide-angle image M1902064, and (b) rendered map of the topography of Mars' North Polar ice cap, from Mars Orbiter Laser Altimeter (MOLA) observations taken during the Mars Global Surveyor (MGS) mission.

though the origin of these structures could also be due to quite different processes related to the glaciology of the water ice (e.g., see Hvidberg and Zwally, 2003).

3.5 SEASONAL VARIATIONS IN MARS' ZONAL MEAN CIRCULATION

As the seasons progress, the subsolar latitude moves between the extremes of around $\pm 23^\circ$, and we would therefore expect the Hadley circulation and other aspects of the large-scale flow in the atmosphere to change in response to the seasonal cycle. We will now look at these seasonal variations in the zonal mean flow, using both observations and GCM simulations.

3.5.1 TES observations

Figure 3.18 (see colour section) shows a sequence of latitude–height maps of temperature in the Martian atmosphere, in which the temperatures measured with the TES instrument on MGS have been averaged in latitude and over a time period of a few tens of days, which is relatively short compared with the annual cycle. These maps show the temperature structure representative of the main seasons (by analogy with the twelve months on Earth), starting with the northern hemisphere summer ($L_S = 90^\circ$, where L_S is the solar longitude (see Appendix A) – see top left of Figure 3.18). At this time of year the subsolar latitude is well north of the equator itself, and the temperature distribution is correspondingly quite asymmetric about the equator. At the autumn equinox, however, around $L_S \sim 180^\circ$, the distribution is more symmetric about the equator apart from a light cold bias towards the South Pole, which is just emerging from winter. A strong equatorward temperature gradient occurs at almost all heights above the North Pole, close to the edge of the retreating polar ice cap, but has yet to form strongly in the north, since the Northern Polar cap has only just started to form. This is a reflection of the thermal inertia of the surface, which takes some time to change its temperature and tends to lag behind the seasonal movement of the subsolar point.

In northern winter (Figure 3.18, see colour section, around $L_S \sim 270^\circ$), this pattern again develops to become very asymmetric about the equator. By this time, the warmest point on the surface is not near the equator, but close to the summer pole. This is surprising at first sight, since the subsolar latitude is only around 23°S by this time. However, the total integrated sunlight at this time actually peaks at a much higher latitude than this, because at latitudes poleward of 65°S (Mars' 'antarctic circle'), the Sun never sets and the pole is constantly in daylight. A similar effect also occurs on Earth, but the response is much less extreme because much of the Earth's surface is covered by oceans, which have an enormous thermal capacity (and hence thermal inertia) preventing the surface temperature from responding quickly to changes in solar heating. On Mars, however, there are no oceans, and so the ground and surrounding air can heat up quickly and produce very strong changes in circulation and temperature. At the winter pole inside Mars' 'arctic circle', on the other hand, there is now perpetual darkness, and so the surface and atmosphere cool rapidly until CO_2 condenses at its maximum rate. The winter pole is typically seen to be shrouded in a thick 'polar hood' cloud of CO_2 ice particles, which presumably precipitate as 'dry ice' snow onto the surface (as seen in both MOLA observations and models; e.g., Colaprete and Toon, 2002). This leads to an intense thermal gradient at the edge of the growing polar ice cap, which extends to high altitude.

In southern spring (Figure 3.18(c), see colour section), we return to equinox conditions, and the temperature pattern is again quite similar to northern spring and is reasonably symmetrical about the equator. Finally, in southern summer (Figure 3.18(d)), the pattern becomes almost the mirror image of northern summer, with the South Pole becoming the warmest site on the planet and with an intensely cold North Pole. There are some differences of detail between the two solstices, which arise partly because of the differing altitudes of the two hemispheres (the northern hemisphere is relatively low-lying and smooth, whilst the south is generally more mountainous having a higher mean altitude), and also because the ellipticity of Mars' orbit leads to the southern summer being shorter but more intense than its counterpart in the north.

3.5.2 GCM simulations of the seasonal zonal mean circulation

In this section we give a brief summary of seasonal change on Mars, by showing the GCM zonal-mean circulation, at three cardinal seasons: northern hemisphere spring equinox, northern summer solstice, and northern winter solstice. Each map has been formed from quantities time-averaged over 30° of areocentric longitude following equinox or solstice, rather than an equal amount of time (see Figure A.1), using a GCM run with the prescribed dust scenario illustrated in Section 2.5.2.3, which gives a reasonable overall match with MGS thermal observations in the lower atmosphere.

The zonal-mean quantities are produced by taking an average in longitude of model fields, which have been interpolated from model σ-surfaces onto surfaces of constant pressure. All the maps are plotted with a vertical log-pressure axis, with an approximate altitude guide axis on the right, labelled "log-pressure altitude". The approximate altitudes are plotted assuming a constant pressure scale height, $H = RT/g$, of 10 km. The approximation is useful in the lower and middle atmosphere, but the scale height becomes smaller at greater altitudes (and lower temperatures), so the values toward the top of the altitude axis are overestimates; the actual geometric height at the top of the region shown is close to 120 km, rather than nearly 140 km as indicated. Atmospheric fields are shown over almost the full current height range of the GCM, cut off just below the upper sponge region. This includes altitudes, particularly between about 40 km and 120 km, where there are few observations currently available with which to validate models.

Three basic fields are shown: zonal-mean temperatures, zonal-mean zonal winds (winds in the longitudinal direction), and mean meridional circulation streamfunction. The meridional streamfunction shows the mean atmospheric mass transport in the latitudinal and vertical directions, with flow parallel to the contours, proportional to the gradient of the streamfunction, and in an anticlockwise (clockwise) sense around positive (negative) regions. A non-linear contour interval has been adopted around the zero contour, in order to show better the full extent of the circulation, which is otherwise biased strongly toward the main cells in the lower atmosphere, which transport most mass.

Figure 3.19 (see colour section) shows the zonal-mean atmospheric state at northern hemisphere spring equinox. The northern hemisphere autumn equinox is not shown, but is similar in appearance, although the seasons are not totally symmetrical because of the ellipticity of Mars' orbit and the hemispherically-asymmetric topography, with a high southern hemisphere compared with the north.

It is worth noting that the equinoctial zonal-mean state (Figure 3.19, see colour section) is rather similar to the annual-mean state (Figure 3.7), although the circulation is only in this pattern for a relatively short part of the Martian year, during the transition from the summer solsticial state to the winter, or vice versa. The temperatures are highest near the surface and close to the equator, with a pattern of westerly jets (eastward wind, or positive u) in the wind field in the mid-latitudes of both hemispheres. These jets are related to the equator-to-pole temperature gradients through the thermal windshear relations:

$$2\Omega \sin\phi \frac{\partial u}{\partial z} \sim -\frac{g}{T}\frac{\partial T}{\partial y} \tag{3.21}$$

$$2\Omega \sin\phi \frac{\partial v}{\partial z} \sim \frac{g}{T}\frac{\partial T}{\partial x} \tag{3.22}$$

with terms as defined earlier. The thermal windshear equations are derived by taking geostrophic balance between the largest Coriolis and pressure gradient terms of Eqs (2.5) and (2.6), the hydrostatic equation (Eq. 2.7) and the perfect gas law (Eq. 2.3), and apply on rapidly rotating planets, except near the equator, where geostrophic balance breaks down (e.g., Andrews, 2001; Holton, 1992).

It is worth noting that the thermal windshear equations do not imply a causal relationship between temperature and wind gradients, rather a consistent balance between the two. For the zonal-mean zonal wind, $\bar{u}$ (Eq. 3.21) indicates that the wind should become more westerly with increasing height if the temperature falls from equator to pole, as is the case in the lower atmosphere. This applies in both hemispheres since $\partial T/\partial y$ is negative (positive) in the northern (southern) hemisphere where $\sin\phi$ is positive (negative), and is seen to hold in Figure 3.19 (see colour section). In the upper atmosphere, where the westerly jets decay (or 'close') with increasing height, the equator-to-pole temperature gradient is reversed.

The meridional circulation shows a pattern of rising motion near the equator with a large Hadley cell in each hemisphere and some evidence of smaller, counter-rotating Ferrell cells at higher latitudes. The Hadley cells are of unequal strength, but this is sensitive to the averaging period chosen and is also true on the Earth. Unlike the Earth, the Hadley circulation on Mars is not constrained by a strong tropopause and extends to above 60 km altitude, although the bulk of the mass flow is in the lowest one or two scale heights, as might be expected.

The Martian equinoctial circulation shown so far is not totally dissimilar from the global circulation of the Earth's atmosphere. Mars, however, becomes much more strongly asymmetric between the hemispheres for much of the year, partly because of the lower thermal inertia of the surface compared with Earth in the absence of oceans. Figure 3.20 (see colour section) shows a set of zonal-mean fields for northern hemisphere summer solstice. The temperature plot shows that the warmest part of the planet moves to high northern latitudes, where horizontal temperature gradients are small. There is a strong horizontal temperature gradient from the equator to the winter pole in the southern hemisphere. This pattern is reflected in the zonal wind which now has a single, strong westerly jet in the winter hemisphere and weaker easterly flow in the summer hemisphere. There is also evidence of a strong easterly mean flow in the equatorial upper atmosphere; the causes of this are complex, but thermal tides and planetary waves both contribute (Lewis and Read, 2003), as discussed in the following chapters.

The mean meridional circulation is now dominated by a single Hadley Cell, spanning more than 90° latitude, which rises in the summer hemisphere and descends in the winter. Such a single cell pattern is well reproduced by all Mars GCMs, although the need to raise the model top above 60 km to capture the full circulation is clear here (Wilson, 1997). This circulation pattern, with descending motion in the winter hemisphere, with air being subject to adiabatic compression, leads to the warming high above the winter pole around 80 km altitude. There is further warming above 120 km altitude in that diagram, in thermal wind balance with the closing westerly jet, which may be a result of deceleration through gravity wave drag, possibly overestimated in this experiment.

Finally, fields for northern hemisphere winter solstice are shown in Figure 3.21 (see colour section). This is the season of most extreme contrasts between summer and winter hemispheres on Mars, largely because of the proximity of the planet to the Sun at this time, with perihelion currently occurring about 20° of areocentric longitude, or about 30 sols, prior to northern hemisphere winter solstice on Mars. For the same reason, southern hemisphere summer is shorter than northern hemisphere summer on Mars, as well as being more intense. Much of the discussion for the northern hemisphere summer case above applies equally here, but the horizontal temperature gradient, the winter westerly jet, the mean meridional circulation and the winter polar warming, are all amplified.

3.5.3 An asymmetric Held–Hou model

The profound changes to the form of the Hadley circulation are perhaps the most striking changes which take place during the changing seasons. Lindzen and Hou (1988) (see also James, 1994) have shown how the simple Held–Hou model of the Hadley circulation can be adapted to allow for solar heating which is not centred on the equator. Non-linearities in even this simple model lead to internal feedbacks which rapidly result in highly asymmetric circulations on either side of the equator, even for relatively small displacements of the subsolar latitude. Figure 3.22, for example, shows a sequence of solutions from a somewhat more sophisticated version of the Held–Hou model (computed for Earth parameters by Lindzen and

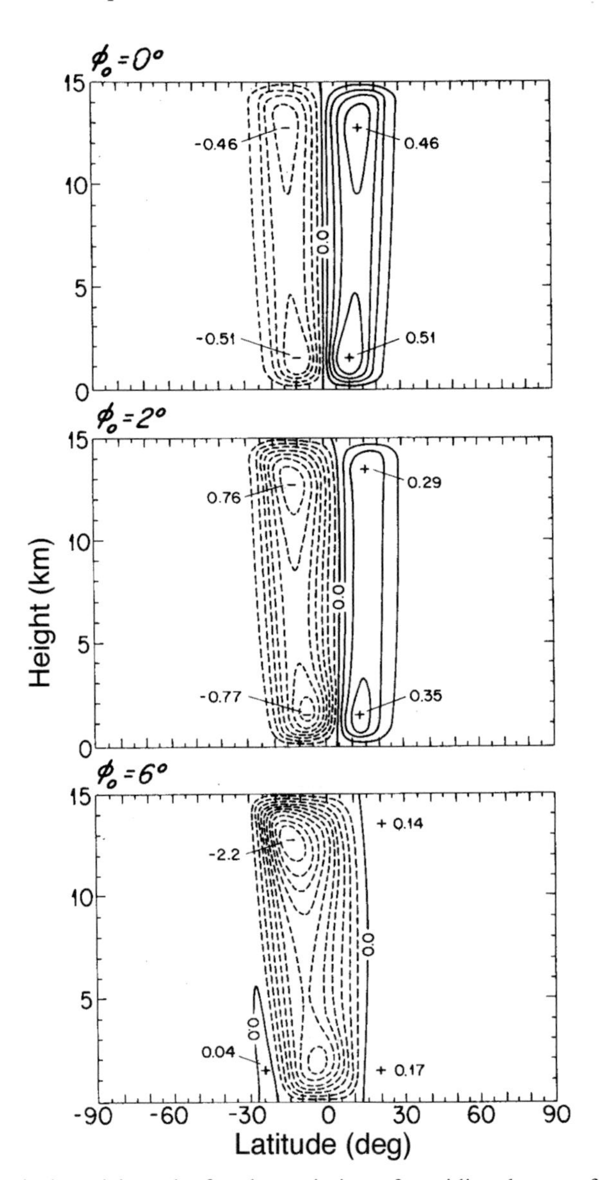

Figure 3.22. Numerical model results for the variation of meridional streamfunction Ψ with subsolar latitude ϕ_0, from the study by Lindzen and Hou (1988), using the numerical model of Held and Hou (1980).

Hou, 1988) which illustrates that, by moving the subsolar latitude to just 6° away from the equator, the initially symmetrically-disposed pair of Hadley cells on either side of the equator at $\phi = 0°$ rapidly gives way to a single, dominant cell which is much stronger than either of the cells at $\phi = 0°$. This results on Mars in the symmetrical equinoctial circulation being a relatively short-lived, transitional phase between the solsticial extremes.

The zonal wind field consistent with the circulations illustrated in Figure 3.22 are predominantly westerly, despite the transition to a single equator-crossing Hadley circulation outside the equinoxes. At more extreme displacements of the subsolar latitude, more typical of solstice conditions on Mars, however, the zonal flow in the 'summer' hemisphere becomes easterly, even in the simple axisymmetric model based on Held and Hou's original formulation (e.g., see Schneider, 1983).

3.6 SUMMARY

We see that many features of the atmospheric circulation of Mars can be understood without needing to take into account non-axisymmetric effects (e.g., due to surface topography, atmospheric waves, and instabilities). The basic low-latitude Hadley-type cells, driven by a varying imbalance with latitude of solar heating and infrared cooling, dominate the zonally-averaged circulation, and account for the atmospheric structure of potential temperature and zonal and meridional winds. The vertical structure of the atmosphere also reflects an adjustment to the vertical and horizontal distributions of net radiative heating and cooling toward increased local static stability. The Hadley circulations change their form in response to the seasonally-varying subsolar latitude, whilst approximately preserving geostrophic (thermal wind) balance throughout the middle atmosphere (away from the ground) and angular momentum conservation in the subtropics.

Even the condensation flow produced by the seasonal condensation and sublimation of the CO_2 ice caps leads to a flow (and surface pressure variation) which can be accounted for (at least qualitatively) by an axisymmetric model, though topography and orbital variations may be necessary to get some of the details right.

There are, however, a number of features which cannot be accounted for with axisymmetric models. These include the presence of easterly or (more especially) westerly flow over the equator itself, especially around equinox when axisymmetric models predict $u \simeq 0$. Such features arise primarily because of the effects of non-axisymmetric tides and travelling waves. Even more so, at high latitudes, axisymmetric models are under-constrained or downright wrong in their predictions for the circulation pattern, predicting (for example) unrealistically strong westerly winds at latitudes approaching the poles. These and many other features of the atmospheric circulation, especially concerning the transport of material tracers such as dust and water, need to take into account non-axisymmetric features of the circulation, most notably due to the substantial topography of Mars' underlying surface, and also to various large- and small-scale wave processes in the atmosphere. In the next few chapters, therefore, we extend our discussion to examine these non-axisymmetric features in turn, starting with an examination of the role of surface topography in the next chapter.

4

Topographical influences on the atmospheric circulation

Mars is far from being a smooth, flat, spherical planet, but exhibits variations in the height of the surface on all horizontal scales, from metre-scale boulders (which can just be seen, for example, in the highest resolution Mars Global Surveyor (MGS) Mars Orbiter Camera (MOC) images) to land masses the size of continents (comparable to that of the planet itself). Furthermore, the amplitude of these variations is considerable, often exceeding that of topographical variations found on Earth. Topographic variations are well known to exert a strong influence on both local- and global-scale circulations in the Earth's atmosphere, and it is no surprise, therefore, to find that the topography of Mars' surface has a substantial effect (at least as strong as on Earth) on the Martian atmosphere too. In addition to topographical variations, the thermal properties of the soil vary significantly from place to place on Mars (e.g., as measured by the surface *thermal inertia*). This can also lead to large-scale 'thermal continents', somewhat similar to (though less dramatically obvious than) the ocean–land contrasts on Earth.

Virtually every topographical influence on the atmospheric circulation found on the Earth seems to have a counterpart on Mars, and in this chapter we examine many of the most important phenomena in turn, starting with features on the largest scales (planetary waves, western boundary currents (WBCs) etc.) and going on to consider smaller-scale dynamics of features such as katabatic drainage flows and mountain waves. We begin, however, with a brief survey of the areography of Mars, as revealed in recent observations from the MGS mission.

4.1 THE TOPOGRAPHY OF MARS FROM MGS/MOLA

Until the MGS mission, the main sources of information on the topography of Mars and the shape of its surface came from a variety of rather indirect sources: ground-based images, which mainly reveal variations in surface reflectance or albedo which are not necessarily closely correlated with topography; space-based images from Mars orbit, which offered higher resolution, and determination of local topographic variations via stereo imagery; radio-occultation measurements (enabling location of the surface in inertial space at a few discrete points); and measurements of Mars' gravity field. These observations led to the derivation of various digital topographic maps (Wu, 1981; Smith and Zuber, 1996) which were used by various groups, including those designing numerical models of Mars' atmospheric circulation, prior to the arrival of MGS at Mars.

4.1.1 Mapping Mars from space

In practice, although these maps represented a great deal of detail on the Martian surface, it is now clear that they were significantly incomplete and subject to some serious systematic errors. From the viewpoint of the influence of topography on the atmosphere, the important attributes of the topography are not simply departures of Mars' planetary figure from a perfect sphere, but departures from a surface of constant gravitational potential (taking into account centripetal effects due to the planetary rotation). The determination of a suitable reference 'areopotential' surface (i.e., Mars' equivalent to 'sea level' on Earth) is therefore quite a complex process, involving not only measurements of the planetary shape or figure, but also detailed variations of the gravity field across the entire planet. Such measurements can only be obtained to high precision from a well characterized platform in fairly low orbit around the planet.

In the case of MGS, this has been achieved using a combination of instrumental techniques. The gravity field has been mapped in considerable detail (Zuber, 2000) by monitoring the instantaneous orbital trajectory of the spacecraft over many months and many orbits. As the spacecraft flies over a region where the subsurface mass density is greater and gravity is slightly stronger than the mean value (representative of the spherically symmetric potential due to the total mass of Mars), it experiences an acceleration which results in a small departure from the purely elliptical orbit expected in a symmetric inverse-square law potential. This acceleration can be measured to a great precision by the inertial navigation systems on the spacecraft (and from Doppler shifts in the frequency of the telemetry signal received on Earth), from which the 'gravity anomaly' can be determined and mapped. The resulting gravity map is shown in Figure 4.1 (see colour section).

The figure of Mars has been measured from MGS to extraordinary precision using a laser rangefinder or altimeter (the Mars Orbiter Laser Altimeter (MOLA) instrument), which can determine the figure to metre-scale accuracy with a 'footprint' on the surface of around 100 m in diameter (Smith *et al.*, 1998). By combining measurements over many orbits, the surface of Mars has now been covered systematically to produce global maps (at the time of writing) with 1/4° resolution or better in latitude and longitude (see Figure 4.2, colour section) and even higher resolution local maps and profiles of various features. These are already proving invaluable to planetary geologists and geophysicists, but are also now available to atmospheric scientists to help establish the role topography has in influencing the local and global meteorology and atmospheric circulation on Mars. So effective has been this campaign of measurements from MGS, that geologists now have maps of the Martian surface topography which are at least as accurate and detailed as are available (at least in the public domain) for the Earth!

4.1.2 Large-scale areography of Mars

There are a number of distinctive features of Mars' surface topography which are remarkable in their own right, and which might be expected to have a strong influence on the atmospheric circulation. The first and most striking aspect of Mars on the largest scale, as clearly evident in Figure 4.2 (see colour section), is the remarkable difference in the mean altitudes of the northern and southern hemispheres (the so-called *hemispheric dichotomy*; e.g., see Zuber, 2000 and references therein). The geophysical reasons for such a dichotomy are still being debated, and presumably relate to events and circumstances prevailing during the very early phases in the evolution of the planet, more than 3.5 Gyr ago. Possible explanations invoke mechanisms such as a planetary-scale, semi-permanent deep convection pattern in Mars' mantle, or the lasting effects of giant impact events in one hemisphere or the other (e.g., see Zuber, 2000 for recent reviews of these issues). The northern plains are comparatively smooth, with a relatively low density of impact craters (which have presumably been eroded away or the surface reprocessed?), whereas the southern uplands are generally significantly rougher with a much higher density of ancient impact features. Whatever the causal explanation of the hemispheric dichotomy may be, the resulting topography which places most of the southern hemisphere some 5 km higher on average than the north is bound to lead to discernible consequences for atmospheric circulation and meteorology.

One particular consequence, highlighted in recent studies using MGCMs (Joshi *et al.* 1995b; Richardson and Wilson 2002b; Takahashi *et al.* 2003b), is a pronounced asymmetry in the strengths of the Hadley circulation cells between the northern and southern hemispheres, even the annual time-mean. The topographic dichotomy evidently leads to the northern winter Hadley cell dominating the time-mean transport circulation in Mars' tropics, regardless of the orbital eccentricity or timing of perihelion within the seasonal cycle (Richardson and Wilson 2002b). Such an assymmetry is also evident in the Oxford MGCM climate (see Figure 3.7, colour section) and is likely to have a significant influence on the long-term transport of material tracers (such as dust and water) between the northern and southern hemispheres. The topographic dichotomy and associated large-scale gradients of topography may also influence the propagation of large-scale waves in the Martian atmosphere (see Section 4.2.2.1).

On a subplanetary scale, Mars has a number of continent-sized upland regions, centred in latitude not far from the equator (see Figure 4.2 (colour section) and the labelled map in Figure 4.3 (colour section)), such as the Tharsis Plateau, Arabia/Syrtis Major, and Elysium. These upland regions are largely of ancient volcanic origin, and several giant extinct volcanoes are clearly evident, especially within and around the Tharsis Plateau, including the gigantic Olympus Mons (rising some 27 km from the surrounding plains to a cone crowned by an immense sunken caldera), the three main Tharsis volcanoes (Arsia, Pavonis, and Ascraeus Montes), the ancient, substantially eroded Alba Patera to the north of Tharsis itself, and Elysium Mons in the far east of the planet. The region around the South Pole itself is also at relatively high altitude (>5 km), whilst the North Pole lies in the middle of a relatively low-lying plain, extending northwards of 40–50°N, which some have speculated could have been the site of a major ocean of liquid water during the early history of Mars (Smith *et al.*, 1998; see also Chapter 8). A remarkable aspect of these continent-scale uplands is their immense vertical scale, rising 6–10 km from the surrounding low-lying northern plains. Such an altitude is comparable with the atmospheric scale height itself (around 10 km), so that the atmosphere in some regions is almost enclosed by continental masses, much as the Earth's oceans are confined to oceanic basins by the terrestrial land masses. One might expect, therefore, to see some circulation features on Mars which emulate some of the effects of continental margins on the circulation of the terrestrial oceans.

On a similar scale, with diameters of 1,000–2,000 km, the southern hemisphere also exhibits two substantial near-circular depressions, the Hellas and Argyre Basins, both around 5-km deep and which are the remnants of massive impact events which took place during the early history of Mars (much like the Mare features on the Moon). Both basins have steep edges on the inside, with raised walls surrounding a relatively smooth plain. These are also on a scale which might have a substantial influence on winds and circulation, and are frequent sites for local and regional dust storms.

On a smaller scale, some regions of the surface are criss-crossed with major systems of canyons, valleys, and channels, some of which resemble immense river outflow channels leading into the northern plains, especially bordering the Chryse Planitia and Acidalia from the Tharsis Plateau. The largest canyon network is Valles Marineris, found to the east of Tharsis close to the equator, and is almost certainly an early tectonic fault-line (much like the Great Rift Valley on Earth). The valley networks and outflow channels, however, are the result of erosion, possibly produced by large releases of liquid water during Mars' early history. Topographic features are thus found on all scales, down to those of individual rocks and boulders (as evident, for example, in the images returned by the Mars Pathfinder (MPF) mission – see Figure 7.1).

The thermal properties of the surface also vary significantly from place to place on Mars, which can lead to the possibility of stationary patterns of atmospheric forcing by the resulting 'thermal continents' in a way which partly parallels that due to the ocean–continent thermal contrasts on Earth. Figure 4.4 (see colour section) shows a map of the surface thermal inertia on Mars, derived from infrared measurements of the surface response to the diurnal cycle of solar heating. This quantity therefore essentially measures the properties of the uppermost few cm of soil. Variations are found over a range of factors up to 5 or more across the planet, and largely reflect differing particle sizes comprising the soil itself. Thus, low thermal inertia regions are commonly found correlated with upland topography, indicative of a relatively rocky, boulder-strewn landscape with little sand. High thermal inertia regions, on the other hand, are more common in low-lying areas, and indicate a more finely divided, granular or sandy soil with plenty of air-filled spaces between the sand grains. The anticorrelation between topography and thermal inertia is far from perfect, however, leading to the possibility of a complicated distribution of mechanical and thermal forcing on the atmosphere.

4.2 STATIONARY PLANETARY WAVES

One of the most substantial effects of planetary-scale topography on the Earth is to force the presence of planetary waves which remain fixed relative to the underlying planet. Such waves are primarily forced by the large-scale wind passing over a large obstacle, such as a mountain range or landmass, and being constrained to rise and fall with the topographic contours. They propagate upwards and may eventually break at high altitude, transporting heat and momentum across the atmosphere. In the Earth's stratosphere, the effect of breaking planetary waves can lead to dramatic phenomena such as sudden stratospheric warmings (Andrews *et al.*, 1987), consequent disruption of the winter polar vortex, and mixing and erosion of the antarctic and arctic ozone holes.

In this section, we examine the forcing of planetary waves on Mars, their structure and evolution with season, and evidence from models and observations for their occurrence and influence on the atmospheric circulation.

4.2.1 Simple models of wave excitation and propagation

The simplest theoretical model which exhibits the response of a rotating, stratified atmosphere to topographic forcing is based on the linearized quasigeostrophic potential vorticity equation (e.g., Andrews, 2001) for an atmosphere of total depth H flowing zonally above a rigid lower boundary of height $h(x)$ (shown schematically in Figure 4.5).

$$\left(\frac{\partial}{\partial t} + U\frac{\partial}{\partial x}\right)q' + \beta\frac{\partial\psi'}{\partial x} = 0 \tag{4.1}$$

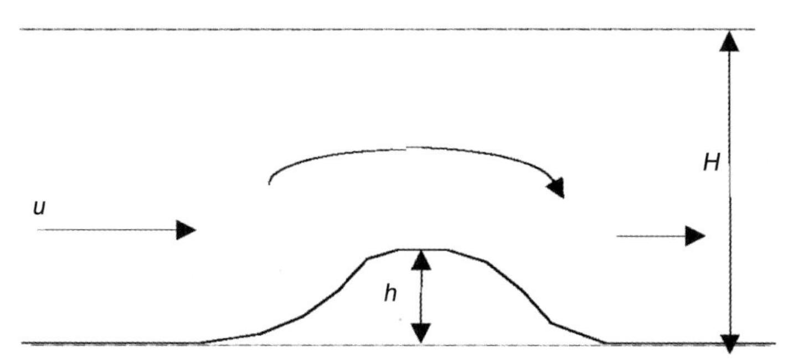

Figure 4.5. Schematic formulation of stratified flow over a pattern of topography $h(x)$ to excite upward-propagating planetary waves.

where U is the mean zonal wind (assumed constant); ψ is the streamfunction for the horizontal flow; β is the planetary vorticity gradient; primes denote departures from zonal mean quantities; and

$$q' = \nabla^2\psi' + f^2 \frac{\partial}{\partial z}\left(\frac{1}{N^2}\frac{\partial \psi'}{\partial z}\right) \tag{4.2}$$

where $N^2(z)$ is the square of the Brunt–Väisäla or buoyancy frequency (assumed to vary only with height z) and f is the Coriolis parameter. Eqs (4.1) and (4.2) apply throughout the atmosphere, and are subject to the boundary condition:

$$\left(\frac{\partial}{\partial t} + U\frac{\partial}{\partial x}\right)q' + \beta\frac{\partial \psi'}{\partial x} = -\frac{fU}{H}\frac{\partial h}{\partial x} \tag{4.3}$$

at $z = h(x)$, where the right-hand side represents the uplift forcing as the air is forced to move vertically in order to follow the topography. Provided that $h(x)/H$ is no larger than the Rossby number (assumed to be small), we can apply this boundary condition as if it were at $z = 0$.

We seek wave-like solutions to Eqs (4.1)–(4.2) of the form:

$$\psi' = A \exp\left[i(kx + ly + mz - \omega t)\right] \tag{4.4}$$

(so that $q' = -(k^2 + l^2 + f^2m^2/N^2)\psi' = -K^2\psi'$, where K is a 'total' wavenumber) which satisfy (4.3) at $z = 0$. For illustration we take $h(x)$ to be a simple sinusoid $h = h_0 \exp\left[i(k_0x + l_0y)\right]$. A steady solution can be found (with $\omega = 0$) of the form:

$$q' = -(k_0^2 + l_0^2 + f^2m^2/N^2)\psi' = -K_0^2\psi' \tag{4.5}$$

from which we can obtain an expression for the amplitude of the streamfunction:

$$A = \frac{fh_0}{H}\frac{1}{K_0^2 - \beta/U} \tag{4.6}$$

This implies a resonant response whenever $K_0 = (\beta/U)^{1/2}$. The corresponding dispersion relation, obtained from substituting our wave solution (Eqs (4.4) and (4.5)) into Eq. (4.1), becomes:

$$\omega = 0 = Uk_0 - \frac{\beta k_0}{K_0^2} \tag{4.7}$$

from which we obtain:

$$m^2 = \frac{N^2}{f^2}\left[\frac{\beta}{U} - (k_0^2 + l_0^2)\right] \tag{4.8}$$

From the condition that m^2 must be > 0 for vertically propagating waves, we obtain the result often known as the Charney–Drazin condition, that stationary planetary waves will only propagate into the upper atmosphere if $0 < U < U_c = \beta/(k_0^2 + l_0^2)$, or for a given U, $k_0 \leq (\beta/U)^{1/2}$.

Thus, propagating planetary waves are only observed if the prevailing zonal flow is eastward (to allow westward-propagating Rossby waves to become stationary in the frame of reference of the underlying planet) and not too strong, and then only for relatively long horizontal wavelengths ($k_0 \leq$ some limit). This means that even when resonant planetary waves are forced, the atmosphere effectively acts as a 'low-pass filter'.

4.2.2 More realistic models

The development of more realistic models of stationary waves on Mars has largely paralleled the equivalent developments for understanding similar features in the Earth's atmospheric circulation. Blumsack (1971) carried out an early study using an analytical model, similar to the simple model described above at mid-latitudes on Mars, to show that regions of high topography would be expected to be correlated with relatively cool temperatures, and vice versa. Later in the 1970s and early 1980s, several studies (e.g., Mass and Sagan, 1976; Webster, 1977; Gadian, 1978; Moriyama and Iwashima, 1980) made use of simplified numerical models to solve for the structure of stationary waves, forced at the surface. Most of these early models used just two layers in the vertical to represent the vertical structure of the waves, and often solved linearized forms of the governing equations (quasigeostrophic or primitive equations). However, they were sufficient to indicate that both mechanical forcing (due to direct uplift over topography) and thermal variations (e.g., due to variations in thermal inertia of the surface) could lead to the generation of large-amplitude stationary waves, at least at certain seasons of the Martian year. Winter-time circulations (either represented heuristically in the variation of zonal wind with latitude (Mass and Sagan, 1976; Webster, 1977) or by simple sinusoidal jets (Gadian, 1978)) showed a strong response (due, for example, to the Tharsis and Noachis Terra Plateaux), with an intensification and either a barotropic response or weak westward tilt with height.

The most detailed simplified mechanistic models of stationary waves on Mars (i.e., short of being full GCMs) solve for steady responses in linearized forms of the governing equations over the full sphere (e.g., Hollingsworth and Barnes, 1996; Nayvelt *et al.*, 1997). Such models make use of generalizations of Eqs (4.1–4.8) above (or their primitive equation equivalents) to take account of spatial variations of $u(\phi, z)$, $f(\phi)$, and vertical stratification. In this case, the dispersion relation for stationary planetary waves in Eq. (4.8) becomes (e.g., see Andrews *et al.*, 1987):

$$m^2 = \frac{N^2}{f^2}\left[\frac{\bar{q}_y}{u(\phi, z)} - \left(\frac{s}{a\cos\phi}\right)^2 - l^2 - \frac{f^2}{4N^2H_p^2}\right] \tag{4.9}$$

$$= \frac{N^2}{4\Omega^2 a^2}\bar{Q}_s\phi, z \tag{4.10}$$

where $\bar{q}_y$ is the latitudinal gradient of zonal mean quasigeostrophic potential vorticity (a generalisation of β to take account of the contribution of the zonal mean flow structure and stratification to the planetary potential vorticity gradient); s is the scaled zonal wavenumber $s = ka\cos\phi$; ϕ is latitude; and H_p is the pressure scale height ($= RT/g$). $\bar{Q}_s$ is often referred to as the quasigeostrophic 'refractive index' (Palmer, 1981, 1982), by analogy with its optical counterpart, so that propagation or evanescence is determined by the sign of $\bar{Q}_s$. From this equation, it is clear that the vertical propagation of stationary waves depends on a number of factors, including the structure of the zonal mean flow (represented by $\bar{q}_y/\overline{u(\phi,z)}$), the stratification (explicitly in $f^2/(4N^2H_p^2)$ and implicitly in $\bar{q}_y/\overline{u(\phi,z)}$), and the zonal and meridional wavenumber of the response. Thus, an equivalent result to the Charney–Drazin condition mentioned above again leads to the expectation that waves forced at the surface will not reach high altitudes unless the zonal flow is westerly or the wavenumbers of the waves are sufficiently small.

4.2.2.1 Forcing at the surface

Since the work of Webster (1977), it has been appreciated that the generation of stationary planetary waves at the surface of Mars can occur through two distinct processes, namely, through mechanical uplift and local heating. Moreover, these may operate in ways which may differ significantly from the equivalent processes responsible for producing large-scale stationary waves in the Earth's atmosphere.

The simplest process common to both Mars and the Earth is straightforward mechanical uplift as near-surface winds move air over mountain ridges and valleys. This is typically represented in linearized models (such as those of Hollingsworth and Barnes, 1996 and Nayvelt *et al.*, 1997) by the boundary condition on vertical velocity w':

$$w' = \bar{\mathbf{u}} \cdot \nabla h' + \frac{v'}{a} \frac{\partial \bar{h}}{\partial \phi} \tag{4.11}$$

Note that this formulation includes a term in the meridional gradient of the zonal mean topographic height $\partial \bar{h}/\partial \phi$, which is usually neglected in comparable models of the terrestrial atmosphere. This is because the zonal mean variations in $\bar{h}$ are much larger on Mars than on the Earth. Figure 4.6 shows a profile with latitude of the zonal mean topography on Mars, taken from the MGS MOLA topographic map, in which it is clearly seen that the height of the surface, even when averaged around an entire latitude circle, varies by several km (i.e., a significant fraction of the atmospheric scale height). This is especially the case at high northern latitudes, where there appears to be a substantial valley extending all around the North Pole. This would be expected to lead to a topographic β-effect southward of around 70°N which opposes the basic vorticity gradient due to the spherical curvature of the planet. This may also be significant in affecting the zonal propagation speeds of any Rossby wave-like disturbance in the northern hemisphere.

The first term in Eq. (4.11) reflects the forcing due to zonally-varying topography. In light of the simple linear model outlined in Section 4.2.1, it is useful to visualize the structure of Mars' topography in the form of a Fourier spectrum of zonal harmonics. Figure 4.7 shows a map of the zonal harmonics of Mars' topography as they vary with latitude. From this, it is clear that at low latitudes, the topography is dominated by wavenumber $m = 2$, reflecting the spacing of the Tharsis and Arabia Terra Plateaux near the equator. At higher northern latitudes, the topo-

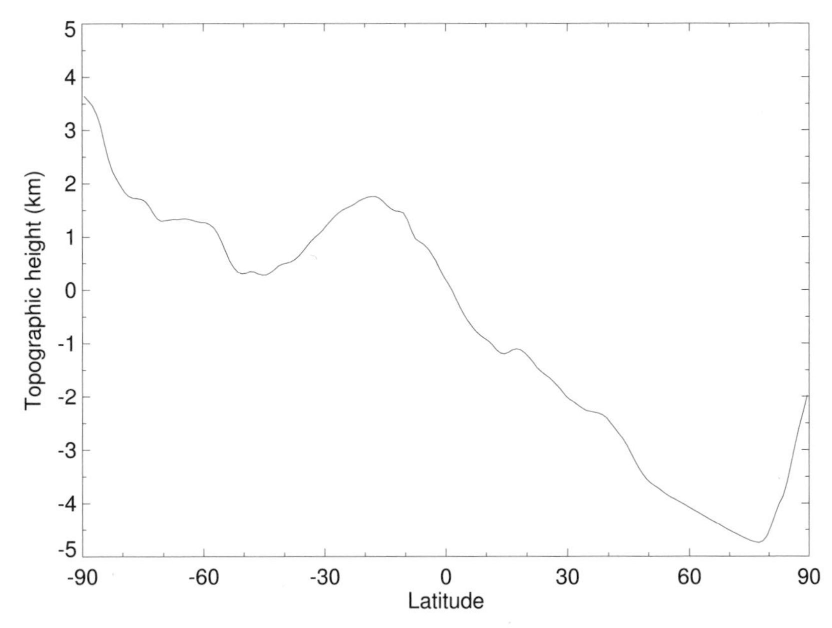

Figure 4.6. Meridional profile of zonally averaged topographic height on Mars, taken from the Mars Global Surveyor (MGS) Mars Orbiter Laser Altimeter (MOLA) dataset and smoothed to a typical model resolution of approximately 3.5° in latitude.

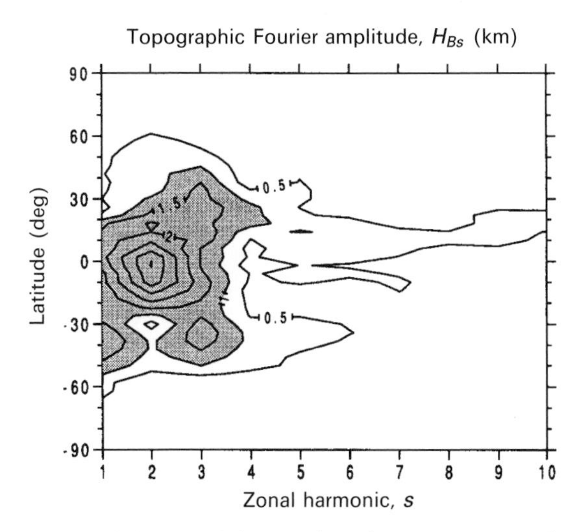

Figure 4.7. Contour map of the zonal harmonics of Mars' topography, derived from the Consortium topography dataset by Hollingsworth and Barnes (1996).

graphy is dominated by $m = 3$, including the effect of the elevated Elysium region around 30°N.

In the south, however, the topography is more complicated, and appears to comprise a mixture of $m = 1$ and $m = 3$ around 40°S. This would indicate the wavenumbers most likely to appear at high altitudes, provided forcing were due primarily to mechanical effects, and that these wavenumbers can propagate according to the Charney–Drazin criterion.

The other main mechanism for exciting large-scale waves on Mars is through local heating. This operates in a way which is quite different from that on Earth. Owing to the strong thermal damping and low thermal inertia of the Martian atmosphere, the ground temperature is determined almost entirely by a local radiative balance between direct solar heating and emission of infrared radiation directly to space, and the altitude of the surface exerts little influence on its equilibrium temperature. This is in complete contrast to the Earth, where the air temperature exerts a strong influence on the ground temperature, which accordingly varies considerably depending on the altitude of the surface. Thus, a mountain on Mars will not only act as a mechanical barrier to the flow, but may also act as a local heat source or sink. This can be seen from the form of the representation of the thermal forcing term in linearized models (e.g., see Nayvelt *et al.*, 1997)

$$S = -\frac{1}{\tau} h' \frac{\partial \bar{T}}{\partial z} \tag{4.12}$$

where τ is a radiative heating timescale. The situation is further complicated on Mars by the frequent presence of a temperature inversion in $\bar{T}$ at low levels, so that $\partial T / \partial z > 0$ in many places near the ground. This tends to make elevated regions act as a heat sink at low levels, while depressions act as a source. If $\partial \bar{T} / \partial z > 0$, then a positive elevation will lead to a negative heating rate S, and vice versa.

This can be seen in Figure 4.8, taken from the paper by Nayvelt *et al.* (1997), which shows the heating rate as a function of position both at higher altitudes ($\sim$2 hPa) (Figure 4.8(a)) and near the surface (Figure 4.8(b)), where the sign of S reverses the correlation with topography. The latter occurs because the heating effect of topography is not confined to the lower boundary, but extends upwards into the free atmosphere. This is because the heating effect arises through radiative heating of the atmosphere via infrared emission from the surface. The effect of such radiative heating leads to the radiative equilibrium temperature of the atmosphere depending approximately just upon the distance from the surface (i.e., on p/p_s – see Goody and Belton (1967), Gierasch and Goody (1968)). A more detailed discussion of these arguments in the context of linear models can be found in Appendix A of Nayvelt *et al.* (1997).

4.2.2.2 *Atmospheric response in linearized models*

In recent work, Hollingsworth and Barnes (1996) considered the atmospheric response to realistic mechanical forcing in the presence of mean zonal flows representative of both northern and southern winter conditions, and also included effects

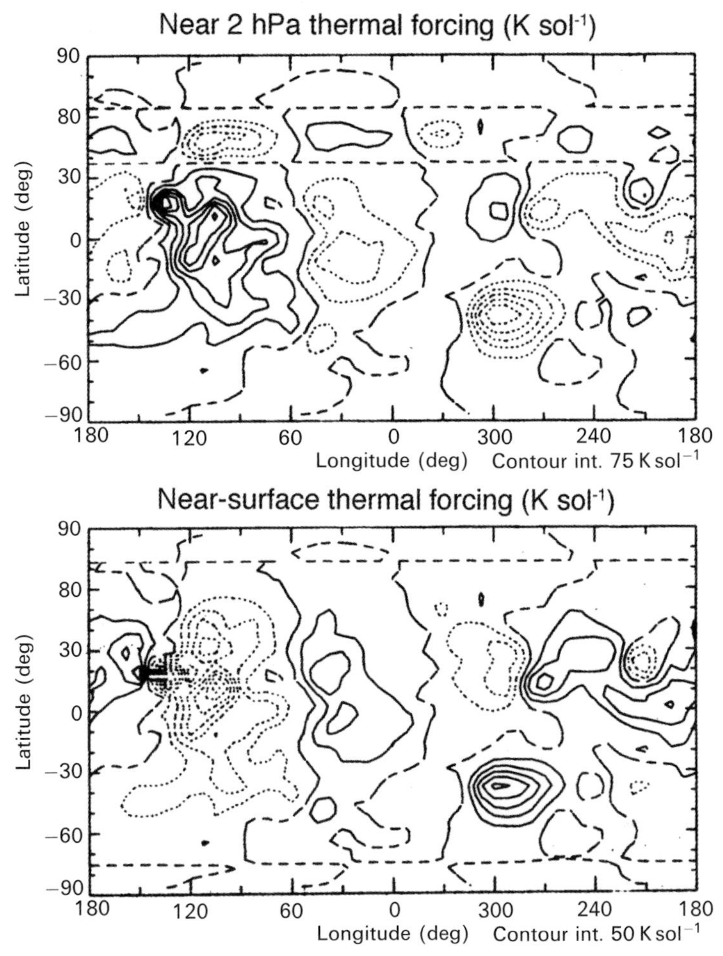

Figure 4.8. Radiative heating rate S (a) at an altitude of approximately 2 hPa, and (b) near the surface, in the atmosphere of Mars, as computed by Nayvelt *et al.* (1997).

of both small and large global dust loadings (derived from corresponding simulations using the NASA Ames Mars GCM). Their radiative forcing/dissipation representation, however, was somewhat simpler than given above in Eq. (4.12), and simply included a radiative damping term. In contrast, Nayvelt *et al.* (1997) used a more complete combination of mechanical and radiative forcing (including Eq. 4.12), but only considered realistic responses under northern winter conditions. Both studies, however, also included investigations using idealized topography to determine the role of various factors which may have influenced their solutions.

From both these studies, it was clear that, at mid- to high-latitudes on Mars (as on the Earth), the mechanical forcing due to direct uplift was the dominant mechanism for generating large-scale stationary waves. Moreover, the atmospheric response in each case was broadly compatible with terrestrial experience, in that the resulting waves at high altitudes were consistent with the spectrum of waves in the distribution of the topography itself subject to the Charney–Drazin criterion as reflected in Eq. (4.10). Thus, as shown, for example, in Figure 4.9, the strongest wave activity occurs in the respective winter hemisphere, and the dominant wavenumber reflects the lowest dominant wavenumber in the topography at the latitude where the refractive index $\bar{Q}_s$ is non-negative. The overall response, however, tends to be more widespread across the planet than would be suggested simply by the regions of positive $\bar{Q}_s$. Wavenumber $m = 2$ dominates at northern latitudes, whereas in southern winter, $m = 1$ is strong around 50–60°S. The forced waves are approximately barotropic in structure, and their zonal phase (in the geopotential field Φ) tends to be correlated with the topography itself with peaks of Φ lying near to elevated topographic features, and vice versa. The correlation is clearer in northern winter than in the south, however, where the deep Hellas Basin evidently leads to a more complex response.

At lower latitudes, however, it was evident that the thermal forcing given by Eq. (4.10) was more important, and became comparable with the mechanical forcing. This evidently leads to a response which contrasts significantly with what happens in the Earth's atmosphere, where latent heating associated with the effects of moist convection mainly over the tropical oceans dominates the forcing at low latitudes. Also, as mentioned above, owing to the temperature inversion in $\bar{T}$ on Mars near the surface, the sign of the thermal forcing at low latitudes is opposite to the equivalent term on the Earth (though this effect has generally been neglected in most studies of topographically-forced waves in the Earth's atmospheric circulation using linearized models). The resulting response is significantly baroclinic, especially in the lowest levels of the atmosphere, with near-surface flow which is southwesterly in the northern hemisphere and with anticyclonic curvature. The response at higher levels is dominated by $m = 3$ and $m = 2$ close to the equator, though with a phase shift in the zonal direction so that peaks in Φ no longer align with regions of high elevation. As discussed by Nayvelt *et al.* (1997), this is a consequence of the rapid thermal response of the atmosphere to local heating, which causes the temperature of the air to adjust quickly to changes in heating as it passes over topographic features. The change in temperature leads to a rapid adjustment of the hydrostatic pressure in this region, which in turn modifies the flow.

4.2.2.3 *Stationary planetary waves in GCMs*

The assumptions made in the linearized mechanistic models discussed above may not be valid at all times on Mars, particularly concerning the neglect of non-linear effects, and also the more detailed seasonal variations in the atmospheric circulation during the Martian year. The possible role of non-linear effects has been highlighted in several studies of stationary waves in the Earth's atmosphere (Chen and Trenberth, 1988; Valdes and Hoskins, 1991; Cook and Held, 1992), in which it is suggested that a non-linear response may become significant if the topographic slope exceeds that of the zonal mean isentropes or potential temperature surfaces. Such a

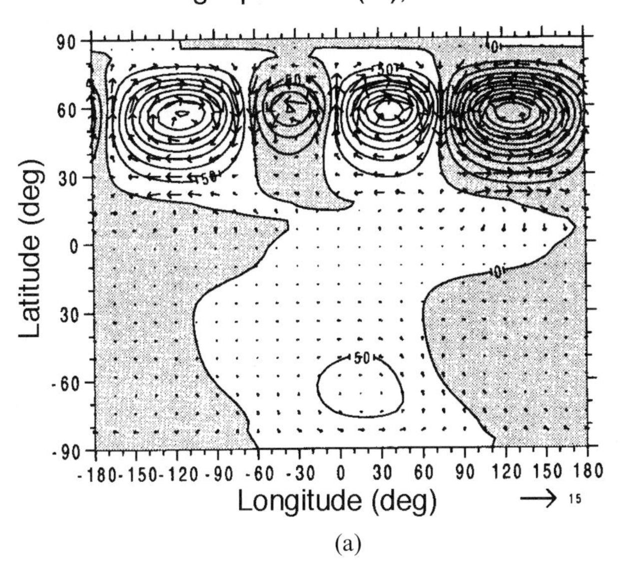

(a)

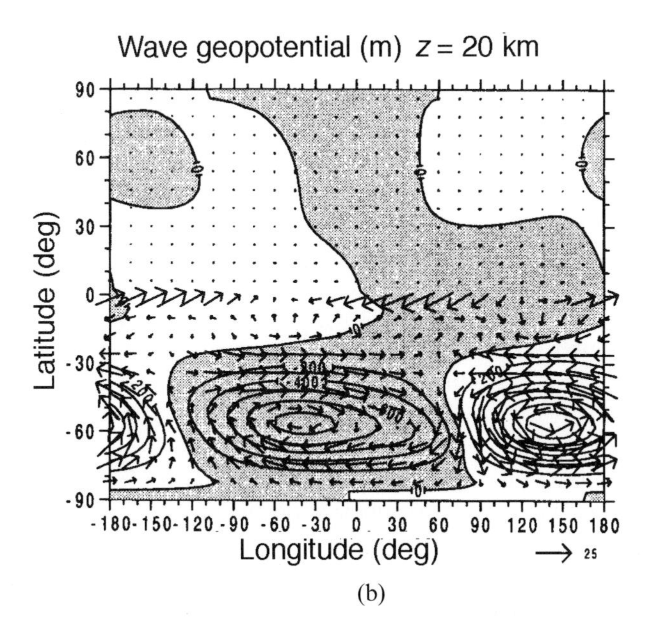

(b)

Figure 4.9. Contour maps of stationary wave activity for (a) northern hemisphere winter and (b) southern hemisphere winter.

From Hollingsworth and Barnes (1996).

situation is found on Earth typically for mountains greater than about 2 km in height, and might well occur frequently on Mars, where the topographic slopes can be considerable. It is preferable, therefore, to check the results of such linearized studies against yet more complex models which do not make linearized approximations. In this respect, comprehensive GCMs of the Martian atmosphere have an important role to play.

An example of the occurence of large-scale stationary waves in a Mars GCM is illustrated in Figure 4.10 (see colour section), in which we show a series of maps of meridional wind at an approximate altitude of 21 km above the mean areoid ($p \sim 1$ hPa), and averaged over an interval of 30° in L_S for each of the 12 'seasons' or 'months' for a typical Martian year. From this sequence, the development of different responses to the surface forcing can be seen as the climatological circulation changes during the seasonal cycle. Thus, we see the strongest response in the northern hemisphere occurring during the peak of northern winter ($L_S \sim 270°$), with a strong $m = 2$ pattern located around 50–80°N. This pattern evidently disappears in northern spring, to be replaced in the southern hemisphere by a strong response in high southern latitudes during southern autumn and winter. The form of this response changes during the onset of autumn and winter, however, developing first a strong $m = 1$ pattern around the equinox period at high polar latitudes, which then evolves to a moderately strong $m = 2 + 1$ pattern around 60°S by the middle of southern winter.

The patterns found in their respective winter seasons appear to be quite well matched to the predictions from the linearized models discussed in the previous subsection, indicating that (at least at these levels) the linear assumptions are reasonably valid. Both the dominant wavenumber and zonal phase seem to agree reasonably well with the linear models, at least at some point in the seasonal cycle, despite differences of detail between the GCM mean zonal state and that assumed in the linear models. This lends confidence in the broad conclusions of those studies, at least for the seasons considered, though the detailed comparison of linear and fully nonlinear simulations of stationary waves remains an area of active current research and a number of open questions remain to be studied in detail.

The strong $m = 1$ response at very high southern latitudes during both equinoxes is an interesting observation which bears further investigation. This may well reflect the occurrence of a phenomenon similar to the stratospheric sudden warming (SSW) on the Earth (e.g., Andrews *et al.*, 1987), in which upward-propagating, stationary planetary waves may reach a large amplitude aloft and 'break' like surf on a beach (e.g., see McIntyre and Palmer, 1983; Andrews *et al.*, 1987). This may result in the disruption of a cold polar vortex, displacing it away from the pole and leading to strong local warming over the pole and a pattern which looks like a strong $m = 1$ response. Such a phenomenon is an important feature of the terrestrial stratosphere around equinox periods, and heralds a major seasonal change in the atmospheric circulation at those levels. More detailed GCM studies for Mars (e.g., Barnes and Hollingsworth, 1987; Wilson, 1997) have indicated that a similar phenomenon may occur on Mars, leading to effects akin to those appearing around the equinoxes in Figure 4.10 (see colour section). This strongly non-linear phenomenon is clearly not a feature of studies using linearized models, and needs the complexity of a full atmospheric GCM to be realized in model simulations.

This would seem to be an example of a more common occurrence at high altitudes on Mars, since GCM simulations extending to very high altitudes ($z > 60$ km) indicate that the strong winter jets may actually be strongly dissipated at high altitudes due (at least in part) to the action of breaking planetary waves. Such effects lead to the decay with height and closing off of the winter jet at high altitude (e.g., see Figures 4.7(b) and 4.19), which is an important feature of the circulation at those altitudes.

4.2.3 Evidence for stationary planetary wave activity

Although model studies suggest that planetary-scale stationary waves may be a strong feature of the Martian atmospheric circulation at various times, direct observations of their occurrence and structure have proved remarkably elusive until quite recently. Conrath (1981) found evidence from Mariner 9 InfraRed Interferometer Spectrometer (IRIS) thermal observations of planetary-scale wave activity in the Martian atmosphere around 60°N during late northern winter. However, because of the sparse sampling of the atmosphere at the time from Mariner 9, which was in a highly elliptical orbit of relatively long period, the identification of the wavenumber and phase speed of the waves was ambiguous. One possible interpretation was of a stationary $m = 2$ wave, but the possibility that there might also be travelling baroclinic waves present (see Chapter 6) could not be ruled out.

More recently, Banfield *et al.* (1996) have analysed aspects of Martian weather in the 15 μm thermal measurements from the Viking orbiter InfraRed Thermal Mapper (IRTM) instrument, which provides information on thermal structures around an altitude of 15 km or so. Though not the primary focus of their study, they identified some stationary planetary-scale structures in this dataset at mid-latitudes in both hemispheres around $L_S \sim 0°$, which they interpreted as being possible stationary planetary waves with wavenumbers $m = 1$ and 2. Some uncertainties remain concerning this interpretation, however, owing to the recent identification of a bias in the original IRTM data caused by surface radiance (Wilson and Richardson, 2000).

Nayvelt *et al.* (1997) made some ingenious use of systematic observations of surface wind streaks (e.g., in the lee of local craters and mountains), which can provide some indication of the prevailing direction of the wind on seasonal timescales. By careful decomposition of the wind fields from their models, Nayvelt *et al.* (1997) showed that stationary wave structures could improve agreement between model predictions and the observed direction of wind streaks in Viking images (Thomas *et al.*, 1984), at least in some locations close to major topographic features.

Some of the most direct observations of stationary planetary waves on Mars have so far come from thermal mapping and radio-occultation measurements of temperature from the MGS mission (Banfield *et al.*, 2000; Hinson *et al.*, 1999;

Hinson *et al.*, 2001). Published results to date refer to observations from the so-called 'science phasing' period of the mission following orbital insertion up to early 1999, during which the orbit of the spacecraft was adjusted into its present sun-synchronous configuration. Initial results from the MGS Thermal Emission Spectrometer (TES) enabled the thermal signature of stationary waves to be mapped directly from Mars' orbit, and led to the identification of $m = 1$ thermal structure in the southern hemisphere. A potential difficulty with these mapping observations, however, was a potential ambiguity which could confuse the identification of a stationary wave with the signature of a thermal tide, which moves westwards with the Sun (see Chapter 5). Such an issue led to similar confusion over the claimed detection of stationary waves at thermospheric altitudes during the aerobraking phase of MGS, during which the possible identification of a stationary $m = 2$ wave was reported from early measurements of atmospheric density structure as the MGS spacecraft flew through the upper atmosphere of Mars (Keating *et al.*, 1998). This was later reinterpreted (Forbes and Hagan, 2000; Joshi *et al.*, 2000) as being due to the way the observations sampled the thermal tide (for operational reasons the spacecraft pass through the atmosphere tended to occur at the same local time on Mars), rather than to a topographically-forced stationary wave, the observation of which at such high altitudes would have been in clear contradiction to the predictions of various models of stationary wave behaviour.

Hinson *et al.* (1999) and Hinson *et al.* (2001) combined a series of radio-occultation measurements of thermal structure in Mars' lower atmosphere around $L_S \sim 75°$ and 66°N with GCM simulations to identify the presence of significant planetary wave amplitude in wavenumbers $m = 1$ and 2. The waves were broadly consistent with those found in the GCM simulations for this season, though (as anticipated from the Charney–Drazin criterion) were largely confined in altitude to levels below 20 km ($p > 1$ hPa).

4.3 LOW-LEVEL JETS AND WESTERN BOUNDARY CURRENTS (WBCs)

Because of the very large amplitude of the topographic variations on Mars, which exceeds the pressure scale height ($H_p \sim 10$ km) in places, not only are large-scale atmospheric waves likely to be generated, but other dynamical effects more akin to those found adjacent to the ocean margins on Earth, might be expected to occur commonly on Mars.

4.3.1 WBCs on Earth

4.3.1.1 Ocean currents

In the Earth's oceans, large-scale currents and gyres are generated primarily by the action of the wind on the ocean surface. In a steady state, this leads to a situation where the horizontal flow is in geostrophic balance almost everywhere, and in which the poleward advection of 'planetary vorticity' is mainly balanced by the generation of vorticity via the so-called 'stretching term' $f\partial w/\partial z$. If the latter is integrated over the entire depth of the ocean, the stretching term is dominated by the effects of the Ekman layers driven by the surface wind stress, leading to:

$$\beta v = f\frac{w_E}{H} \tag{4.13}$$

where H is the depth of the ocean layer. From the theory of Ekman layers (e.g., see Gill (1982), this leads to:

$$\beta v = \frac{1}{\rho}\mathbf{j} \times \nabla \times \boldsymbol{\tau}_w \tag{4.14}$$

where $\boldsymbol{\tau}_w$ is the surface stress due to the winds, and $\mathbf{j}$ is the northward unit vector. The balance represented by Eq. (4.14) is known as the Sverdrup balance (e.g., Gill, 1982), in which the northward mass transport in the ocean is balanced by the curl of the surface wind stress. Note, however, that this balance cannot hold in the presence of a boundary at the western (eastward-facing) edge of an ocean basin. In this case a steady solution cannot exist, but instead a boundary layer forms. In the absence of friction, this boundary layer will get thinner and thinner with time, though the presence of friction (e.g., either in the form of a linear drag or diffusive viscosity) can lead to an alternative steady state with a western boundary layer of finite thickness. This boundary layer leads in practice to a strong current which flows along the coastline of the western boundary of large-scale ocean basins, such as the Gulf Stream along the east coast of North America, the Kuroshio Current along the east coast of south-east Asia, or the Agulhas Current along the east coast of southern Africa.

4.3.1.2 The East African Jet

Such a confinement of north–south flow to the vicinity of a western boundary is not restricted to the Earth's oceans, but may also be observed (in modified form) in the atmosphere. A clear and spectacular example of such a phenomenon was discovered just over 30 years ago close to the coast of East Africa, roughly between northern Madagascar and Saudi Arabia. The land along this stretch of coastline is typically quite mountainous, rising to over 3000 m in places. Moreover, the wind pattern in this region is strongly affected by the seasonal monsoon circulation over southern Asia, with easterly flow across India in winter (January–March) and westerly in summer (July–September) (and vice versa in the southern hemisphere). Much of the air involved in this monsoonal circulation ends up crossing the equator, which happens at low levels in a concentrated jet close to the East African highlands at an altitude of around 2000 m.

This jet was not predicted theoretically, but was discovered empirically by an operational meteorologist, John Findlater (Findlater, 1969, 1977), working for the UK Meteorological Office in the 1950s and 1960s, who noticed consistent reports from pilots flying aircraft in this region of unexpectedly strong north–south winds at levels of less than 10,000 ft. This lead to a systematic mapping of the structure of the

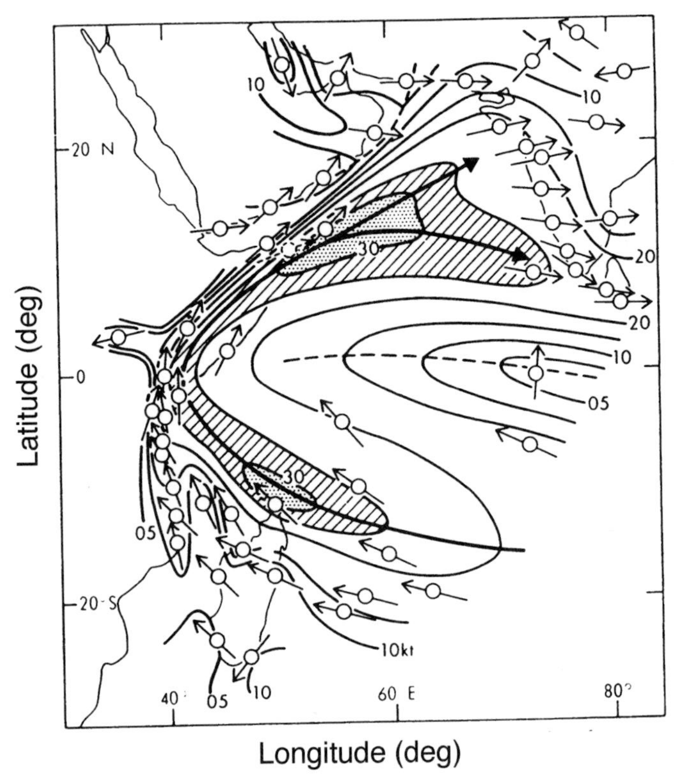

Figure 4.11. Mean monthly airflow at an altitude of 1 km in July, showing the cross-equatorial flow concentrated into a narrow boundary layer of width $\sim 5^\circ$ in longitude. Bold arrows indicate the major streamlines and axis of maximum flow, and contours represent isotachs at intervals of 5 kt. The dashed line indicates the axis of minimum wind speed.

From Findlater (1977).

low-level East African Jet, which demonstrated its link with the Asian monsoon (see Figure 4.11), and showed its concentration near the equator into a tight jet 'leaning against' the East African highlands (see Figure 4.12).

Subsequent work (e.g., Anderson, 1976) showed that the essential dynamical balance in this jet is similar to that in a western boundary current, where the East African mountains play a key role in enabling air to flow freely across the equator, and hence maintain the structure and intensity of the Asian monsoon.

4.3.2 A simple model of an equator-crossing WBC

The previous subsection has indicated the form of western boundary currents as they are found to occur on the Earth, especially with reference to the East African Jet. In the context of Mars, it is interesting to note that, under solstice conditions (as we have seen in Chapter 3), the basic zonal flow is easterly in one hemisphere and

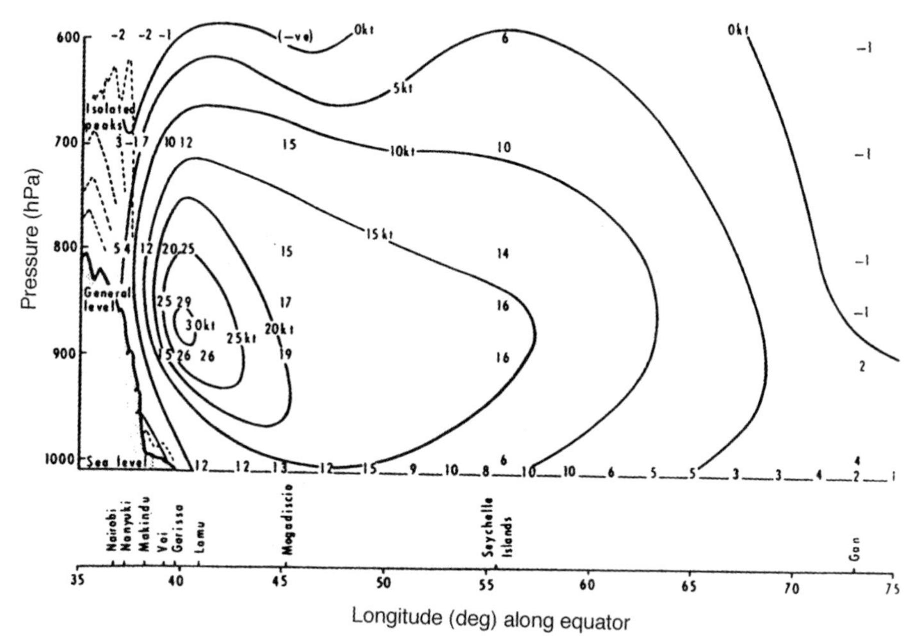

Figure 4.12. Longitude–height sections along the equator in the vicinity of the East African low-level Jet, showing the concentration of the flow close to the East African mountains in both (a) January and (b) July.

From Findlater (1969).

westerly in the other, much as during the monsoon periods when the East African Jet is most prominent on Earth. Also, Mars has continent-scale topographic features close to the equator, with steep, east-facing slopes aligned roughly north–south, again rather like on Earth. It is likely, therefore, that similar western boundary current structures might occur on Mars in association with the large topographic slopes on the flank of the Tharsis and Syrtis Major Plateaux. Before we examine the evidence for this circulation, however, it is useful to develop a simple theoretical model to explore the typical structure and dynamics of equator-crossing WBCs.

Some of the essential features of an equator-crossing WBC can be clearly seen in a remarkably simple theoretical model, if we make some simplifying assumptions about the flow and the shape of the boundaries. For simplicity, we suppose there to be a high mountain ridge, aligned north–south and completely blocking any east–west flow (see Figure 4.13). This is assumed to lie across the equator and, again for simplicity, we use a cartesian coordinate system with the x-direction towards the east and the y-direction towards the north. The ridge is then assumed to extend from $y = -L/2$ to $y = +L/2$.

The background large-scale circulation is taken to represent schematically northern summer solstice conditions close to the surface, so that the basic zonal flow varies from westerly at northern mid-latitudes to easterly in southern

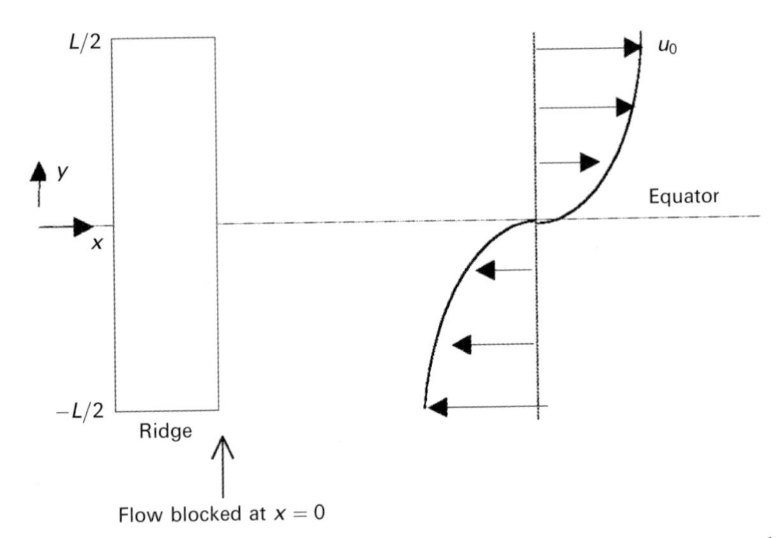

Figure 4.13. Schematic layout of the simple model of an equator-crossing western boundary current (WBC). A high ridge is aligned N–S at $x = 0$ and extends between $y = \pm L/2$. The background circulation is assumed to tend towards a basic zonal flow u_0, with eastward flow in the northern hemisphere and westward in the south – see text for further details.

mid-latitudes.* Accordingly, we set the zonal flow to tend to a simple sinusoidal form $u_0(y) = U_0 \sin \pi y/L$, where $y = 0$ corresponds to the equator.

In his simple model of the East African Jet, Anderson (1976) looked for solutions to the shallow water equations on an equatorial β-plane – an idealization of the region close to the equator as a plane tangential to the equator in which the Coriolis parameter, f, is assumed to vary linearly with y (i.e., $f = \beta y$). Here we simplify the problem even further and consider the vertically-integrated, linearized, barotropic, non-divergent flow, also on an equatorial β-plane. In a steady state, the x and y equations of motion then become:

$$-\beta y v = -\frac{\partial \Phi}{\partial x} - r(u - u_0) \tag{4.15}$$

$$\beta y u = -\frac{\partial \Phi}{\partial y} - r v \tag{4.16}$$

together with the continuity equation for incompressible flow:

$$\frac{\partial u}{\partial x} + \frac{\partial v}{\partial y} = 0 \tag{4.17}$$

* Note that, in the real Martian atmosphere, the flow varies strongly with height, and actually changes direction at higher altitude (e.g., see Figures 4.16 and 4.17). The present model is intended as a simplified idealization to illustrate the effects of the topography on horizontal flow.

where r represents the effects of friction (and forcing) by a linear relaxation towards the large-scale, purely zonal flow $u_0(y)$ and Φ is the geopotential function (representing hydrostatic pressure gradients in the horizontal). By taking the curl of Eqs (4.15) and (4.16) and making use of Eq. (4.17) we obtain a vorticity equation:

$$\beta v = -r\left[\frac{\partial v}{\partial x} - \frac{\partial u}{\partial y}\right] - r\frac{du_0(y)}{dy} \tag{4.18}$$

or, in terms of the horizontal streamfunction ψ and rearranging the terms:

$$\beta v + r\nabla^2\psi = -r\frac{du_0(y)}{dy} \tag{4.19}$$

Using the form of u_0 given above, Eq. (4.19) becomes:

$$\beta v + r\nabla^2\psi = -\frac{r\pi U_0}{L}\cos\frac{\pi y}{L} \tag{4.20}$$

This has the general (separable) solution:

$$\psi = \left[A_1 \exp(ax) + A_2 \exp(bx) + \frac{U_0 L}{\pi}\right]\cos\frac{\pi y}{L} \tag{4.21}$$

where a and b are the two roots of a quadratic equation and are equal to $-\beta/2r \pm [\beta^2/4r^2 - \pi^2/L^2]^{1/2}$.

The boundary conditions on ψ in x are $\psi = 0$ on $x = 0$ to represent the blocking effect of the topographic ridge, and ψ must remain bounded as $x \to \infty$. In the y direction, we assume the zonal flow must match onto the unblocked zonal flow at $|y| \geq L/2$, which implies that $\partial\psi/\partial y = U_0$ at $y = L/2$ and $\partial\psi/\partial y = -U_0$ at $y = -L/2$.

Finally, we can simplify the solution yet further by taking the low friction limit $r \to 0$, so we can neglect the exponent a relative to b and take $b \to -\beta/r$. Hence, we arrive at the simple solution:

$$\psi = \frac{U_0 L}{\pi}[1 - \exp(-\beta x/r)]\cos\frac{\pi y}{L} \tag{4.22}$$

This form of the streamfunction captures many features of the vertically-integrated structure of an equator-crossing western boundary current, as may be seen in Figures 4.14 and 4.15(a) and (b).

Figure 4.14 shows a cross-section on the 'equator' ($y = 0$) through the solution given by Eq. (4.22), which shows how the gradient of the streamfunction in x is concentrated into a narrow region (of thickness $\sim r/\beta$) close to the western boundary. This is further illustrated in Figure 4.15, which shows contour maps of ψ for two different values of r. Note that the smaller we make r, the thinner the boundary layer becomes. This is something of an idealization, since there are clear physical limits on how small r can become. In reality, for small enough r the linear assumption of the model will become invalid and/or some other form of friction will become important, and change the character (and complexity) of the solution considerably. However, this simple model is useful for getting some insight into

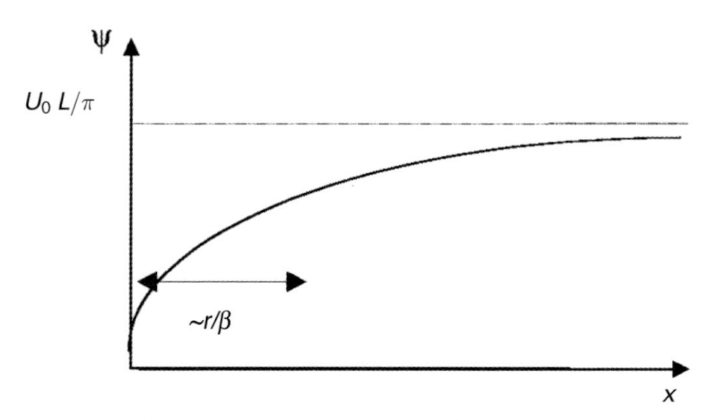

Figure 4.14. Cross-section at $y = 0$ through the solution for ψ represented by Eq. (4.22).

why and how a western boundary current might be expected to form across the equator.

4.3.3 WBCs – more realistic models

The idealized model in the previous subsection gives some indication of the basic mechanisms which lead to the concentration of north–south flow close to the eastward-facing flanks of major topographic features. The details of this effect on Mars can be seen more clearly, however, in numerical models which include an accurate representation of the major topographic features on Mars, and also represent more realistically the effects of 'friction' in the atmospheric boundary layer. Figure 4.16, for example, shows a cross-section of north–south wind along the equator from one of the earliest attempts to investigate the formation of WBCs in a Mars GCM (Joshi *et al.*, 1994). This represents the mean north–south flow across the equator under conditions typical of the northern summer solstice, and two runs are shown (a) at fairly low horizontal resolution (approximately 6°) and (b) at much higher resolution (approximately 3° in longitude). Both simulations clearly show the northwards flow at low levels to be strongly concentrated into low-level jets pressed up against the eastward-facing flanks of the Tharsis and Syrtis Major Plateaux, in ways which clearly resemble the observed cross-sections of the East African Jet on Earth (cf Figure 4.12). In later work, this effect has been clearly confirmed in much more sophisticated GCMs, such the NASA Ames (Joshi *et al.* 1997), Geophysical Fluid Dynamics Laboratory (GFDL) (Wilson and Hamilton, 1996), and Euromars models (Forget *et al.*, 1999). Note that for a frictional timescale of around 1 sol, representing the gross effects of boundary layer friction, the width of the low-level jet in longitude is around 5°, which compares very well with the expectation from the simple model of the previous subsection that jets will have a width $\sim r/\beta$.

When the full seasonal and diurnal cycles are represented in these comprehensive

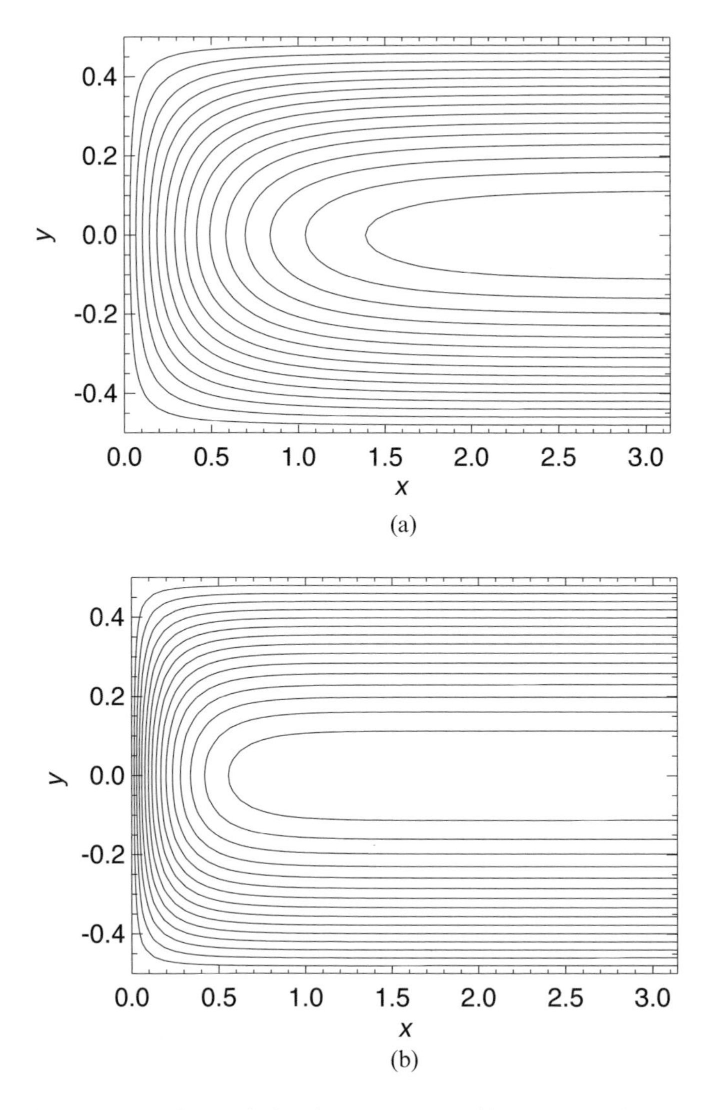

Figure 4.15. Contour map of the solution for Ψ represented by Eq. (4.22) for (a) $\tau = 1$ sol, and (b) $\tau = 5$ sol.

GCMs, the change in direction of the WBCs is clearly seen as the circulation changes from northern winter to northern summer (see Figure 4.17). But the same pattern remains, confirming that the solsticial Hadley circulation across the equator is strongly controlled by the large-scale topography. The low-level jets virtually disappear as the circulation approaches an equinoctial pattern, though the topography always exerts a strong modifying effect on the flow. Diurnal variations also modify the form and strength of the WBCs, though do not disrupt them altogether.

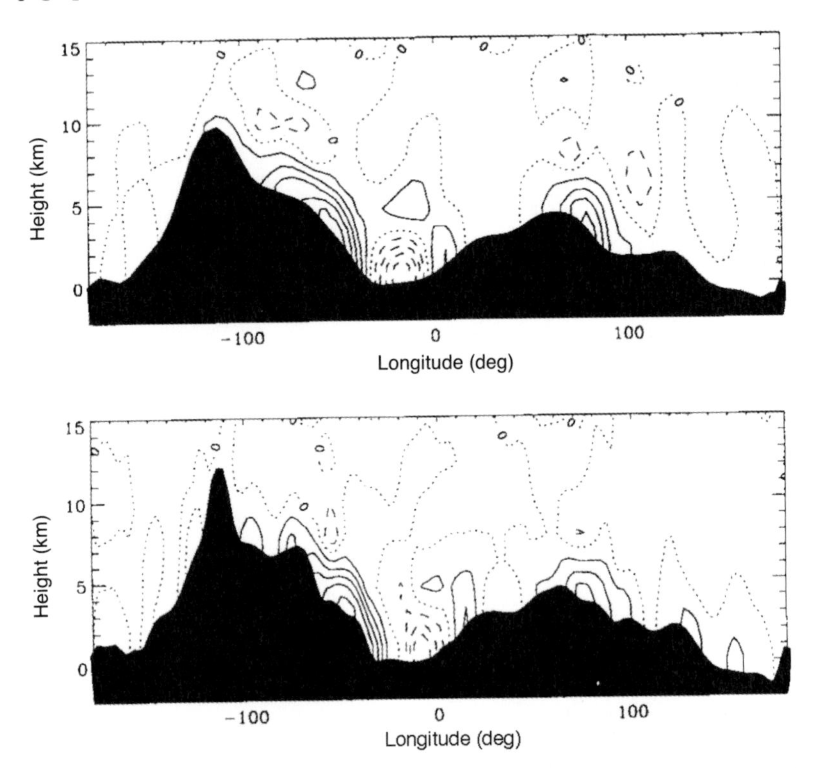

Figure 4.16. Cross-section in longitude and height of the time-averaged north–south wind v across the equator under northern summer solstice conditions, from simplified GCM simulations by Joshi *et al.* (1994). The topography at this latitude is represented by the black filled region: (a) shows a simulation at T21 resolution (approximately 6° in longitude) while (b) shows the equivalent flow at T42 resolution (approximately 3°).

Indeed, during southern summer the diurnal temperature variations lead to strong low-level slope winds (e.g., Savijärvi and Siili, 1993 – another topographically controlled phenomenon, though discussed in more detail under diurnal effects in Chapter 5), which can reinforce the strength of the winds in the WBC during the day. This can lead to some locally very strong winds in some regions of the southern hemisphere during the summer, which may have a strong effect on the lifting of dust and initiation of dust storms at this time (see Chapter 7).

The strong flow across the equator may also lead to the cross-hemispheric transport of other tracers, notably potential vorticity q (Joshi *et al.*, 1997). The equator is a rather special place for a dynamical variable such as q, since this is the latitude at which it normally changes sign (so northern hemisphere q is positive and southern hemisphere q is negative). As q is forced to cross the equator as a conserved tracer in the strong flow in the WBCs, however, it may find itself at a latitude where it has the opposite sign to its surroundings. This situation is usually

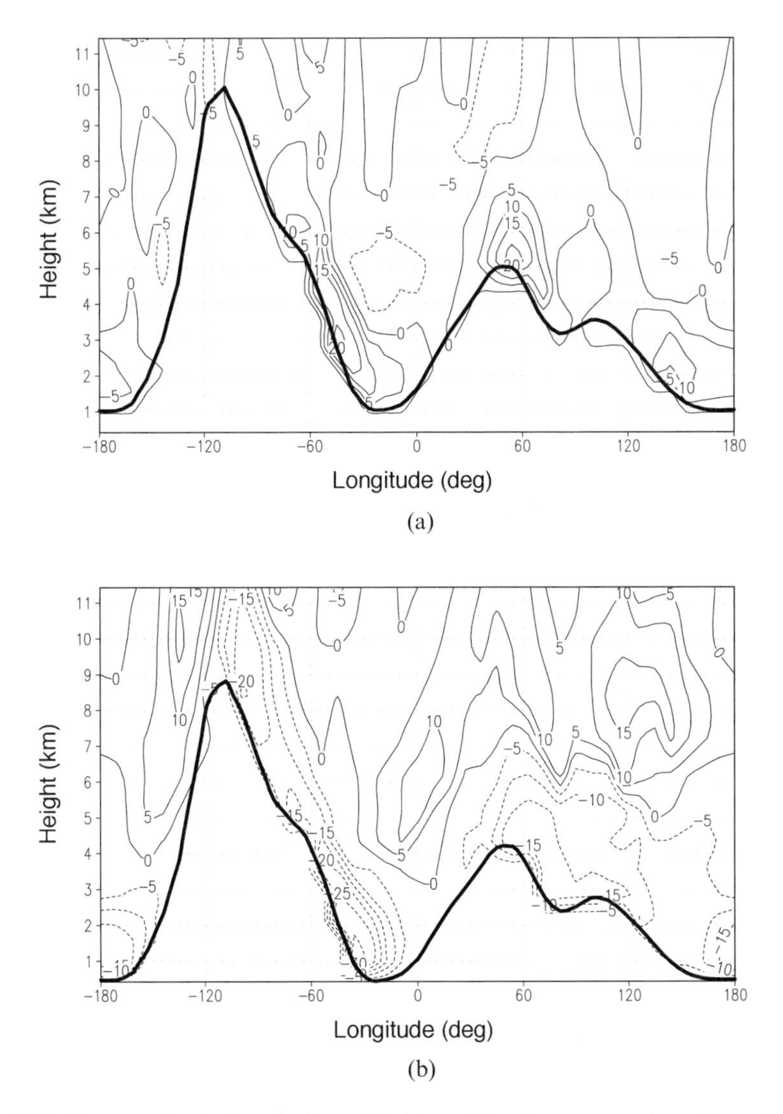

Figure 4.17. Cross-section in longitude and height of the time-averaged north–south wind v across the equator under (a) the northern summer solstice conditions, and (b) the northern winter conditions, from the comprehensive NASA Ames GCM simulations by Joshi *et al.* (1997).

unstable to inertial instabilities (e.g., see Andrews *et al.*, 1987), and might be expected to lead to the break up of the low-level jet. In practice, however, frictional effects near the ground appear to be sufficiently strong on Mars (at least in the most realistic GCMs) to suppress these instabilities, and so the low-level WBCs persist as strong features of the cross-hemispheric circulation.

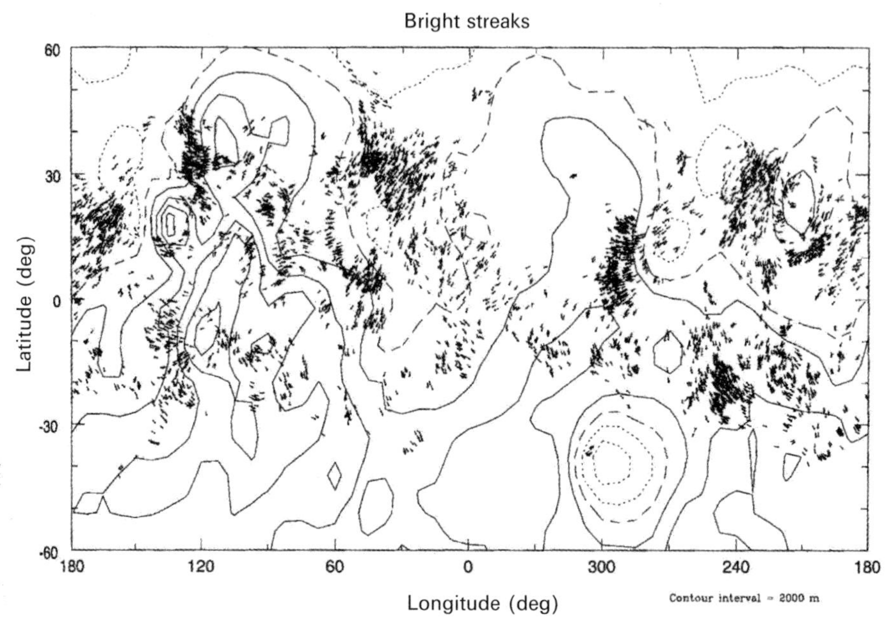

Figure 4.18. Map in longitude and height of the prevailing wind direction obtained from bright streaks on Mars (data from Peter Thomas – see Nayvelt *et al.*, 1997), superposed upon a low-resolution map of Martian topography.

4.3.4 Observations of WBCs

Although models suggest that these topographically-controlled low-level WBC jets are a major feature of the Martian atmospheric circulation, especially during the summer and winter seasons, the direct observational evidence for them on Mars is quite hard to come by. Because they are concentrated near mountainous regions at low levels, they cannot be detected easily from remote-sensing observations from space, and the terrain in which they occur most strongly is too mountainous for lander spacecraft to have explored, at least as yet.

The main sources of indirect evidence are discussed by Joshi *et al.* (1995), who emphasize the patterns seen in maps of dust streak alignments (e.g., see Figure 4.18). Dust streaks occur as a result of persistent winds blowing over local obstacles such as mountains or craters, and can either lead to systematic erosion or deposition of dust in the lee of these features (depending on the details of the surface and local flow patterns). The presence of these streaks can provide some indication of the prevailing direction of the wind during the preceding season, but not its strength – hence the information can be quite ambiguous or even misleading. Such maps do, however, suggest the presence of persistent north–south flow along the eastward-facing flanks of the Tharsis and Syrtis Major Plateaux, as suggested by the models, though the strength of such flows is unknown from these maps.

Joshi *et al.* (1995) also note that the Viking Lander (VL) 1 spacecraft landed in a relatively flat region around 28°N, but in which the prevailing winds may have been affected by the outflow from the Tharsis low-level WBC jet. This flow may have produced a bias in the prevailing wind at low levels, which could account for some systematic differences between the winds observed by VL 1 and those predicted by a 1-D boundary layer model (Haberle *et al.*, 1993), though this could also be due to other, more local effects which were not taken into account in that model. It would seem, therefore, that detailed confirmation of the existence of these low-level WBC jets will have to await further observations from new missions, perhaps including landers and rovers which can venture directly into mountainous terrain.

4.4 SMALL-SCALE GRAVITY WAVES

The flow over mountains, especially when the thermal stratification is statically stable, may lead to the excitation of other phenomena, typically on a much smaller scale than the planetary Rossby waves discussed above. In particular, so-called gravity waves may be generated by individual mountain peaks or obstacles on scales of a few tens or hundreds of km. Such waves occur frequently on Earth, and are responsible for the formation of stationary wave-like cloud features (such as lee wave clouds). In extreme cases these waves can lead to violent windstorms on the lee side of mountains, and strong turbulence which can be hazardous to aircraft flying close to large mountain ranges (e.g., see Smith, 1979).

Gravity waves may also propagate upwards to high altitude in the atmosphere before breaking and dissipating. At these altitudes, this can lead on the Earth to 'clear air turbulence' (CAT) experienced by aircraft, giving a bumpy ride for passengers. The momentum carried by such waves can also exert a strong influence on the large-scale atmospheric circulation at high altitudes, leading to a significant effective drag force acting on the atmosphere. It is increasingly recognized in terrestrial weather prediction and climate models that the effects of topographically-forced gravity waves must be taken into account in computing the momentum budget of the atmosphere, even if the waves themselves are not resolved directly by the model (e.g., see Palmer *et al.*, 1986).

Given that Mars has both a stably-stratified atmosphere and very large amplitude topography on all scales, it is hardly surprising that small-scale gravity waves are generated in the vicinity of mountains and craters. Figure 4.19, for example, shows trains of lee wave clouds from some mountain features on Mars, as seen from the vantage point of an orbiting spacecraft. It is natural, therefore, to think about what role such small-scale gravity waves might have in the Martian atmospheric circulation.

4.4.1 A simple model for internal gravity waves generated by mountains

Although the dynamics of gravity waves in a real atmosphere such as on Earth or Mars may be quite complicated, because of the effects of non-linearities and 3-D

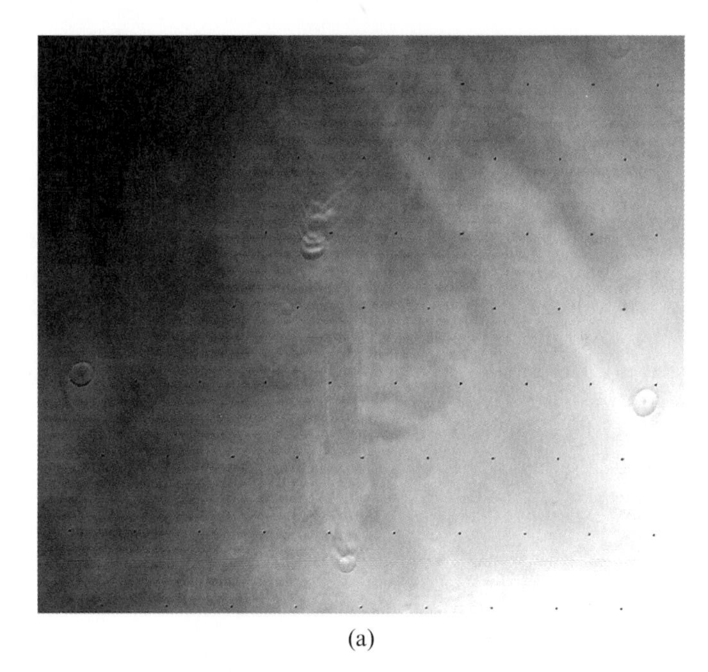

(a)

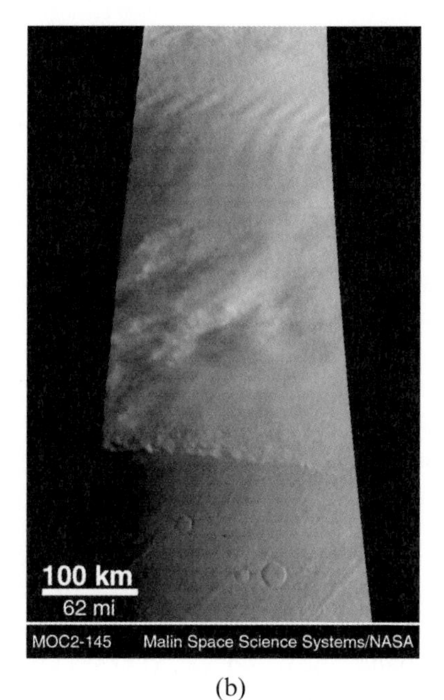

(b)

Figure 4.19. Two examples of trains of lee waves generated at low levels on Mars by flow over mountains and craters, as seen (a) by the Viking orbiter camera, and (b) by the Mars Orbiter Camera (MOC) on Mars Global Surveyor (MGS).

From NASA/JPL/Malin Space Science System.

complexities due to the irregular shape of the mountains and other topographic features which launch them, the basic principles are relatively simple and can be understood using a straightforward theoretical model. Gravity waves occur in a stably-stratified atmosphere and depend on buoyancy forces to restore vertically displaced parcels back to their equilibrium level. They occur on the scale of individual mountains or groups of mountains (10–1,000-km wavelength), and the oscillation periods involved are typically much shorter than a day (from a few minutes–1 sol), and so our simple theory can neglect the effect of planetary rotation and spherical curvature from the outset. It is convenient to start from the linearized, hydrostatic primitive Eqs in 2-D (the horizontal x-direction and the vertical z-direction), neglecting planetary rotation, and assuming an atmosphere at rest with uniform stratification (represented by the buoyancy frequency $N =$ constant), (e.g., see Andrews *et al.*, 1987, chapters 3–4):

$$\frac{\partial u'}{\partial t} + \frac{\partial \Phi'}{\partial x} = 0 \tag{4.23}$$

$$\frac{\partial u'}{\partial x} + \frac{1}{\rho}\frac{\partial \rho w'}{\partial z} = 0 \tag{4.24}$$

$$\frac{\partial^2 \Phi'}{\partial z \partial t} + N^2 w' = 0 \tag{4.25}$$

where Φ is the hydrostatic geopotential variable and primes ($'$) indicate perturbation quantities (assumed to be of small amplitude). From these equations, u' and Φ' can be eliminated to obtain a single equation in w':

$$\frac{\partial^3}{\partial t^2 \partial z}\left[\frac{1}{\rho}\frac{\partial}{\partial z}(\rho w')\right] + N^2 \frac{\partial^2 w'}{\partial x^2} = 0 \tag{4.26}$$

If the density is assumed to vary as $\rho = \rho_0 \exp -z/H_p$, Eq. (4.26) is satisfied by a wave-like solution of the form:

$$w' = W_0 \exp z/2H_p \exp i(kx + mz - \omega t) \tag{4.27}$$

where W_0 is a constant; k and m are the horizontal and vertical wavenumbers; and ω is the wave frequency, satisfying the dispersion relation:

$$\omega = \pm \frac{Nk}{\left[m^2 + \frac{1}{4H_p^2}\right]^{1/2}} \tag{4.28}$$

The sign on the right-hand side of Eq. (4.28) determines the direction of propagation of the gravity wave.

Mountain-induced gravity waves are produced when air flows over the mountains, and a coherent response is obtained when the horizontal propagation speed of the waves relative to the ground is zero. We can represent this in the simple model given above by now assuming a constant flow (say) in the $+x$-direction of

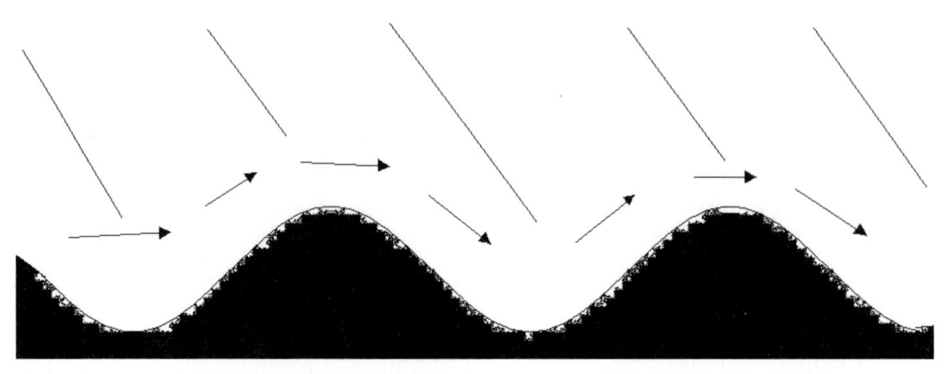

Figure 4.20. Schematic illustration of the excitation of gravity waves by a mountain range, given for simplicity by a sinusoidal variation in height with x.

speed U_s over a set of sinusoidal mountains of horizontal wavenumber k (see Figure 4.20).

If we now Doppler shift to a reference frame moving with the flow at $u = U_s$ (so the local flow relative to us is now zero) and look for solutions with a horizontal phase speed $\omega/k = -U_s$, this will give us the appropriate solution for gravity waves which are stationary relative to the mountains. Thus, we need the negative root of Eq. (4.28), which leads to an expression for m of the form:

$$m^2 = \frac{N^2}{U_s} + \frac{1}{4H_p^2} \tag{4.29}$$

Note that m^2 must be greater than zero for the waves to propagate in the vertical direction, which in turn means that $U_s^2 < 4N^2H_p^2$. This means that, in order to generate strong gravity waves behind mountains, we need both stable stratification and flow U_s near the surface which is not too strong, otherwise the waves will become evanescent and trapped close to the ground. We can also see that if the waves can propagate in the vertical, the vertical group velocity:

$$c_{gz} = \frac{\partial \omega}{\partial m} = +\frac{Nkm}{\left[m^2 + \dfrac{1}{4H_p^2}\right]^{3/2}} \tag{4.30}$$

is positive-definite, confirming that energy will be transported upwards by the waves under these conditions. This is important whenever strong gravity wave activity is generated near the ground, since these waves can then carry energy and momentum to high altitudes, until the waves break and dissipate. If the background flow at higher levels becomes sufficiently strong to render the waves evanescent, then the wave activity may become confined in the vertical and propagation is restricted to the horizontal direction. In this case the waves may become horizontally ducted and form intense lee wave structures, which may then be visualized by clouds forming in the upward-moving crests of the waves (cf Figure 4.19).

4.4.2 More realistic models of mountain lee waves

The simple model outlined above conveys the essence of gravity wave generation by mountain ridges, but is too simplistic to account in detail for the formation of localized lee and mountain waves. In practice, mountain waves are not typically generated by quasi-infinite linear ridges of mountains (assumed for convenience above to make the problem 2-D). Also, observed lee wave cloud features imply that gravity waves generated near the ground must become confined in the vertical into a horizontal duct, which suggests that the basic flow in which the waves propagate must be more complicated than the simple uniform flow with constant stratification assumed above.

Pirraglia (1976) made the first attempt to develop a more realistic model of atmospheric lee waves on Mars, based on the early theoretical treatment of Scorer (1956), Scorer (1956), and Scorer and Wilkinson (1956). Though linear, Pirraglia's model allows for some vertical structure in the atmospheric flow by using a two-layer formulation with different wind speeds (though in the same direction) in each layer. The model also considers waves excited by a single, isolated crater, so is effectively 3-D in form. The mathematical treatment is quite complicated, and will not be reproduced in detail here. But the essential features can be summarized as follows: (a) a background zonal velocity which is subcritical in the lower layer (so the waves generated at the surface can propagate freely in the vertical) but supercritical in the upper layer (so the upward propagating waves become evanescent, confining activity to a finite height close to that of the interface between the two layers); and (b) use of a 2-D Fourier expansion of gravity waves excited by an isolated crater source (represented in Fourier space as an ensemble of periodic ridges oriented at all angles $\pm\pi/2$ relative to the airstream). The resultant wave pattern effectively comprises an interference pattern of many waves, propagating horizontally at all angles relative to the airstream. Pirraglia (1976) computed maps of the lines of constant phase within such wave patterns, in order to illustrate the forms which would be expected, which resembled diverging 'ship-wake' patterns, rather like those observed in the lee wave cloud images in Figure 4.19.

This approach was developed further by Pickersgill and Hunt (1979), who used a more sophisticated approach than Pirraglia (1976), and allowed the static stability to vary with height as well as the wind speed, such that the so-called Scorer parameter l, defined by:

$$l^2 = \frac{gB}{U^2} - \frac{1}{U}\frac{d^2U}{dz^2} + \frac{1}{2}\frac{d\bar{R}}{dz} - \frac{\bar{R}^2}{4} \tag{4.31}$$

where B is the static stability parameter $B = (1/\theta)\, d\theta/dz$; $U(z)$ is the basic flow profile; and $\bar{R} = (1/\bar{\rho})d\bar{\rho}/dz$ (and where $\bar{\rho}(z)$ is the basic density profile), is constant with height in each layer. They also carried out a more complete analysis including the continuous spectrum of evanescent waves instead of the more restricted, discrete normal-mode spectrum assumed by Pirraglia (1976). Finally, in their numerical computations they calculated fields of vertical velocity and parcel displacement, to determine more clearly where clouds might be expected to form

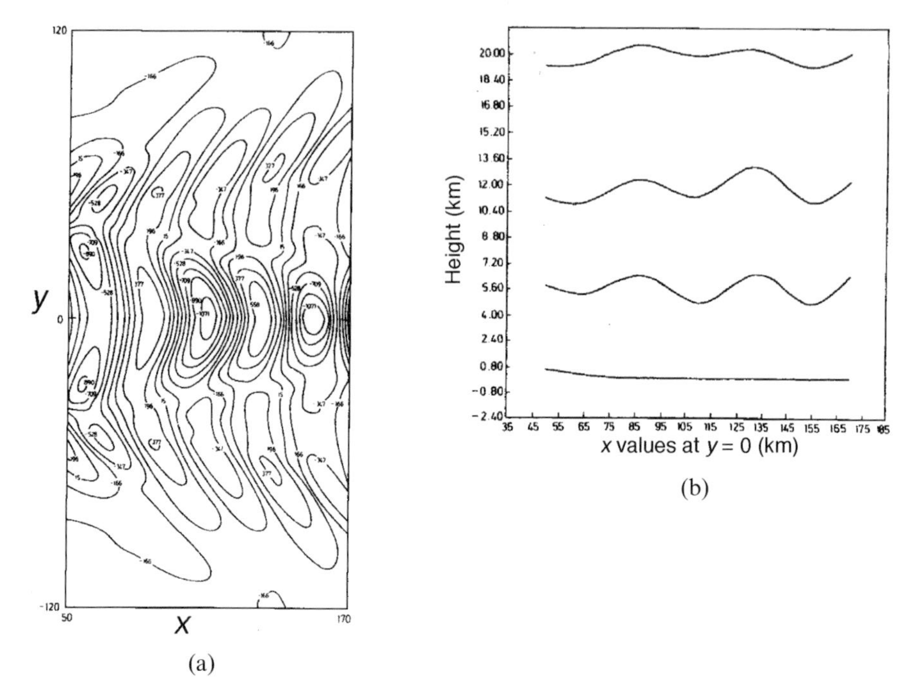

Figure 4.21. (a) Contours of vertical velocity at an altitude of 6 km in the lee of a crater of diameter 100 km, as computed by Pickersgill and Hunt (1979), and (b) comparison of the streamline displacements at various levels in the same flow, downstream of the crater.

within the flow. An example of such a calculation is shown in Figure 4.21, which clearly illustrates the ship-wake pattern of waves generated by a crater of diameter 100 km under conditions suitable for lee-wave formation, with the largest vertical excursions at altitudes of around 5–15 km above the surface.

Pickersgill and Hunt (1979) found the occurrence of the 'ship-wake' lee-wave pattern to be quite sensitive to the parameters of the flow, but such waves predominated when the shear satisfied the subcritical/supercritical criterion in the lower/upper layers and the static stability decreased with height. When they occurred, such flows could launch some wave activity into the upper atmosphere. Also, the presence of the continuous spectrum of evanescent wave modes allowed some influence to propagate upstream of the crater.

Pickersgill and Hunt further extended their analysis to the formation of lee waves generated by single, isolated mountains such as the giant volcanoes on the Tharsis Plateau (Pickersgill and Hunt, 1981), but following the single-layer approach previously used by Sawyer (1962), Crapper (1959), and Crapper (1962). Flow around isolated peaks can be significantly more complicated than over mountain ridges, because the static stability may restrict vertical motion at low levels so that air flows horizontally *around* the mountain rather than *over* it (e.g., see Baines, 1995). Near the mountain summit, however, the air may flow over the mountain, leading to the generation of vertically-propagating gravity waves, as before. The resulting lee-wave patterns may end up looking quite different from those produced by craters, in some cases forming isolated crescent-shaped clouds rather than a 'ship-wake' (e.g., see Figure 4.22).

A similar approach as used to model the crater 'ship-wake' patterns works just as successfully for the case of an isolated mountain (Pickersgill and Hunt, 1981). In this case, however, Pickersgill and Hunt (1981) use a single-layer atmosphere, in which the wind and static stability vary but the Scorer parameter (see Eq. 4.31) is kept constant. Again by considering an isolated mountain as a Fourier decomposition of linear ridges at all orientations relative to the upstream flow direction, Pickersgill and Hunt (1981) obtain solutions which clearly show a series of crescent-shaped patches in vertical velocity, with downwards streamline displacement over the mountain itself and a substantial region of uplift just downstream (see Figure 4.23).

The flows illustrated in Figure 4.23 correspond to a Froude number ($Fr = U/Nh$, where N is the buoyancy frequency and h the topographic height scale) of between 0.22 and 0.55. This would indicate that this solution is outside the limits of validity of linear theory. In practice, however, it implies that the low-level flow around the flanks of the mountain would probably be blocked and decoupled from the gravity waves near the mountain summit. The shape and size of the crescent-shaped pattern is, however, quite similar to the cloud patterns observed near Ascraeus Mons on occasions. Cloud patterns near the larger Olympus Mons, however, are often observed to be more irregular and turbulent than suggested from Pickersgill and Hunt's (1981) model, which may be due to a number of factors, including the larger amplitude of the latter volcano, and differences in the wind profile at lower altitudes.

4.4.3 Observations of mountain waves on Mars

We have already mentioned several times the quite common occurrence of wave cloud features associated with lee-wave structures generated by mountains and craters on Mars. Such features are almost certainly tenuous clouds of water ice condensing in regions of strong uplift reasonably close to the surface downwind of the topographic obstacle. Their occurrence, wavenumber, and extent, can provide some indication of the flow parameters in the lower atmosphere in the region of the mountain feature, though it has not yet proved possible to gain much detailed quantitative information, either about the background flow or the waves themselves, from such observations.

Profiles of temperature and density, obtained during the descent of the VL and MPF spacecraft to the Martian surface, showed various wave-like features which are suggestive of gravity waves. The identification of these structures as internal gravity waves, whether or not they are associated with topographically-generated features, is quite difficult, however. This is partly because the actual trajectories followed by the spacecraft during descent are often far from being a simple vertical descent, but typically involve a shallow downward-inclined trajectory which travels a large

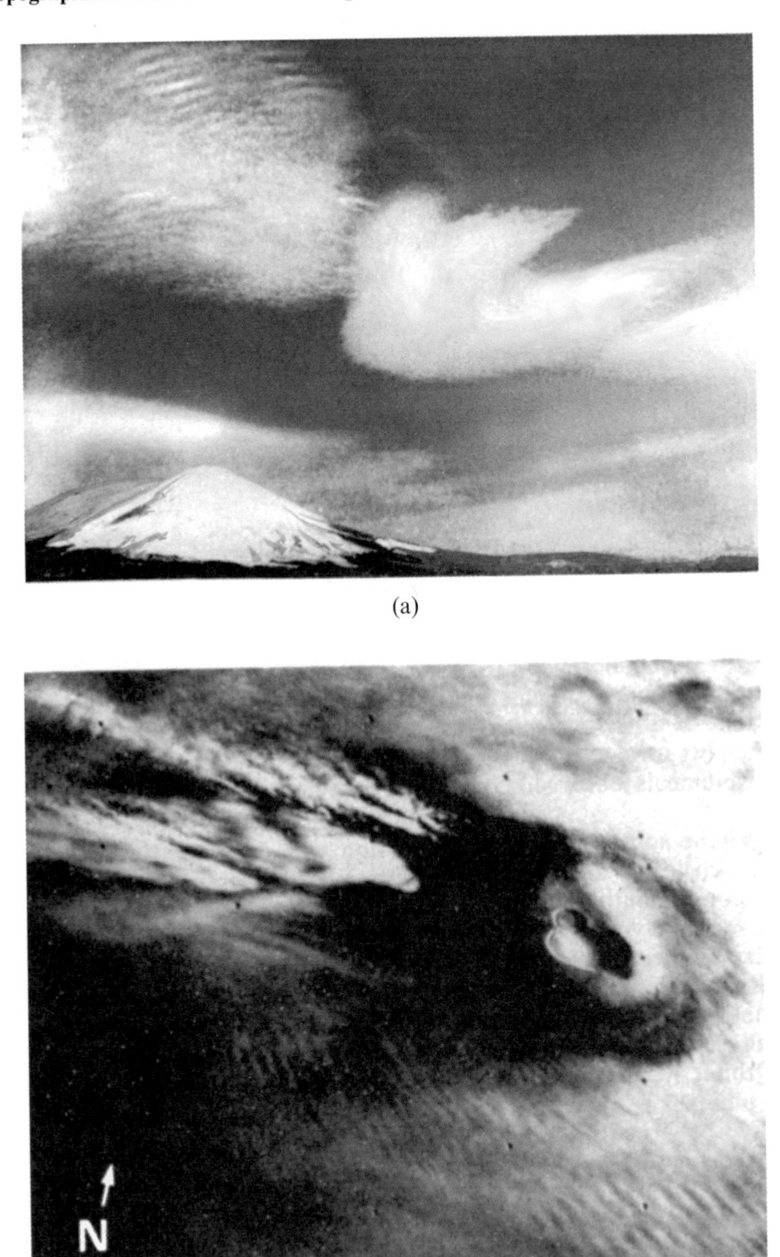

Figure 4.22. Photographs of crescent-shaped lee-wave clouds: (a) Turusi generated by Mount Fujiyama on Earth (photograph taken by Prof. Masaneo Abe in 1938), and (b) generated by Ascraeus Mons on Mars, taken from a Viking orbiter image.

Courtesy of (a) the Japanese Meteorological Agency and (b) NASA/JPL.

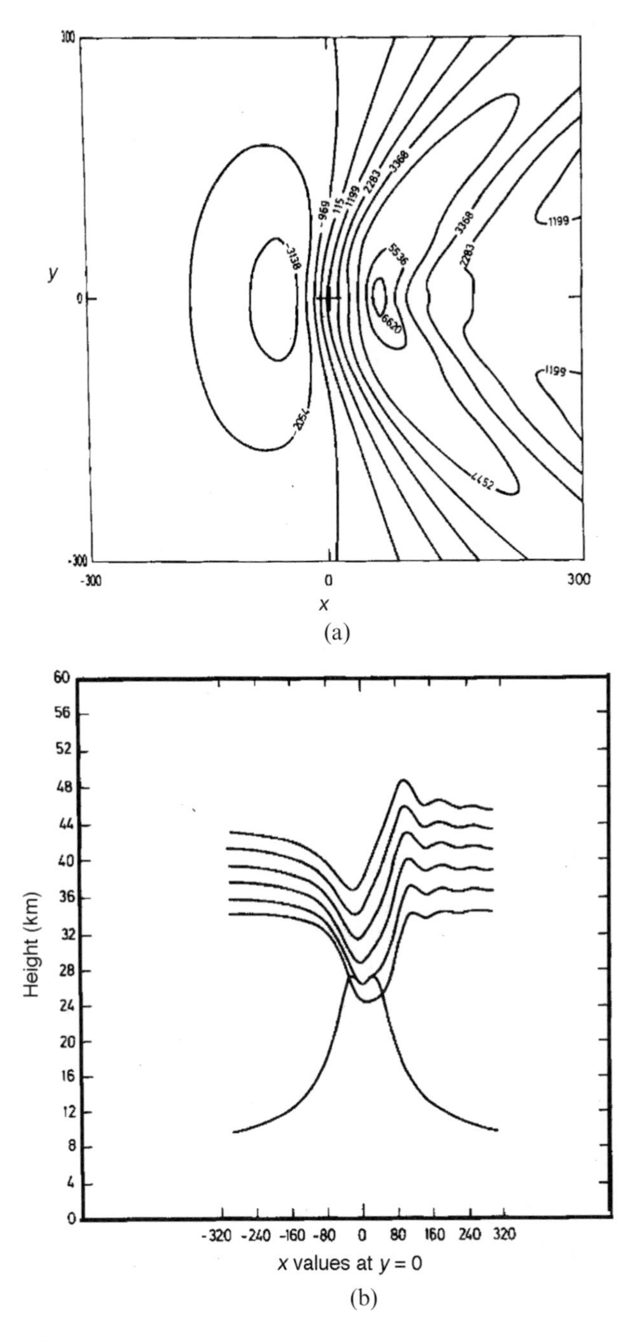

Figure 4.23. (a) Contour map of streamline displacement at an altitude of 40 km in a stationary lee wave, and (b) a vertical cross-section showing a series of displaced streamlines, generated by an isolated mountain similar to Ascraeus Mons on Mars, in a flow computed by Pickersgill and Hunt (1981).

distance horizontally before parachutes are deployed for the final descent. The current concensus would suggest that the strongest wave-like features seen in these descent profiles are more likely to represent strong diurnal thermal tides than small-scale gravity waves.

Radio-occultation measurements, especially from MGS (Hinson *et al.*, 1999) may have sufficient resolution and sensitivity to detect gravity waves. Recent measurements (Hinson *et al.*, 1999) indicate the occurrence of gravity waves, especially at northern and equatorial latitudes in work published to date (which apply only to a period around southern summer). Such waves evidently were propagating to high altitude, with some evidence of wave breaking taking place (where the waves were reaching a sufficiently large amplitude to reduce the local static stability close to zero). Such an effect would suggest that these gravity waves might exert a significant influence on the large-scale circulation and flow structure (see Section 4.4.4 for further discussion).

Some further evidence of waves in the lower atmosphere of Mars was found recently from laser altimeter measurements by the MOLA instrument on MGS over the polar hood CO_2 ice cloud at high northern latitudes during northern winter (Pettengill and Ford, 2000). Some of these waves appear to represent gravity waves generated by surface topographic features which Pettengill and Ford (2000) found were consistent with a static stability expected from the properties of the vapour pressure of dry ice.

Mountain waves are expected to propagate to much higher altitudes before breaking than covered by the observations outlined above. However, they are unfortunately very difficult to detect by direct means from remote-sensing observations at altitudes greater than about 40–50 km (though are occasionally visible in images of limb hazes from Mars' orbit (e.g., see Anderson and Leovy, 1978). This remains an aspect of the Martian atmosphere which deserves greater attention than has been the case thus far, but will need to await future spacecraft missions, probably using descent probes, in order to make further significant progress in characterizing mid-tropospheric gravity waves. In the next subsection, however, we consider the indirect effects of gravity waves on the large-scale circulation. Such effects can be studied by comparing model simulations with observations of the large-scale thermal structure of the atmosphere, and is what we concentrate on next.

4.4.4 Large-scale gravity-wave drag

We have already alluded to the possibility that gravity waves may be important, not just because of the local effect of perturbations to the thermal structure and winds in the atmosphere where the waves happen to be passing by, but also on a larger – even global – scale. As waves propagate upwards from the surface, their amplitude increases exponentially with height (cf Eq. 4.27). This applies not only to the amplitude in vertical velocity, but also to other fields such as the temperature. At very large amplitudes, the waves become non-linear and steepened in places, which

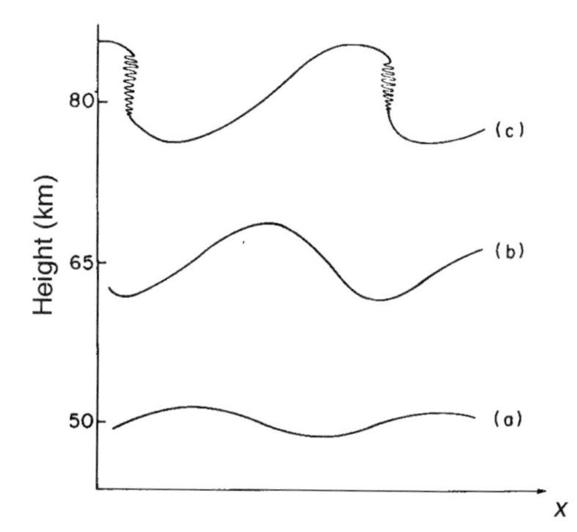

Figure 4.24. Schematic diagram showing the increase in amplitude of gravity waves with height, leading to breaking.

After Andrews *et al.* (1987).

can lead to the local static stability becoming very small or even negative (see Figure 4.24). Where this happens, local convective instability can break out, leading to turbulence and irreversible mixing of temperature, momentum, and other tracers. Even though this may be a local phenomenon within each wavelength of the gravity wave, the widespread nature of the wave propagation can lead to this strong mixing occurring on a large scale.

On Earth, this effect is now well known in the stratosphere and mesosphere (e.g., Houghton, 1978; Lindzen, 1981; Holton, 1982; Fritts, 1984) and even in the troposphere (Palmer *et al.*, 1986), where the consequences for the large scale can be substantial. The mixing of momentum when the waves break leads to an effective drag on the atmospheric circulation, forcing it towards the phase speed of the gravity waves themselves (e.g., Killworth and McIntyre, 1985), that is, towards zero flow relative to the ground in the case of mountain-induced gravity waves. Such a drag can be significant both for the momentum budget of the atmosphere, and also for the thermal structure. In particular, gravity-wave drag is widely considered to be responsible for the 'closing off' of the mesospheric zonal jets in both winter and summer hemispheres and the associated reversal of the equator–pole thermal gradient (so that the poles may become warmer than the equator at mesospheric heights (e.g., Andrews *et al.*, 1987).

On Mars, several observational studies of the Martian middle atmosphere have suggested that, at least during certain seasons, the polar regions of Mars are significantly warmer than expected on the basis of local radiative equilibrium (Martin and Kieffer, 1979; Deming *et al.*, 1986; Jakosky and Martin, 1987; Santee and Crisp, 1993; Théodore *et al.*, 1993). This is generally thought to owe its origin to a dynamical phenomenon, which might include the direct forcing of descending flow over the poles (e.g., under conditions of enhanced dust loading in the atmosphere

(Haberle *et al.*, 1993), breaking planetary waves (Hollingsworth and Barnes, 1996) or gravity waves). The latter suggestion has been examined in several modelling studies in recent years (Barnes, 1990; Théodore *et al.*, 1993; Joshi *et al.*, 1995; Collins *et al.*, 1997), in which the impact of various ways of parameterizing the effects of gravity-wave drag in large-scale dynamical models have been investigated. Most studies have used schemes based on the Lindzen 'saturation hypothesis' (Lindzen, 1981), in which a gravity-wave generated at the surface propagates freely upwards until it reaches its 'breaking level' z_b, at which the minimum local gradient Richardson number ($Ri = N^2/(\partial u/\partial z)^2$) in the wave decreases to a critical value of 1/4. Above this level, the flow is adjusted to retain a minimum value for Ri of 1/4, which implies a diffusive drag force on the flow which depends strongly on $(\bar{u} - c)$ (where $c = 0$ for topographically forced waves), and also on the wavelength of the gravity wave and the background profile of N^2. Some allowance is also typically made for sporadic breaking events which may occur below z_b, and the strength of the drag is also adjusted by a tuneable parameter $\alpha < 1$ to allow for effects such as wave transience (Holton, 1983), which may act to reduce the effect of gravity waves in a time-averaged sense.

Some of the most sophisticated applications of this approach have been investigated by Collins *et al.* (1997) and Forget *et al.* (1999), who have used a form of the scheme developed for terrestrial GCMs by Lott (1997) in the Oxford and LMD Mars GCM. This includes a complex surface stress parameterization which takes into account both the amplitude and orientation of unresolved mountain features and 'blocking effects' to predict the large-scale drag and lift forces at low levels, together with the Lindzen scheme for gravity wave saturation at upper levels of the atmosphere. In common with the earlier studies which used simpler dynamical models (e.g., Barnes, 1990; Théodore *et al.*, 1993; Joshi *et al.*, 1995a), the effects of parameterized gravity-wave drag were found to be important, especially at high latitudes. However, on comparison with available observations of the thermal structure of Mars' atmosphere over the poles it was apparent that quite a lot of the dynamically-induced warming over the pole was attributable to planetary-wave breaking. Though also significant, the contribution due to gravity waves was smaller than expected based on the experience of terrestrial GCM modellers. Indeed, Collins *et al.* (1997) found they had to reduce the tuneable parameter α to a lower value than typically used in terrestrial GCMs in order to prevent the scheme exerting an unrealistically large drag on the circulation. The reasons for this are still not well understood, but may be partly due to the much stronger effects of radiative damping on gravity waves in Mars' atmosphere compared with the Earth. Radiative damping has not so far been taken into account in the parameterizations implemented in Mars' GCMs, but would be expected to reduce the strength of gravity wave activity, especially at upper levels where the thermal inertia of the atmosphere is very low. Schemes to date have also represented the topographically-generated gravity waves by a single, monochromatic wave, rather than a more realistic spectrum of waves. Schemes have been proposed by terrestrial modellers to represent a spectrum of gravity waves in GCMs (e.g., Fritts and Lu, 1993), although such schemes are computationally expensive and have not yet been widely used for the Earth, let alone for Mars, for which there is very little information available to constrain such models.

Finally, we note that gravity waves in Mars' atmosphere may also be generated by phenomena other than obstacles at the surface, such as local convection, dust storms, and other meteorological systems (e.g., fronts). In that case, the gravity waves would propagate at horizontal phase speeds $c \neq 0$, with a corresponding tendency to accelerate the flow towards $u = c$. The generation and influence of such waves on Mars have not been considered seriously in any studies so far, and their importance remains unknown at present.

5

Diurnal phenomena

5.1 THE DIURNAL CYCLE ON MARS

It is very commonly the case on Mars that the local atmospheric conditions are governed more by time of day than by any day-to-day variability; often the best short-term weather forecast is simply that conditions will follow a repeatable diurnal pattern. This applies particularly in tropical regions and in the summer hemisphere, where transient weather systems are expected to be absent or small, as described in the following chapter. The diurnal cycle is often very strong, with day–night contrasts of as much as 100 K, owing to the low thermal inertia of the atmosphere and the strong solar heating per unit mass during the day, reminiscent of extreme desert-like conditions on Earth. This chapter reviews components of the atmospheric circulation which are driven by contrasts in heating and cooling associated with the diurnal cycle on Mars.

Throughout this chapter it should be noted that it has been convenient to divide the Martian mean solar day (sol) into twenty-four 'hours', each of 3,699 s in contrast to terrestrial hours of 3,600 s, and each of which can be divided into 60 'minutes'. The difference between Martian and terrestrial hours and minutes is, of course, small because of the similar rotation rates of Mars and Earth, but the use of hours and minutes based on the solar day means that 1200 local time (LT) still means the sun is at its zenith, and avoids awkward conversions for days and fractions of days. Seconds are only used in the context of a universal, S.I., time unit.

5.1.1 Mars Pathfinder (MPF)

Data returned by the MPF lander, during its 90 day lifetime in 1997, provides a good example of the dominance of the diurnal cycle in the Martian atmosphere, at least in equatorial regions in late summer. Pathfinder landed near 19°N, 33°W, at $L_S = 142°$. The Atmospheric Structure Investigation/Meteorology experiment (Schofield *et al.*, 1997) provided observations of temperature at three heights on a 1-m mast (whose base was around 40 cm above the local surface), surface pressure and wind direction, at least 51 times per sol (and often much more frequently) throughout the landed phase of the MPF mission. Data are shown here for the first 30 sol of the mission.

Figure 5.1 (see colour section) illustrates the diurnal air temperature variations on the MPF meteorological mast and the strong contrast between night-time temperatures, falling as low as 190 K at 0600LT, and mid-afternoon temperatures, reaching over 260 K near the surface between 1400LT and 1600LT. It is also of interest to note the extreme vertical temperature gradients implied by these data, especially closer to the ground when the lowest sensor is included. A simple, linear fit to the mean curves implies that temperature falls by around 5–10 K m^{-1} near the surface during the day, as the surface is heated by solar radiation. In other words, an astronaut might find ambient temperatures at head-height almost twenty degrees cooler than at his or her feet! At night, the situation is reversed with a more moderate gradient between the upper sensors, but a strong inversion develops close to the ground as it cools by emitting infrared radiation. Such temperature contrasts and inversions are familiar from terrestrial deserts, but generally much less extreme in magnitude.

Figure 5.2 shows a comparison between the upper temperature sensor on MPF and LMD and Oxford GCM results taken from the Mars Climate Database (MCD) (Lewis *et al.*, 1999, see Appendix A). The GCM seems able to reproduce the overall

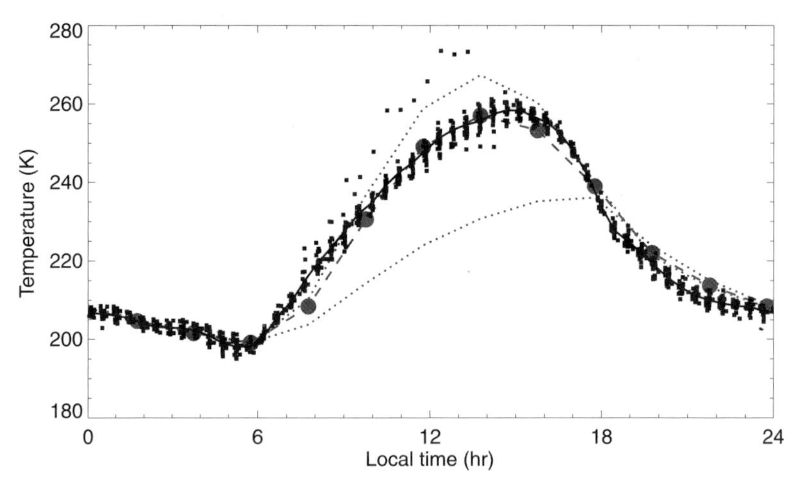

Figure 5.2. A comparison of the Mars Pathfinder temperature at the top of the 1 m meteorological mast with GCM results. As in Figure 5.1 (see colour section), the smaller dots show the individual measurements, and the solid line is the mean over 30 sol. The dotted lines are Mars Climate Database (MCD) GCM surface temperatures (top line at midday) and the MCD lowest atmospheric level temperature (bottom line) which is at a height of about 4–5 m. The circles connected by dashed lines show a linear interpolate between the two at a height of about 1.4 m, corresponding to the top of the Pathfinder mast for comparison.

cycle shape and a large temperature contrast (up to 40 K) between the surface and the lowest atmospheric level at around 4–5 m. The GCM results are a mean over the $L_S = 120°–150°$ period, but the model displays highly repeatable temperatures at this location and time with almost imperceptible variations between different days, other than a slow seasonal trend. The repeatability of the cycle of atmospheric temperatures measured by MPF is also clear. The few outlying warmer temperatures are from a period at the start of the mission when the mast was lying horizontally and being directly heated by the sun, and so should be ignored, or compared with surface temperatures. The remaining spread in the observations also contains a component of seasonal trend, as the Martian atmosphere passed from summer toward autumn, and the true day-to-day variability is small, on the order of 1 K.

Not only temperature, but also surface pressure and wind direction, vary in a consistent way with time of day, as shown in Figure 5.3, which again compares observations from the first 30 sol of MPF operations with model results. The diurnal pattern of heating induces a strong diurnal (24-hour period) and semidiurnal (12-hour period) response in the surface pressure, as shown in the upper panel of Figure 5.3. This variation is a result of large-scale atmospheric tidal modes, to be discussed in more detail later. It is apparent that the surface pressure signal is highly repeatable, again with a spread that can largely be explained as a seasonal trend as the polar CO_2 ice caps freeze and sublime. The surface pressure every day peaks close to 0800LT and falls to a minimum around 1800LT. This swing, of order 40 Pa peak-to-peak, is much larger than any variability at a fixed time of day during the MPF primary mission period. Such a tidal surface pressure response of the atmosphere to the thermal forcing from solar heating is also present on Earth, although, in contrast, it is generally small compared to other sources of variability close to the ground. It has been known for many years that there is a diurnal and semidiurnal signal in the surface pressure record, and the question of why the semidiurnal response ($\sim$100 Pa = 1 mb) is larger than the diurnal response on Earth was noted by Lord Kelvin in 1882. This is now understood as the response of the whole atmosphere to heating, primarily via ozone absorption in the middle atmosphere, and will be discussed further in the following section. Thermal tides become clearly significant in the terrestrial stratosphere, mesosphere, and thermosphere, growing as they propagate upward (e.g., Andrews *et al.*, 1987).

The wind direction, shown in the lower panel of Figure 5.3, does show more scatter, perhaps with more small-scale gustiness and short-period variations. The mean, however, still presents a consistent pattern, in accordance with the smoother, large-scale winds predicted by the GCM. The overall picture is of almost steady, southerly winds during the night, followed by a repeatable rotation through a full circle during the day, again a result of the thermal tide passing over the MPF site.

5.2 PLANETARY THERMAL TIDES

Atmospheric thermal tides are global-scale waves directly forced by the day–night contrast in solar heating. Unlike for ocean tides on Earth, lunar and solar

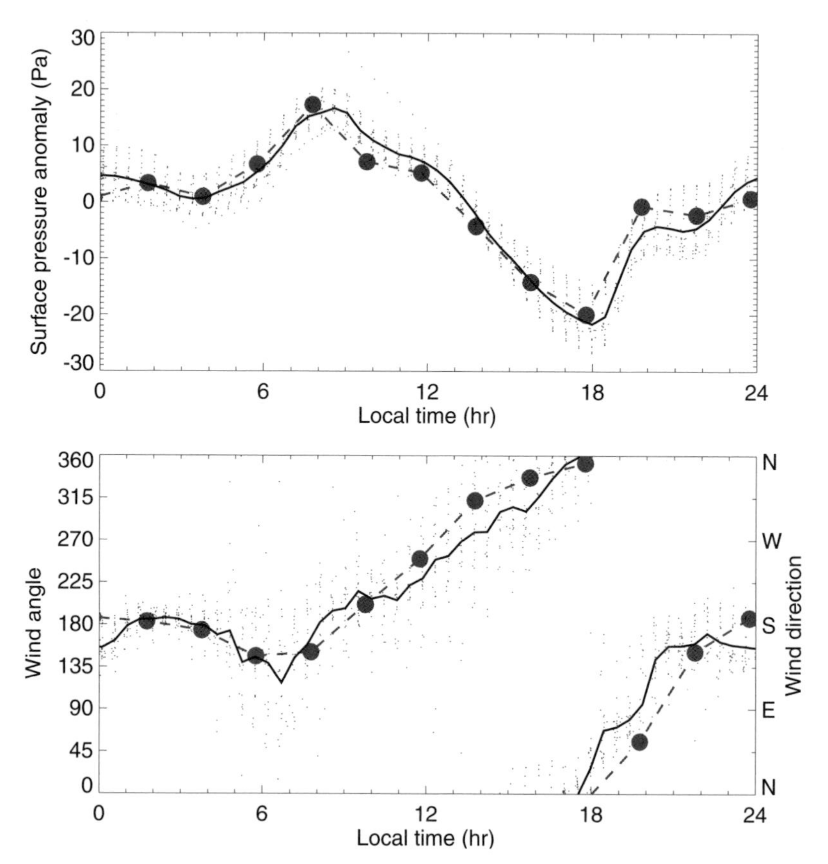

Figure 5.3. Upper panel: A comparison of the observed Mars Pathfinder surface pressure diurnal cycle with the Mars Climate Database (MCD). As in Figure 5.2, dots show individual measurements and the solid line a mean. In both cases the diurnal-mean surface pressure has been subtracted for clarity; there is not an exact match in absolute surface pressure because of the limited horizontal resolution of the model. Lower panel: A comparison of wind direction obtained from the Mars Pathfinder sensors with the 4–5-m wind direction in the MCD.

gravitational forces are of little significance in both the terrestrial and Martian atmospheres; the Martian moons, Phobos and Deimos, are much less massive than the Earth's moon.

Atmospheric tides are a major component of the variability in the Martian atmosphere as a result of the strong solar heating per unit mass of the atmosphere. Most of the time, the Martian atmosphere is relatively clear, and a significant fraction of incoming solar radiation falls on, and rapidly heats, the surface. This heat is exchanged with the atmosphere, both via radiative and convective processes, the latter in particular during the day when the planetary boundary layer is most active. The solar radiation which is absorbed by dust in the atmosphere, either a small fraction in a clear atmosphere, or the bulk of the radiation in a high-dust state, also heats the atmosphere directly and provides strong tidal forcing over a greater range of altitudes, according to the dust distribution. Tides can also be forced in the Martian upper atmosphere by direct gaseous absorption of solar radiation.

The response of the Mars GCM surface temperature to solar forcing is shown in Figure 5.4 for several times of day at northern hemisphere spring equinox. The results shown are from the default prescribed dust assumption for Mars Global Surveyor (MGS)-like conditions, described in Section 2.5.2.3, which is relatively clear at this time of year. The warm day-side, with maximum temperatures at about 1300LT, can been seen to move around the planet, following the subsolar point.

The response of the atmosphere to this diurnal heating pattern is not solely in atmospheric temperature; as shown in the MPF observations earlier (Figure 5.3) the tidal modes which are forced also have surface pressure and wind components. On Mars, the diurnal wind signal tends to be dominated by local topographic and thermal effects, to be described in following sections, and is consequently complex when viewed on global scales. The surface pressure, on the other hand, is related to the sum of the atmospheric mass above any point and so, as a vertically-integrated quantity, tends to reflect features of larger horizontal scale as well. Figure 5.5 shows the surface pressure anomaly, as a function of time of day, for times corresponding to the temperatures shown in Figure 5.4. The time-mean pressure has been subtracted from each plot in order to remove the dominant topographic signature from the surface pressure field for clarity. A pattern can be seen to propagate westwards with a period of one day, again following the sun and the temperature anomaly, and with a similar peak-to-peak amplitude to that seen at the MPF site. It is also noticeable, particularly near the equator, that the surface pressure response contains significant longitudinal wavenumber one and two components (the diurnal and semidiurnal tides). At times, two maxima or minima in surface pressure can be seen on a line of constant latitude.

5.2.1 Migrating tides

Migrating, or sun-following, tides are those which have a constant phase with respect to the solar heating, as it moves around the planet once per day, and have been studied extensively on the Earth. The most important two examples of migrating tides are the diurnal and semidiurnal modes, which propagate westward, have longitudinal wavenumbers one and two, and periods of twenty-four and twelve hours, respectively.

The classical theory of tides dates back to Laplace in the 1780s and is described in detail by Chapman and Lindzen (1970) and Andrews *et al.* (1987). Classical theory treats linearized disturbances about a stationary background state, with no horizontal temperature gradients, and has been applied to Venus and Mars, as well as to the Earth (e.g., Lindzen, 1970; Zurek, 1976, 1986; Hamilton, 1982; Wilson and Hamilton, 1996). Progress has also been made in extending the terrestrial theory to account for non-zero mean winds and for latitudinal temperature gradients.

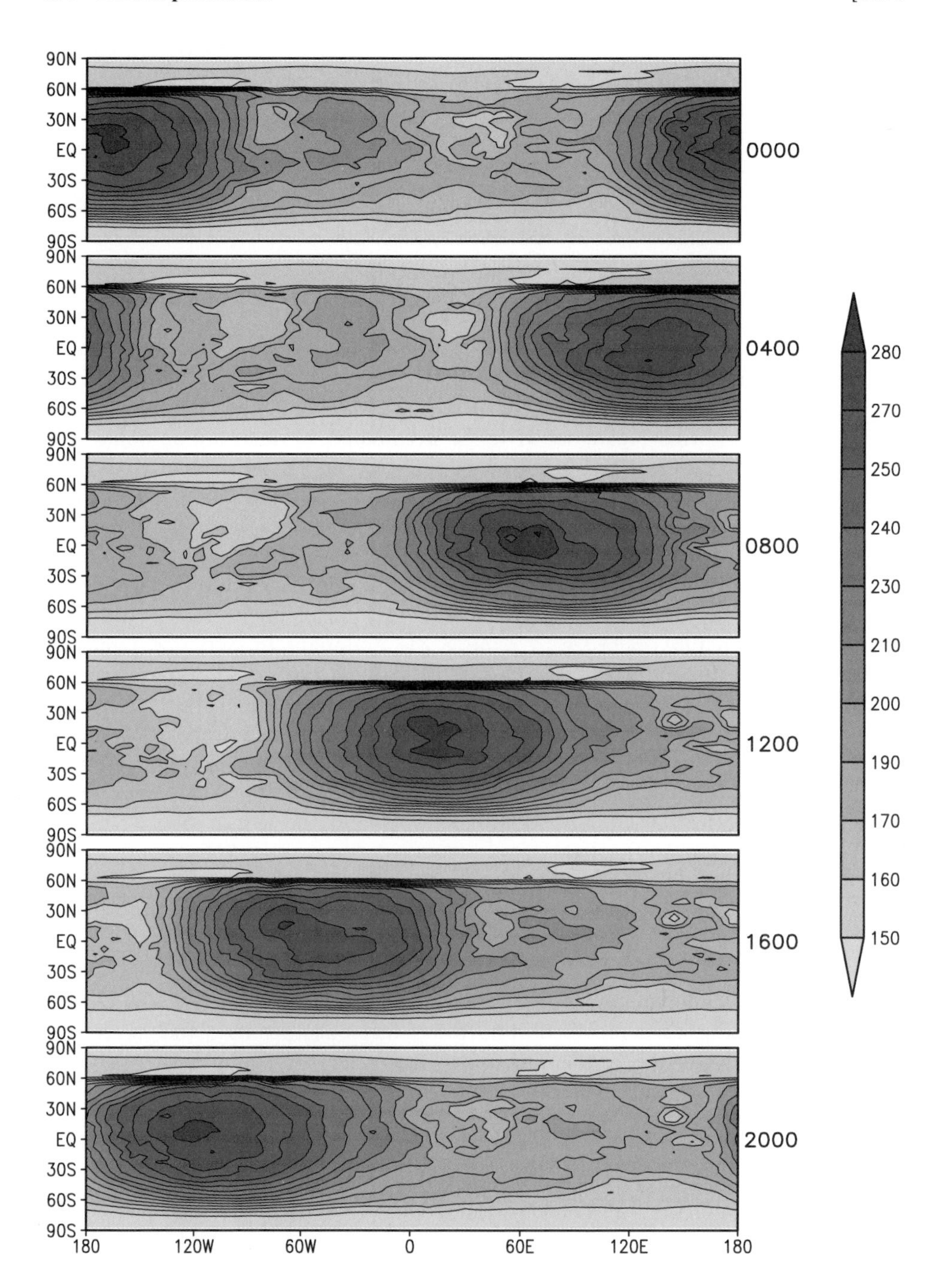

Figure 5.4. Surface temperature at six times of day, with universal time (UT, the local time at 0° longitude on Mars) labelled on the right. Results were taken for the Oxford Mars GCM at northern hemisphere spring equinox, $L_S = 0°$.

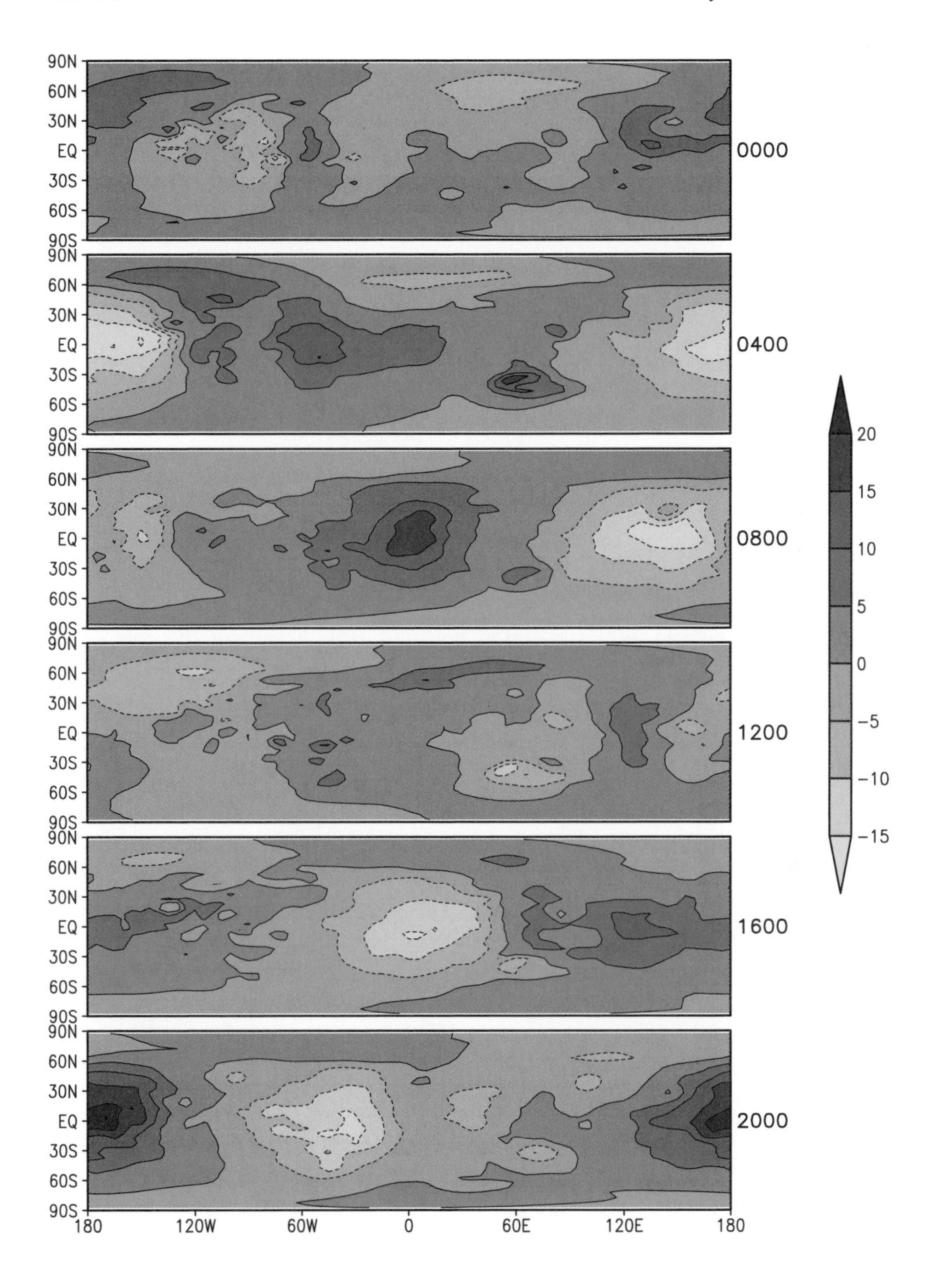

Figure 5.5. Surface pressure anomaly at six times of day, corresponding to Figure 5.4. The time-mean pressure has been subtracted from each plot.

The procedure for describing tidal modes begins with forms of the hydrostatic primitive equations (Eqs 2.2–2.7), which have been linearized for disturbances about a resting state. For simplicity, viscous and forcing terms may be at first neglected. Following Andrews *et al.* (1987), solutions in pressure coordinates may be found for perturbation quantities of the form:

$$(u, v, \Phi) = \exp(z/2H_p)U(z)(\tilde{u}, \tilde{v}, \tilde{\Phi}) \tag{5.1}$$

and

$$w = \exp(z/2H_p)W(z)\tilde{w} \tag{5.2}$$

where $\Phi = \int g\,dz$ is the geopotential; $H_p = RT/g$ is the scale height; and $\tilde{u}$, $\tilde{v}$, $\tilde{\Phi}$, and $\tilde{w}$ are all functions of longitude, λ, latitude, ϕ, and time, t, only. This gives Laplace's tidal equations:

$$\frac{\partial \tilde{u}}{\partial t} - f\tilde{v} + \frac{1}{a\cos\phi}\frac{\partial \tilde{\Phi}}{\partial \lambda} = 0 \tag{5.3}$$

$$\frac{\partial \tilde{v}}{\partial t} + f\tilde{u} + \frac{1}{a}\frac{\partial \tilde{\Phi}}{\partial \phi} = 0 \tag{5.4}$$

$$\frac{1}{a\cos\phi}\left(\frac{\partial \tilde{u}}{\partial \lambda} + \frac{\partial(\tilde{v}\cos\phi)}{\partial \phi}\right) + \frac{1}{gh}\frac{\partial \tilde{\Phi}}{\partial t} = 0 \tag{5.5}$$

where gh is introduced as a separation constant with h playing the role of an equivalent depth. Eqs (5.3–5.5) are the same as those derived by Laplace for disturbances to a thin layer of fluid on a sphere, with mean depth h and the disturbance departure from the mean $\tilde{\Phi}/g$. Solutions can be found in the form;

$$(\tilde{u}, \tilde{v}, \tilde{\Phi}) = \mathrm{Re}\,(\hat{u}(\phi), \hat{v}(\phi), \hat{\Phi}(\phi))\exp(i(s\lambda - 2\Omega\sigma t)) \tag{5.6}$$

which have longitudinal wavenumber s and period $1/(2\sigma)$ days, with negative σ used to indicate westward propagation (the definition of σ with the factor of two is conventional in tidal work). $\hat{u}(\phi)$, $\hat{v}(\phi)$, and $\hat{\Phi}(\phi)$ take the form of Hough functions, which have been tabulated (e.g., Longuet-Higgins, 1968).

In the vertical, the form of the solutions is given by the vertical structure equation:

$$\frac{d^2W}{dz^2} + \left(\frac{N^2}{gh} - \frac{1}{4H_p^2}\right)W = 0 \tag{5.7}$$

with $U(z) = dW(z)/dz - W(z)/2H_p^2$, the relation between the vertical structures of different variables. Hence, solutions of this form can either be trapped in the vertical and decay exponentially away from the forcing region, or, if $0 < h < 4N^2H_p^2/g$, they can have sinusoidal form in the vertical, with wavelength:

$$\lambda_v = 2\pi\left(\frac{N^2}{gh} - \frac{1}{4H_p^2}\right)^{-1/2} \tag{5.8}$$

according to the equivalent depth, h, which should not be confused with the wavelength.

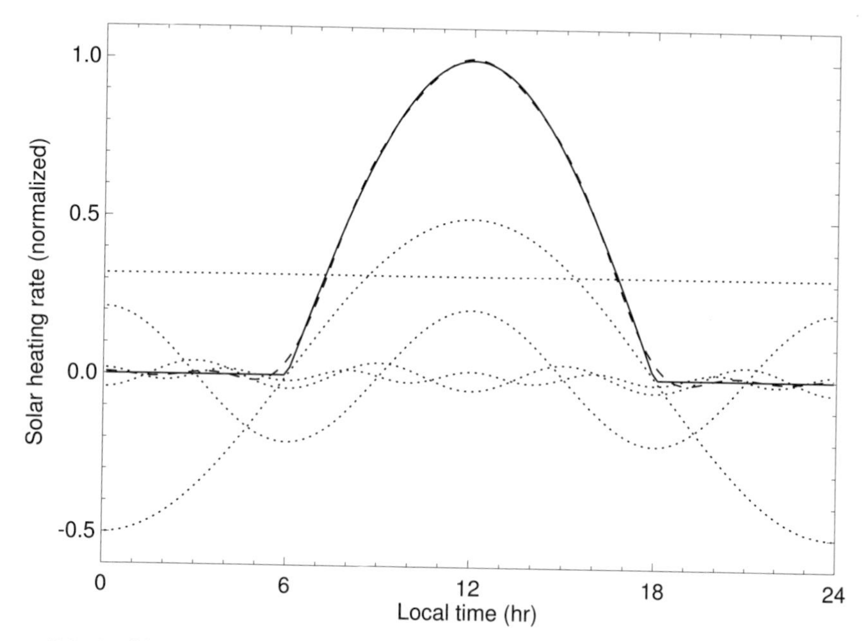

Figure 5.6. An idealized plot of solar thermal forcing (solid line) versus local time of day. The main components are shown as a mean, diurnal, semidiurnal, 6-hour and 3-hour period waves (dotted lines). The sum of these first five components is shown as a dashed line, indicating how they combine to almost reconstruct the original signal.

Thermal forcing can be included in the theory above. Chapman and Lindzen (1970) and Andrews *et al.* (1987) give more complete details of the derivation, with the forcing decomposed into Fourier harmonics in longitude, and then Hough functions in latitude. This results in a modified form of Eq. (5.7), adding a forcing term $Q^{(s)}(z)$ to the right-hand side of an inhomogeneous version of the equation for the vertical structure of each mode of longitudinal wavenumber s.

It is possible to see how even a simple solar heating term, of cosine form during the day, peaking at local noon, and zero at night, can give rise to tides with several different longitudinal wavenumbers. Figure 5.6 shows how such an idealized forcing function can be broken down into a combination of a steady component (of $1/\pi$ times the peak amplitude of the variable forcing), a diurnal period mode (of the same peak-to-peak amplitude as the forcing) and a series of modes of period $1/2n$ sol, where the semidiurnal is the largest. It is a straightforward exercise in Fourier analysis to show that the amplitudes of these modes falls as $2/\pi(4n^2 - 1)$ with increasing n. Responses on all these timescales would be expected, even in the simplest, linear case.

For sun-synchronous tidal modes, the classical procedure is to choose a wavenumber and period ($s = 1$, $\sigma = -1/2$ for the diurnal tide, or $s = 2$, $\sigma = -1$ for the semidiurnal) and then to solve Laplace's tidal equations for an equivalent depth, h, which gives a vertical wavelength (Eq. 5.8) for that mode. For the major migrating tides on both Earth and Mars the results are remarkably similar. The wavenumber one, diurnal component propagates vertically in the tropics, with a vertical wavelength of about 30 km, and tends to be vertically trapped at low levels in the extratropics (Andrews *et al.*, 1987; Wilson and Hamilton, 1996). The wavenumber two, semidiurnal tide has a much longer vertical wavelength, >100 km on both planets, with little vertical phase propagation.

It is the difference in the vertical wavelength of these two modes which accounts for the relative amplitudes observed. When the Martian atmosphere is clear, heating is concentrated near the surface and the diurnal tide is the largest single mode excited, as in Figure 5.6. If the heating is spread over a deeper region of the atmosphere, comparable with the wavelength of the diurnal mode, or greater, then the diurnal mode may suffer from destructive interference. This is indeed the case when the Martian atmosphere is dusty: the diurnal tide is seen to become weaker, with a growth in the semidiurnal tide. Under these circumstances, the semidiurnal tide, which has a vertical wavelength much greater than the depth of the forcing and therefore effectively integrates the total thermal forcing in the atmosphere, is most strongly forced. In fact, the correlation on Mars between the dust optical depth and the amplitude of the semidiurnal tide is thought to be good enough to make estimates of dust opacities based on Viking Lander (VL) 1 pressure data (Zurek, 1981), when the optical depth could not be observed directly.

On Earth, this answers the question highlighted by Lord Kelvin in 1882. There is normally thermal forcing over a range of heights in the atmosphere, particularly from heating through ozone absorption, centred around the stratopause, and this leads to destructive interference in the diurnal tide and enhanced forcing of the semidiurnal mode, which is seen at the surface. The situation on Earth is permanent rather than temporary, unlike that on Mars, where very dusty states are exceptions to the more common moderately dusty or clear atmosphere, and where there is very little ozone or other gaseous absorption in the middle atmosphere.

5.2.2 Interaction with topography

In addition to the migrating tides described above, the diurnal thermal forcing of the atmosphere can give rise to a range of non-migrating tides. These modes still have $1/n$-sol periods, with $n = 1, 2, \ldots$, but they do not maintain a constant phase with respect to the sun, which would require westward propagation with a phase speed of $2\sigma/s = -1$ in the notation adopted earlier. The non-migrating tides are driven by the interaction of solar forcing with surface variations, primarily in topography, but also in albedo and thermal inertia. Most notably, in the case of Mars, the diurnal tide and wavenumber two longitudinal topographic variations can give rise to a westward-propagating wavenumber three tide ($s = 3, \sigma = -1/2$) and an eastward-propagating wavenumber one tide ($s = 1, \sigma = 1/2$), both non-migrating and with diurnal periods.

The later diurnal mode, a Kelvin wave (e.g., Gill, 1982; Andrews *et al.*, 1987), has maximum amplitude at low latitudes and its frequency is close to atmospheric

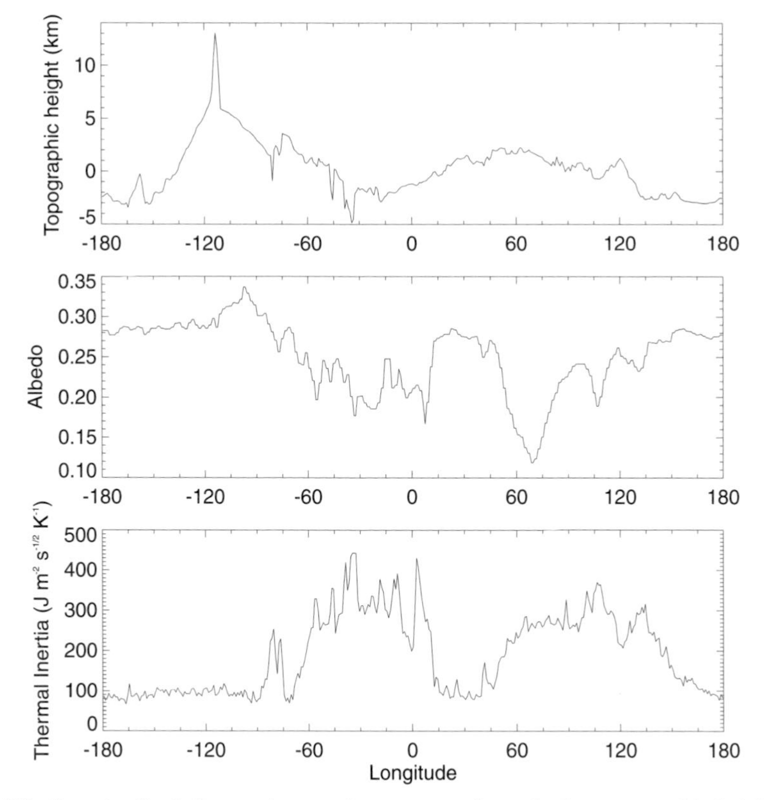

Figure 5.7. Longitudinal slices, taken at the equator, through the topographic height (upper panel), surface albedo (middle panel), and thermal inertia (lower panel) surface fields.

resonance on Mars, as discussed in the following section. Kelvin waves are eastward-propagating, equatorially-trapped inertia-gravity waves which, in an atmosphere at rest, consist effectively of a simple high and low pressure system, centred on and symmetrical about the equator, with zero meridional velocity. In an ocean (and in the atmosphere under some circumstances), in addition to the equatorially-trapped wave, it is possible to have a Kelvin wave which 'leans' against a vertical solid boundary to support its pressure fluctuations (Gill, 1982).

The likelihood of a strong Kelvin mode response on Mars has been known for some time (e.g., Zurek, 1976). Wilson and Hamilton (1996) discuss their presence in Mars GCM simulations and Wilson (2000) identifies a diurnal Kelvin mode in MGS temperature data. Such a strong atmospheric Kelvin mode response, especially near the surface, is particular to Mars, although Kelvin modes have been shown to play important roles in the quasibiennial oscillation in the Earth's middle atmosphere (Andrews *et al.*, 1987) and in equatorial ocean dynamics (Gill, 1982).

Figure 5.7 shows equatorial cross-sections through surface topography, albedo, and thermal inertia fields for Mars. In addition to the smaller-scale variability, all three fields possess large-scale signals at low longitudinal wavenumbers. In particular, a wavenumber two pattern is present, with two broad maxima and minima discernible in each cross-section. In the case of the topography, the maxima are formed by the Tharsis Ridge and Syrtis Major, almost on opposite sides of the planet to each other. It appears that the topographic signal is the most important for generating Kelvin waves on Mars. This can be demonstrated using a GCM by removing all topographic and thermal variations, which results in no non-migrating thermal tides being produced, and then by adding back topographic and thermal variations separately, comparing the non-migrating tide response to that of the full GCM. The topographic influence on thermal forcing occurs because the solar radiation primarily heats the surface, when the atmosphere is not very dusty, which then transfers heat to the atmosphere via convection and thermal radiation. Thermal forcing of the lower atmosphere will, therefore, roughly follow the terrain, inducing a stationary, wavenumber two modulation to the forcing as viewed on a constant-pressure level in the atmosphere.

A simple illustration of the way in which the surface modulation can give rise to a Kelvin mode, and a non-migrating diurnal wavenumber three mode, is shown in Figure 5.8. The upper panel shows a simple westward-propagating, diurnal wavenumber one mode. This is the principal component of the migrating thermal tide discussed in the previous section. In the lower panel is the same wave, modulated by a stationary $\cos(2\lambda)$ pattern. The result is equivalent to a wavenumber one pattern propagating eastward and a wavenumber three pattern propagating westward, both with diurnal periods. In other words, two waves are produced with the sum and difference of the wavenumbers ($s = 1$ and $s = 2$) and frequencies ($2\sigma = -1/\text{sol}$ and $2\sigma = 0$) of the diurnal tide and modulating pattern; this can be verified with basic trigonometric identities.

The illustration shown here is the primary combination of modes which gives rise to a Kelvin mode on Mars, but it is easy to see that with the full spectrum of migrating tides, and all the wavenumbers present in the surface fields, a complex set of interactions may produce a rich spectrum of responses. Another significant interaction, for example, is that of the semidiurnal tide ($s = 2$, $\sigma = -1$) with the wavenumber two surface components, which results in a wavenumber four tidal mode ($s = 4$, $\sigma = -1$) and a semidiurnal modulation to the zonal mean flow ($s = 0$, $\sigma = 1$).

5.2.3 Planetary free modes

The discussion so far has concerned tides which are direct responses to thermal forcing. It is also possible to use Eqs (5.3–5.5) and (5.7) to search for free, or normal, modes. These are natural resonant modes of the atmosphere, which should emerge if the atmosphere is forced by any means, including essentially stochastic processes over a range of frequencies. The structure and frequency of the free modes are determined by the background state of the atmosphere, rather than by the forcing process, and are analogous to the normal modes of vibration of a string which is fixed at both ends. The procedure to find such modes is almost the reverse of the classical tidal theory, in which the period and wavenumber are fixed,

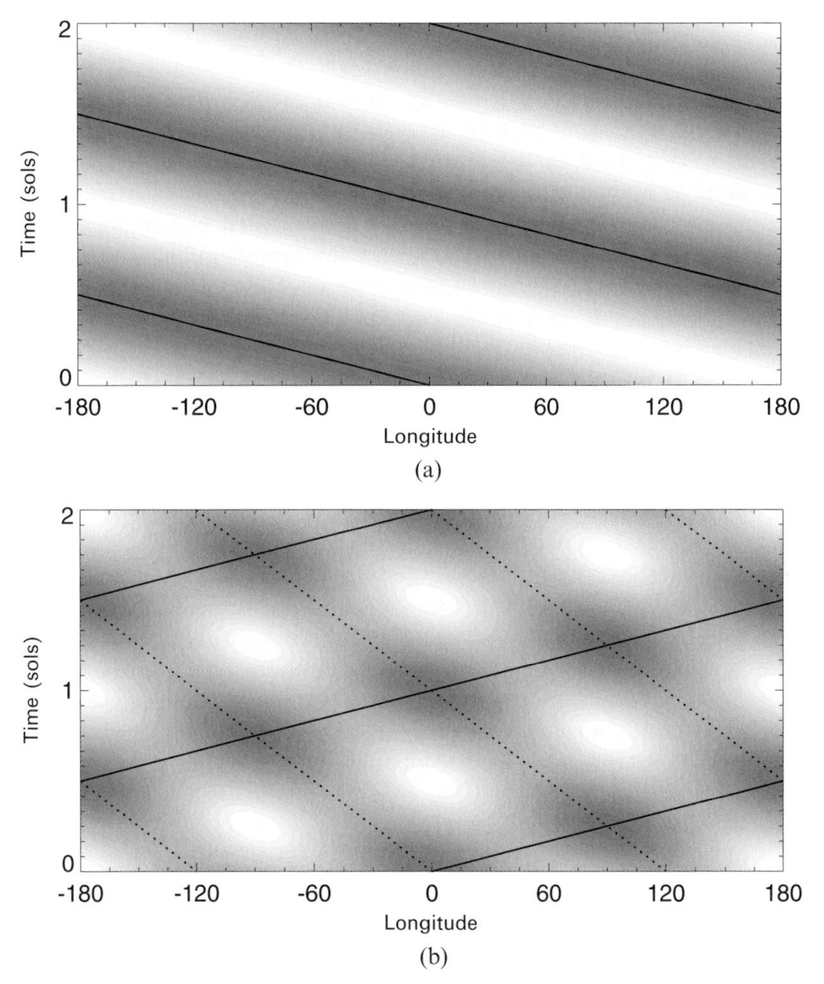

Figure 5.8. Upper panel: A single, diurnal period, longitudinal wavenumber one wave, propagating westward, shown as a function of longitude and time over 2 sol (with phase lines marked). Such a wave might be a large component of the direct response to thermal forcing. Lower panel: The same wave, modulated by a cos (2λ) stationary pattern. This has the effect of producing a wavenumber three, westward propagating wave (dotted phase lines) and a wavenumber one, eastward propagating wave (solid phase lines), both with diurnal periods.

and then the equivalent depth and hence vertical wavelength found. Rather, firstly Eq. (5.7) is solved, with suitable boundary conditions, to find eigenvalues which determine a set of equivalent depths for the free modes. Each equivalent depth can then be used to determine the frequency and wavenumber of the free mode by solving Laplace's tidal equations (Eqs 5.3–5.5).

On Earth, the most prominent free modes tend to be at longer periods than the tidal modes, a well known example being the five-day wave in the stratosphere (e.g.,

Andrews *et al.*, 1987). On Mars, in contrast, the possible range of free mode frequencies varies greatly (Zurek, 1988), and some free mode frequencies are very close to those of the most strongly driven atmospheric tides. There is the possibility, therefore, of exciting very large-scale resonant responses by tidal forcing of the natural modes of oscillation of the Martian atmosphere.

Several short periods of a few sol of the VL surface pressure record appear to show the atmosphere passing through resonant states with the growth of large wave modes with periods very close to diurnal and semidiurnal. Tillman (1988) suggests that these are evidence of the excitation of free Kelvin modes with periods which are not exactly one or one-half a sol. Two occasions when the appearance of these free modes seem clearest are prior to the 1977b and 1982 great dust storms. Tillman (1988) identifies the phase shift seen in the VL 1 pressure record, just before the 1977b dust storm, as the rapid decay of the diurnal tide and the growth of a 1.1-sol period free mode. He then hypothesizes further that the generation of this large, resonant response enables the atmospheric circulation to initiate a great dust storm. The details of such a mechanism are not clear; Zurek (1988) calculates that resonance is more likely in a clearer, colder atmosphere, such as in the northern rather than southern hemisphere summer. In southern hemisphere summer, the preferred time for great dust storms, the raised background dust levels favour a directly forced, thermal tide. Winds associated with a large amplitude, resonant free mode, if excited, might combine with the global circulation, other tidal modes, and topographic flows to produce wind speeds great enough to raise dust from the Martian surface at some locations. This is essentially a particular case of the dust storm mechanism proposed by Leovy *et al.* (1973), which requires the superposition of different flows, including the general circulation and thermal tides, to generate winds large enough to lift dust in some regions.

5.2.4 Tides in a Mars GCM

The section on planetary thermal tides concludes with an overview of tides as simulated by the full Oxford MGCM. This briefly illustrates many of the features of migrating and non-migrating thermal tides which have been discussed in previous sections.

One convenient way in which to summarize the tidal behaviour is to Fourier transform the surface pressure at the equator, in both time and longitude, in order to find the amplitudes of each tidal surface pressure mode as a function of frequency and longitudinal wavenumber, Figure 5.9. Four cases are shown, with typical background dust levels (with which the GCM broadly matches temperatures as observed by the MGS spacecraft) at both northern summer and winter solstice, with a heavy global dust storm of $\tau = 5$ at northern winter solstice and a repeat of the global dust storm in an idealized GCM with a smooth surface which has the topography, albedo, and thermal inertia fields all replaced by the Martian global-mean values. The background dust loading, used to represent a year similar to the first one observed by MGS without dust storms, was described in Chapter 2 (Figure 2.2). At northern

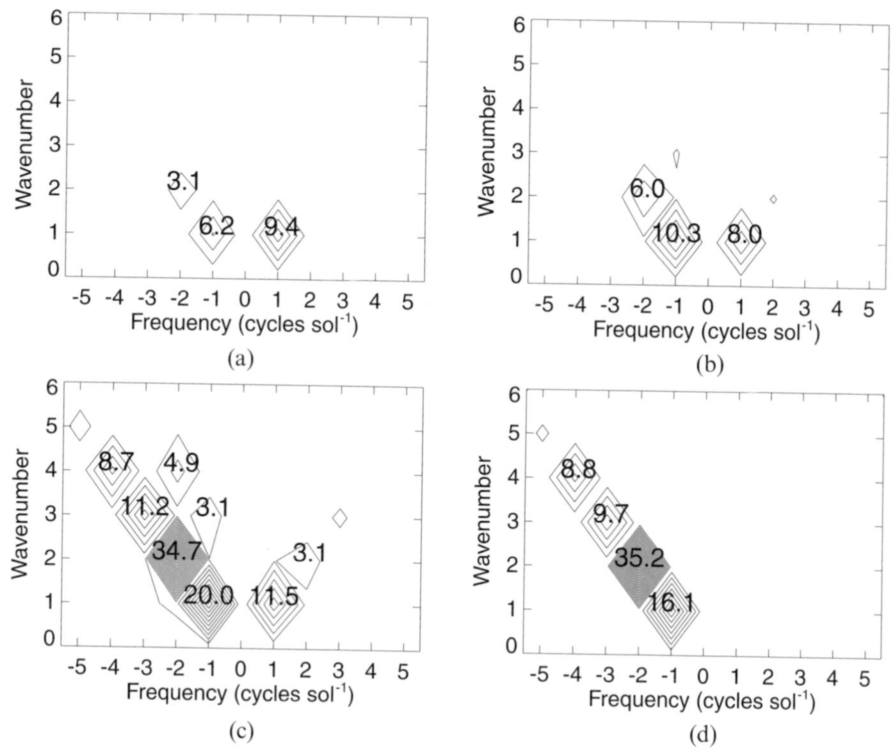

Figure 5.9. Thermal tide modes as seen in the surface pressure signal at the equator in the LMD and Oxford Mars GCM. Surface pressure data, as a function of time and longitude, was Fourier transformed in both dimensions to show the spectrum as a function of frequency in cycles/sol, 2σ, and longitudinal wavenumber, s. Negative frequencies indicate westward propagation and the sun-synchronous modes all lie on a line passing through the origin with slope -1. Contours are plotted at intervals of 2 Pa and modes with amplitude greater than 3 Pa are labelled. (a) Northern hemisphere summer solstice with background Mars Global Surveyor (MGS) dust loading. (b) Northern hemisphere winter solstice with background MGS dust loading. (c) Northern hemisphere winter solstice with a dust storm of optical depth, $\tau = 5$. (d) Northern hemisphere winter solstice with a dust storm of optical depth, $\tau = 5$, but smooth topography and thermal properties.

summer solstice the equatorial visible dust opacity used by the GCM is $\tau \approx 0.2$ and at northern winter solstice $\tau \approx 0.45$, referenced to a pressure level of 700 Pa.

Figure 5.9(a) shows the typical tidal response in the colder, clearer northern hemisphere summer. Both diurnal and semidiurnal (of half the magnitude of the diurnal) tides are visible, but the strongest mode present is a diurnal Kelvin mode, eastward propagating with longitudinal wavenumber one, which is half as strong again as the sun-following diurnal tide. This mode is close to resonance in the relatively clear Martian atmosphere, as discussed above, and is strongly excited.

Figure 5.9(b) shows a similar picture at northern hemisphere winter solstice, but here the direct thermal forcing is stronger, both because of the increased dust loading in the atmosphere and, to a lesser extent, because of the increased solar flux close to perihelion. This results in an increased diurnal and semidiurnal tide, with the semidiurnal tide slightly stronger relative to the diurnal in the presence of increased dust. Notably, the Kelvin mode has not grown in proportion, but is actually weaker than in the clear case, consistent with the idea that the normal mode is less close to resonance in the dustier atmosphere.

Figures 5.9(c) and 5.9(d) show the tidal modes in the case of heavy, global dust loading, $\tau = 5$ at 700 Pa. The increased strength of the tides, with the clear presence of more activity at higher harmonics, reflects the strong internal heating which the atmosphere is subject to, as a result of direct absorption of solar radiation by the dust. The amplitude of the semidiurnal mode now exceeds the amplitude of the diurnal mode, consistent with the ideas presented earlier that heating over a deeper range in the atmosphere may cause destructive interference in the diurnal tide, but will only enhance the semidiurnal tide with its longer vertical wavelength. A Kelvin mode which, relative to the amplitudes of the diurnal and semidiurnal modes, is actually much weaker than in the clearer atmosphere cases, and several other non-migrating tides are also present in Figure 5.9(c). Figure 5.9(d) demonstrates the role of topographic and thermal forcing in the production of these complications; on a smooth planet, without longitudinal variations in thermal properties, only a set of migrating tides are produced.

5.3 MORNING CLOUDS, FOGS, AND FROSTS

The appearance of clouds, fogs, and frosts on Mars are another phenomenon often related to the diurnal cycle. As shown later in Chapter 8, where clouds will be discussed in more detail, these phenomena were regularly observed by the Viking and MPF landers and from orbiting spacecraft. The presence of condensates in the Martian atmosphere, and on the surface, has long been known from Earth-based telescopic observations, with the Hubble Space Telescope (HST) adding to this record. It is even possible to see cloud features on Mars using a good amateur telescope, usually enhanced with a blue or violet filter.

Clouds and hazes which form over the winter poles, known as polar hoods, may contain significant amounts of carbon dioxide ice, but the white clouds observed in the Martian tropics, often over regions of high terrain such as Tharsis and Elysium in northern summer, and in the subpolar regions in winter, are predominantly composed of water ice. In contrast to the terrestrial atmosphere, the latent heat release of water condensation is probably not very important on Mars because of the tiny amounts involved; typically there are a few precipitable microns of water in total in an atmospheric column, see Chapter 8. The radiative effects of ice clouds, fogs, and frosts, however, can become large, with changes in surface albedo and water clouds contributing significantly to the observed optical depth of the Martian atmosphere at certain regions and times of year and day (Smith, 2002).

Observations of clouds related to the diurnal cycle include morning and evening or afternoon clouds. Morning clouds often recorded by astronomers are bright, isolated patches, which may actually be surface fog or frosty ground, rather than true clouds, near the morning limb. Both fogs and frosts form overnight, as the surface temperature drops below the condensation point for water (Figures 8.18 and 8.19, see colour section, show good examples). Fogs usually dissipate by mid-morning, as the atmosphere is heated and the daytime convective boundary layer develops, while the frosts may burnoff or persist for most of the Martian day, depending on the season. Early morning clouds have been recorded, both by the MPF lander (Figure 8.17, see colour section), and from satellite images, especially near the large Tharsis Ridge volcanoes (Figure 8.15, see colour section) (Pickersgill and Hunt, 1981).

Evening clouds have a similar appearance to morning clouds, though they are often larger, and form in the late afternoon on Mars as the atmosphere begins to cool, and grow into the evening. Both morning and evening clouds can be seen in Figure 5.10 (see colour section), an image of Mars taken in northern hemisphere summer, when the atmosphere is most often clear of dust and at its coolest overall.

5.4 NOCTURNAL JETS

Low-level winds are known to form in relatively flat, desert landscapes on the Earth as night falls and turbulence in the convective boundary layer dies away. Such an effect is likely to be at least as strong on Mars, under the usual clear sky conditions, and can produce a wind several hundreds of metres above the surface which is significantly greater than might be expected from geostrophic balance.

It is possible to understand the processes which might form a nocturnal jet, away from the equator, following the treatment of Blackadar (1957, 1962). The ageostrophic component of the wind, in balance with the stress in the planetary boundary layer during the daytime, undergoes an undamped inertial oscillation at night when the turbulent stresses decay.

Linearized equations for the eastward and northward wind components, in the boundary layer, may be written, based on Eqs (2.5) and (2.6):

$$\frac{\partial u}{\partial t} - fv + ru = -\frac{1}{\rho}\frac{\partial p}{\partial x} = -fv_g \tag{5.9}$$

$$\frac{\partial v}{\partial t} + fu + rv = -\frac{1}{\rho}\frac{\partial p}{\partial y} = fu_g \tag{5.10}$$

with $f = 2\Omega \sin\phi$, the Coriolis parameter at latitude ϕ; the boundary layer stress modelled as linear, Rayleigh friction, with coefficient r; and the geostrophic velocity, (u_g, v_g), which is the steady-state wind in balance with the horizontal pressure gradients outside the boundary layer.

During the day, the friction is large, and steady-state solutions to Eqs (5.9) and (5.10) are:

$$\left(1+\frac{r^2}{f^2}\right)u = u_g - \frac{r}{f}v_g \tag{5.11}$$

$$\left(1+\frac{r^2}{f^2}\right)v = v_g + \frac{r}{f}u_g \tag{5.12}$$

During the night, the boundary layer collapses ($r = 0$), and the flow becomes time dependent, governed by:

$$\frac{\partial u}{\partial t} - fv = -fv_g \tag{5.13}$$

$$\frac{\partial v}{\partial t} + fu = fu_g \tag{5.14}$$

which have solutions in the form of geostrophic flow plus a superimposed inertial oscillation:

$$u = u_g + u_{io}\cos(-ft+\phi) \tag{5.15}$$

$$v = v_g + u_{io}\sin(-ft+\phi) \tag{5.16}$$

where the amplitude of the inertial oscillation, u_{io}, and its phase, ϕ, are found by matching the nighttime solutions, Eqs (5.15) and (5.16) to the daytime winds, Eqs (5.11) and (5.12), at sunset.

Figure 5.11 illustrates this model for the case where the geostrophic wind is taken to be $u_g = 10\,\mathrm{m\,s^{-1}}$, with $v_g = 0\,\mathrm{m\,s^{-1}}$, at a latitude of 30°N ($f = 7.1\times10^{-5}\,\mathrm{s^{-1}}$), with frictional parameter $r = 3\,\mathrm{sol^{-1}} = 3.4\times10^{-5}\,\mathrm{s^{-1}}$ for the daytime boundary layer. During the day, the friction maintains a wind, deflected northward compared to the geostrophic wind, with $u = 9.0\,\mathrm{m\,s^{-1}}$ and $v = 4.3\,\mathrm{m\,s^{-1}}$ in the boundary layer. At 30°N, the inertial oscillations have a period of 1 sol. Assuming an equal night and day length, the wind will then rotate clockwise during the night so that it is blowing with a southward component of $v = -4.3\,\mathrm{m\,s^{-1}}$ by 0600LT. At higher latitudes the wind will rotate more during the night, and in the southern hemisphere the rotation will be in the opposite direction. A notable feature is the effect on the eastward wind component, which reaches $u = 14.4\,\mathrm{m\,s^{-1}}$ at about 0100LT, more than fifty per cent in excess of the slightly sub-geostrophic day-time eastward wind.

The process of the formation of a nocturnal jet can also be demonstrated in a more complicated numerical model, where the winds vary as a function of height above the surface. Figure 5.12 shows the eastward wind, u, as a function of time and height from a 1-D model based on the full physical parameterization schemes taken from the LMD and Oxford Mars GCMs and described in Chapter 2. The dynamical part of the model is represented by imposing an effective meridional pressure gradient which drives a constant $u = 10\,\mathrm{m\,s^{-1}}$ geostrophic wind throughout the model domain, and integrating equations for the wind with a Coriolis parameter

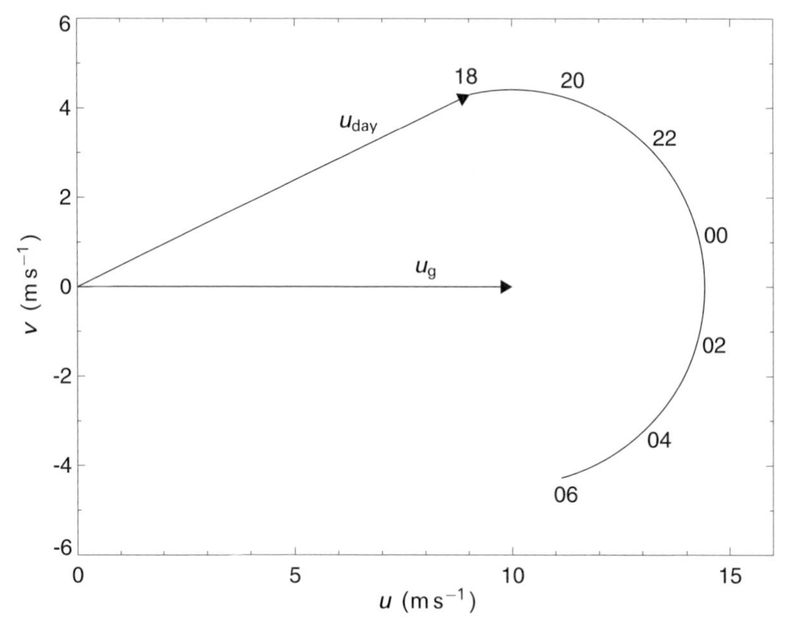

Figure 5.11. The wind vector in a classical model of a nocturnal jet. The geostrophic velocity, u_g, of the imposed pressure gradient is shown as an eastward flow of $10\,\mathrm{m\,s^{-1}}$. The day-time, steady wind produced in the frictional boundary layer is shown as u_{day}. Over the course of the night, this wind vector rotates as shown on the curve, with the night-time local time labelled in hours. Assuming that the sun rises at 0600LT, the wind will rapidly return to its day-time value. Parameters are appropriate for Mars at 30°N latitude and with r taken to be $3\,\mathrm{sol^{-1}}$ during the day.

appropriate to 30°N, adding the tendencies generated by the physical parameterization schemes. Radiative heating is calculated in full assuming a relatively clear atmosphere, with total dust optical depth set at 0.13 at the surface pressure of 610 Pa and with dust well-mixed in the lower atmosphere, under northern hemisphere spring equinox conditions.

The model domain extends in the vertical to above 80 km, but over most of this depth, away from the planetary boundary layer, the wind stays close to the imposed geostrophic eastward velocity of $10\,\mathrm{m\,s^{-1}}$. Figure 5.12 focuses upon the lowest 2 km of the atmosphere, where the wind varies strongly during the day. It can be seen that the eastward wind speed at heights of a few hundred metres varies between significantly subgeostrophic flow of less than $6\,\mathrm{m\,s^{-1}}$ during the day up to a maximum of about $15\,\mathrm{m\,s^{-1}}$ in the early hours of the morning, in a similar way to the simple analytical model described previously. There is even evidence of a slight tilt in the maximum winds, with the collapsing boundary layer releasing the constraint on the wind earlier in the evening at greater altitudes as it begins its rapid descent. Wind speeds at heights below about 50 m are always smaller, as there is a boundary layer coupling the atmosphere to the surface, even though it is shallow at night.

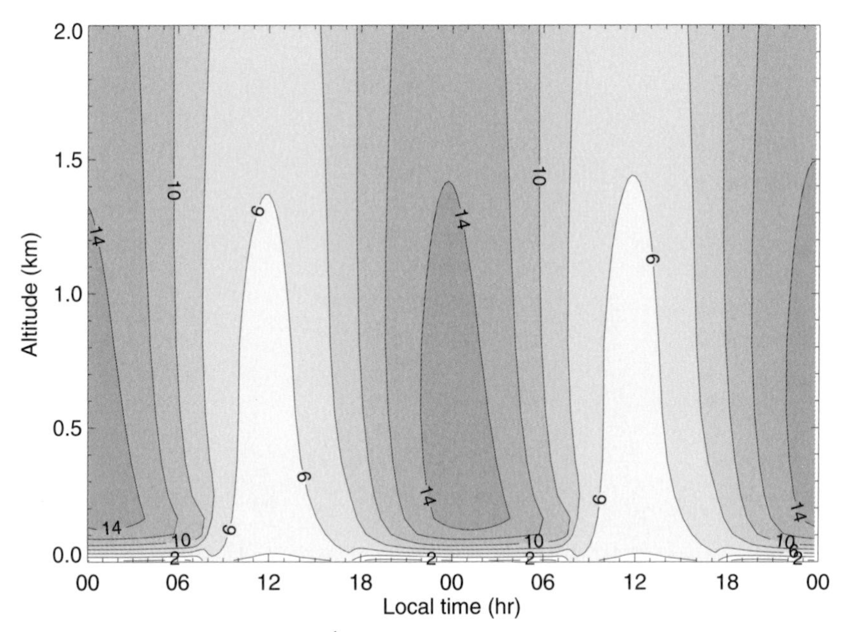

Figure 5.12. Eastward wind, u ($\mathrm{m\,s^{-1}}$), as a function of height and local time of day (over 2 sol) in a 1-D, column version of the full LMD and Oxford GCM physics, with an effective horizontal pressure gradient imposed to drive a $10\,\mathrm{m\,s^{-1}}$ eastward wind over the whole height of the atmosphere. The column parameters were set for 30°N at $L_S = 0°$. A full GCM vertical range and resolution were used, but only the lowest 2 km above the surface are shown.

Figure 5.13 shows the subgrid-scale wind standard deviation from the boundary layer scheme, q from Eq. (2.19), over the same region. This illustrates the strong vertical growth of the turbulent region during the morning, which in the full GCM may be even stronger and reach altitudes well above 10 km at its maximum, and the decay at sunset to a confined nocturnal boundary layer, perhaps 200-m thick. During daylight hours, the region of turbulent stress provides an effective drag on the wind, which is released at night as the turbulence almost vanishes except very close to the surface.

The examples of nocturnal jets shown were based on purely inertial oscillations. It is difficult to find such a 'clean' case in a full Mars GCM, as a result of a host of factors which can affect the strength of the near-surface wind. These include sloping topography, horizontal gradients in surface thermal properties, inhomogeneities in dust, tidal modes, baroclinicity, and temporal changes in the geostrophic wind. The essential mechanism of a strong boundary layer during the day providing a frictional drag to the wind and the lack of a boundary layer at night releasing the flow still applies, however. Despite the complicating factors, it is still the case that, in many locations, night-time wind speeds may exceed those during daytime in the lowest few kilometres of the atmosphere.

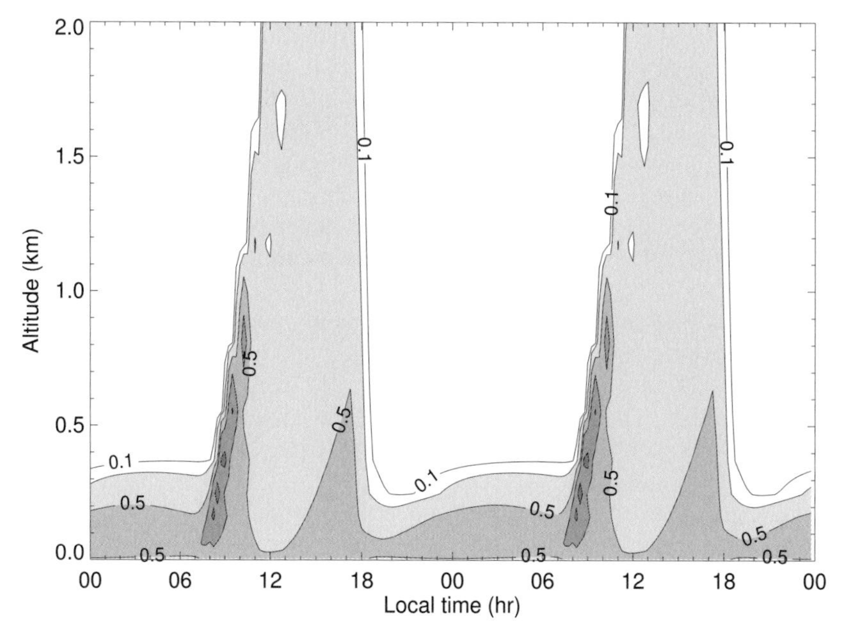

Figure 5.13. The boundary layer scheme subgrid-scale wind standard deviation, q (m s^{-1}), from the same experiment as Figure 5.12. A pseudologarithmic contour interval is used (0.1, 0.2, 0.5, 1.0, 2.0 m s^{-1}). The explosive growth in the boundary layer during the day and its rapid fall to a thin nocturnal layer can be clearly seen.

5.5 SLOPE AND THERMAL CONTRAST WINDS

On the Earth, the presence of large-scale sloping terrain is a major factor in understanding low-level wind phenomena, for example, the nocturnal jet over the Great Plains in the USA and slope winds over Antarctica or Greenland (e.g., Garratt, 1994). On smaller scales, the concepts of rising thermal circulations over heated slopes during the daytime, and of strong downslope, katabatic flows at night, are familiar. These slope winds are forced by buoyancy as air warmed by a heated slope becomes more buoyant than its surroundings and rises, or, conversely, night-time cooled, dense air forms a drainage flow downhill.

Contrasts in surface thermal properties can also produce diurnal wind effects, notably sea-breezes and land-breezes (e.g., Atkinson, 1981). Sea-breezes are winds directed from the sea on to the land, generally during the afternoon, caused by the land warming faster than the sea-surface, creating a horizontal temperature gradient. Air rises over the land, and a circulation forms with low-level horizontal winds blowing from sea to land and a weaker return flow higher in the atmosphere, with downwelling over the sea. Land-breezes are a similar night-time phenomenon in reverse, with the land cooling more quickly than the sea.

Similar phenomena can occur in the Martian atmosphere, sometimes in an even more pronounced way because of the atmosphere's low density and thermal inertia.

The surface is strongly heated by incoming solar radiation and can cool efficiently to space at night, when the atmosphere is optically thin, which is the case except under the heaviest dust storm conditions on Mars. This gives the clear possibility of strong Martian slope winds (Savijärvi and Siili, 1993). Winds caused by surface thermal contrasts are equally possible, either through albedo and thermal inertia variations (Siili, 1996), or across the edges of the polar ice caps, for example. Winds from thermal property contrasts over land sometimes occur on Earth, but are generally swamped by other effects, unlike the more extreme contrasts between land and sea, or ice.

Figure 5.14 (see colour section) shows the winds from the lowest GCM layer (about 4 m above the surface) over a limited region of a full Mars GCM experiment, at six times of day at northern spring equinox. The plots are shaded according to surface temperature. The rotation of the near-surface winds with time of day can be clearly seen, along with the development of some strong slope winds, up to $20\,\mathrm{m\,s^{-1}}$ only 4 m above the surface. In this region, slope effects tend to dominate any surface thermal effects, as appears to be the case over large regions of Mars, with the main exception being close to the polar cap edge, not shown in this diagram. Particular attention should be directed to regions around the edge of the Tharsis Ridge, and to the slopes of the major volcanoes, such as Olympus Mons, where the wind is dominated by strong upslope flow when the surface is warm and downslope flow at night when the surface is cold.

A vertical cross-section through the wind at one point, over the course of 2 sol, is shown in Figure 5.15. The point chosen for the plot, at 52.5°N, 100°W, is near the

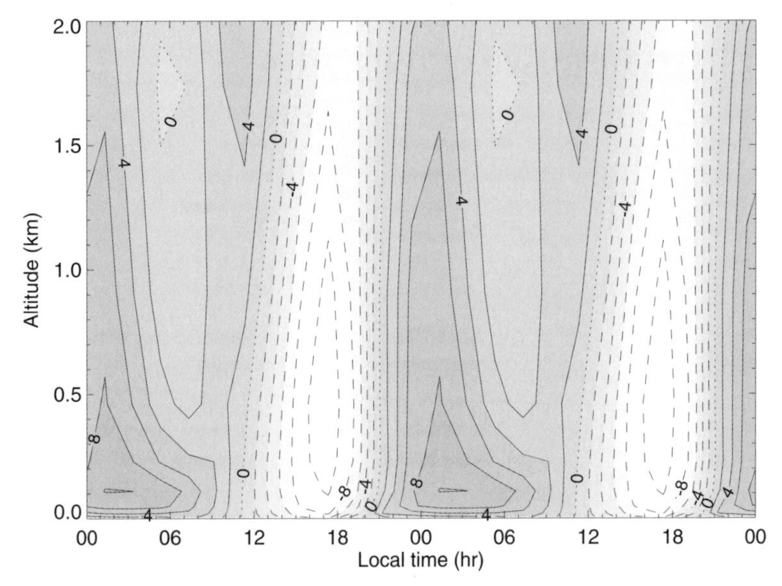

Figure 5.15. A vertical cross-section over the lowest 2 km of the atmosphere of the northward component of the wind, v, plotted against local time of day over 2 sol, at 52.5°N, 100°W, near the centre-upper part of Figure 5.14 (see colour section), at the northern edge of the Tharsis Ridge.

northern edge of the Tharsis Ridge, with a mean slope running downward in a northward direction. Figure 5.15 shows the northward component of the wind at this point, within 2 km of the surface (i.e., positive windspeeds are downslope (northward) and negative are upslope (southward) in this diagram). This is not the region of greatest slope in Figure 5.14, but it is clear that the meridional wind is reversing with a diurnal period and reaching maximum windspeeds of over $10\,\mathrm{m\,s^{-1}}$, both upslope and downslope, with a peak in the lower atmosphere between 100-m and 1-km altitude. It is noticeable that the downslope wind forms over a similar range of altitudes to the upslope wind, at the start of the night, but then becomes concentrated into a powerful jet, peaking in the lower part of the height range, close to 100-m altitude, through the middle of the night. This is typical of downhill, drainage flows and forms a contrast to the upslope wind during the day, which remains spread over a height range of more than 1-km. The reason for this is the stronger turbulent mixing in the planetary boundary layer during the day, see Figure 5.13, which mixes the flow over a deeper region. At night, the atmosphere is stable and mixing is weak, which allows the more intense and vertically-confined jet to develop.

6

Transient weather systems

6.1 VARIABLE WEATHER ON EARTH

Engage almost anyone in the UK in casual conversation and the topic of discussion is almost bound to converge eventually upon the weather. The highly variable, and notoriously unpredictable, nature of weather at mid-latitudes on Earth, especially in places like the UK on the boundary between regions dominated respectively by large land masses and oceans, makes it a favourite subject about which to complain or think about – to the extent that some say that the UK does not have a climate... only weather!

Although the weather at mid-latitudes is indeed highly variable and difficult to predict more than a few days in advance, its chaotic nature is nevertheless organized by a clear pattern of circulation features consisting of alternating low and high pressure centres (or cyclones and anticyclones). These are major features in the troposphere, especially prominent at low levels (e.g., in the mean sea level pressure patterns shown on weather maps), though also important at upper levels near the subtropical jetstreams, and show the weather to be organized on scales of around 1,000–3,000 km. On the Earth, however, these patterns seldom form particularly regular patterns, especially in the northern hemisphere, where the generally east–west alignment of trains of cyclones and anticyclones are typically broken up by the pattern of continents and mountains at middle latitudes. In the south, however, where there are fewer topographical obstacles at mid-latitudes, the pattern of cyclones can sometimes appear somewhat more regular with typically between 5 and 7 cyclones arranged around a latitude circle (e.g., see Figure 6.1). Even so, the behaviour of such terrestrial weather systems is seldom coherent for more than a few days, with individual centres undergoing characteristic life cycles of birth, development, and decay (e.g., see James, 1994).

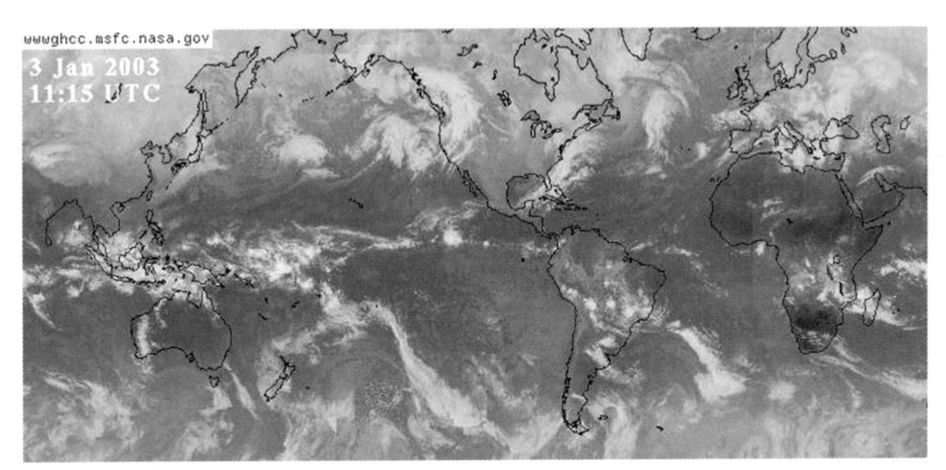

Figure 6.1. Satellite images of cloud patterns on Earth, showing the organization of mid-latitude flow into trains of cyclones (with cloudy regions) alternating with anticyclones (relatively clear).

Image from NASA.

Variable weather organized by large-scale eddies is not restricted to the middle latitudes, but also occurs from time to time in the tropics. The so-called African Wave disturbances (e.g., Holton, 1992), which develop in the tropical easterly flow close to the equator and can form the precursors to tropical cyclones, hurricanes, and typhoons, are also highly variable, but develop on scales of a few thousand km.

6.1.1 Origins as instabilities?

The origin of almost all of these weather systems is widely understood to arise from a basic instability of the large-scale, zonally symmetric pattern of winds and temperature which is produced by the differential heating between the tropics and polar regions. Historically, the first kind of instability to be studied in the context of large-scale weather systems (in the 1930s and 1940s) was what is now known as *barotropic instability*. In this case, eddies are envisaged as developing in a region where a zonal flow (deriving from other large-scale processes) varies strongly with latitude, leading to strong shear. Eddies which develop with the property that their perturbations in u and v are correlated can lead to the transport of zonal momentum in latitude. If this transport is in the appropriate sense as to lead to a reduction in the shear of the basic zonal flow, the kinetic energy of the latter will also be reduced. Since energy is conserved (in the absence of any significant friction), the kinetic energy lost by the basic zonal flow will not disappear from the system but will be converted into an increase in the kinetic energy of the eddies. Hence the original perturbations can grow in strength and the instability can develop spontaneously.

With more detailed study based on the greater availability of better observations of the atmosphere in the late 1940s and 1950s, however, it became clear that the mid-latitude weather systems have a more complicated structure than the simple barotropic eddies envisaged above (which don't change their structure with altitude, for example). In particular, they have a thermal structure which leads to the systematic transport of heat (by having perturbations in v and w correlated with those in T), as well as momentum, in both latitude and height. In practice, it was found that the energy changes produced by the heat transport within these weather systems was much more significant than their momentum transport, leading to the conclusion that mid-latitude weather systems typically derive their energy mainly by releasing *potential* energy from the basic zonal flow. Such an instability has some features (though not all, since it occurs in a stably-stratified environment, and background planetary rotation is important) in common with the basic mechanism of simple thermal convection (e.g., see Section 3.1), and is known as *baroclinic* instability or *sloping convection* (e.g., Hide and Mason, 1975). In fact the eddies developing into cyclones and anticyclones via baroclinic instability on Earth are the dominant means by which heat is transported from the equator toward the poles in the mid-latitude atmospheric circulation.

Given the importance of these major, large-scale instabilities for the Earth's atmospheric circulation, its climate, and its variability day by day, it is natural to consider what role such instabilities might have on Mars, and how they might manifest themselves. In the next section, we look at the observational evidence for organized weather patterns on Mars, and examine what form they may take.

6.2 OBSERVATIONS OF MARTIAN 'WEATHER' SYSTEMS

An observer outside the Earth would have little difficulty in detecting the large-scale organization of weather systems from the ubiquitous patterns of water cloud associated particularly with the upward moving air in cyclones (Figure 6.1). On Mars, however, condensate clouds are comparatively rare (except, perhaps the polar hood cloud in winter), and so (with a couple of exceptions discussed below) other means of detecting weather systems need to be used.

6.2.1 Cloud features and weather systems

Although condensate cloud features are less common on Mars, there have been a few observations which are suggestive of weather systems somewhat like those on Earth. Some 'front-like' cloud structures with linear organization were seen in some Mariner 9 images by Briggs and Leovy (1974), though were not extensive enough to visualize the weather pattern which was organizing the flow.

The other main category of cloud formation relevant here is that of the organized spiral (Gierasch *et al.*, 1979; Hunt and James, 1979; James *et al.*, 1999). These are comparatively rare features on Mars, and seem to occur most frequently during northern summer, close to the edge of the residual polar cap. They can take a well-developed spiral form (see Figure 6.2), with tenuous clouds of water ice winding around a centre over a region of diameter 200–500 km. They are also sometimes seen

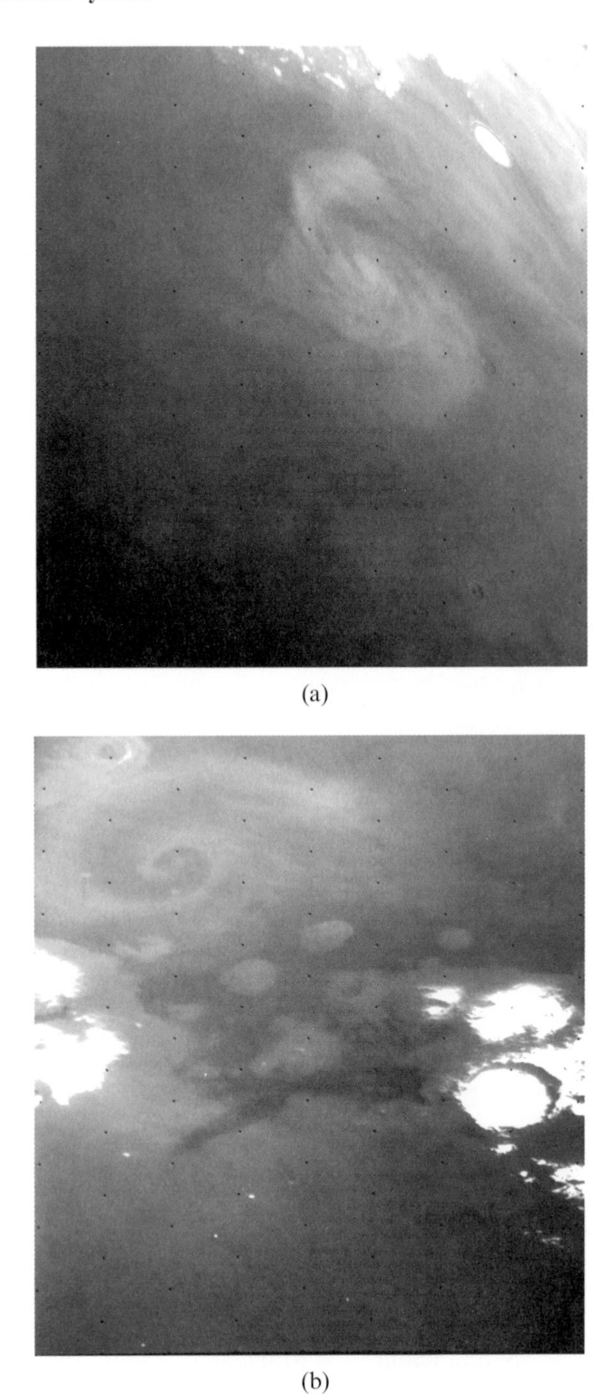

(a)

(b)

Figure 6.2. Viking Orbiter images of small-scale spiral water ice cloud features on Mars. From Gierasch *et al.* (1979); images from NASA/JPL.

to entrain dust in a developing wave structure (James *et al.*, 1999), suggesting that they are associated with the origin of certain kinds of localized dust storm (see Chapter 7).

They are comparatively small in size, and appear to be quite shallow features organizing clouds just at low levels.

6.2.2 Viking Lander (VL) time series

The first real evidence of atmospheric variability due to travelling weather disturbances on Mars came from the VL spacecraft in the late 1970s. Equipped with meteorological sensors to measure local wind speed and direction, temperature and pressure at about 1.5 m above the surface of Mars, these landers provided a nearly continuous record of weather variables for more than two Mars years (more than three in the case of VL 1). As can be seen from Figures 3.12, A.2, and 6.3, the variations of surface pressure during this period were quite different at the two landing sites. The VL 1 series, at a latitude of just 22.5°N, seems comparatively smooth throughout the year, apart from a thickening of the line during northern summer, reflecting the enhancement of the signal due to the diurnal thermal tide. In the VL 2 series, however, at a latitude of 48.0°N, the line is much more 'noisy' during winter and the seasons immediately before and after. The 'noise' apparent at this time is not due to instrumental effects, but is rather due to the passage of meteorological systems past the spacecraft. In fact a similar signal is also present in the VL 1 series, but of much smaller amplitude, indicating that the meteorological systems are

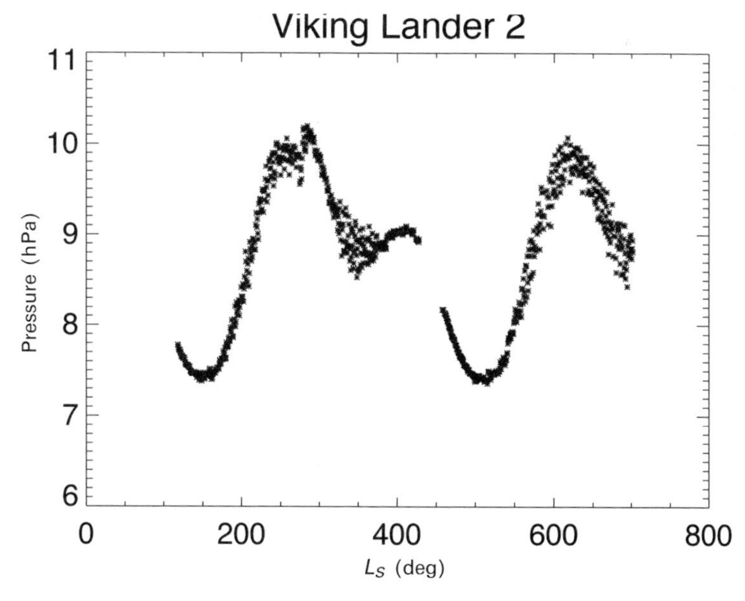

Figure 6.3. Variation of surface pressure at the Viking Lander 2 site at 48.0°N, 225.7°W measured over nearly two Mars years.

concentrated at relatively high latitudes compared with that of the VL 1 spacecraft. The 'noisy' signal disappeared altogether during the summer season from late spring to early autumn.

More detailed analyses of these time series (including the other variables, u, v and T) were carried out by the VL teams (Tillman, 1977; Ryan *et al.*, 1978; Tillman *et al.*, 1979; Sharman and Ryan, 1980; Barnes, 1980, 1981), which enabled many features of the meteorological systems to be deduced. Spectral analysis of the VL 2 data from the first northern winter–spring season showed prominent peaks with periods of 6–8 sol and around 3 sol (Tillman, 1977; Tillman *et al.*, 1979; Sharman and Ryan, 1980; Niver and Hess, 1982). Other periodicities were found at other times, however, ranging from around 8–10-sol periods found following the 1977b dust storm to less than 3–4 sol during the second Viking year when the atmosphere was significantly less dusty.

By looking at the correlation between pressure, wind, and temperature variations during the passage of the weather systems past the landers, it was also possible to deduce various properties of the wave-like patterns moving around the poles on Mars. Ryan and Henry (1979) noted that the north–south wind tended to be from a southerly direction when the pressure was falling, and vice versa when the pressure was rising at both landers. As can be seen from the schematic diagram in Figure 6.4, if the pressure variation is assumed to be due to a pressure wave of constant amplitude, moving at zonal velocity c with the meridional wind v in geostrophic balance with p, it is clear that the sense of the correlation found by Ryan and Henry (1979) is consistent with an eastward-moving wave. A more sophisticated analysis allowed Ryan *et al.* (1978) and Ryan and Henry (1979) to deduce that the zonal wavenumber of the wave train was around $m = 4$–6 with a phase speed of 5–15 m s^{-1}, and that the waves were concentrated somewhat to the north of the VL 2 site.

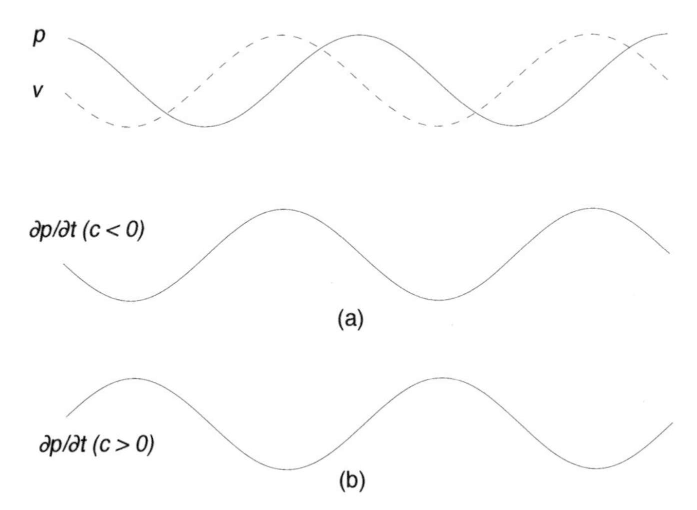

Figure 6.4. Schematic relationships between p and v in a travelling geostrophic pressure wave in the northern hemisphere, showing that if (a) $c > 0$, $\partial p/\partial t$ is anticorrelated with v, while a positive correlation of $\partial p/\partial t$ with v (b) would imply $c < 0$.

Jeff Barnes of Oregon State University carried out the most complete cross-spectral analysis to date of the VL data (Barnes, 1980, 1981), and made some remarkable discoveries about the Martian weather systems passing the spacecraft. He noted that the variations in p, u, v, and T were mutually very coherent, and that the spectra were strongly periodic for intervals of several tens of sols at various times. The latter property was especially remarkable since they suggested that the weather disturbances passing the spacecraft could be much more orderly and regular than any comparable phenomenon found on Earth. Figure 6.5 shows several examples of coherent peaks in frequency spectra taken over intervals of around 100 sols, clearly visible in all the main variables. The early period (VL 2 sols 350–440) shows clear peaks with frequencies of $\sim$0.13 and 0.22 cycles sol^{-1}, while the later period (VL 2 sols 871–999) shows a spectrum dominated by a peak at $\sim$0.42 cycles sol^{-1}. He also calculated more accurately than before the wavenumbers and phase speeds of the disturbances, and generally found phase speeds of around 10–20 m s^{-1}. Longer period oscillations (6–8 sol) were associated with lower zonal wavenumbers ($m \leq 2$), mostly during the autumn and winter seasons of both Viking Mars years, while the shorter period oscillations (2–4 sol) were generally associated with $m \simeq 3$.

Most recently, Collins *et al.* (1996) used a more sophisticated form of statistical decomposition (singular value decomposition or EOF* analysis) of the VL 2 pressure time series to show that the irregular behaviour of the data could be represented as an erratic flipping between two or more coherent states. Each one of these states corresponded to an oscillation with a particular period, suggesting that the flow was jumping back and forth between two different wavenumbers (cf Figure 6.6(b)), which shows the varying amplitudes of the two dominant periodic patterns in the time series other than the thermal tide.

Such behaviour is quite unlike that of weather systems on Earth, which tend to be much more erratic and seldom lead to time variations which can be represented as the interaction of just a few periodic patterns.

6.2.3 Travelling thermal waves

The VL spacecraft enabled scientists to acquire measurements of daily weather variations at two isolated sites on Mars. To understand the nature of Mars' weather systems, however, it is highly desirable to obtain more complete (and, ideally, at least daily) 'weather maps' of the atmosphere. On Earth, this is done by making use of a vast and dense network of daily (or usually more frequent) meteorological observations at weather stations covering the entire globe, supplemented by measurements from ships, aircraft, and orbiting spacecraft. Mars' weather network is as yet very much smaller and less well developed, though spacecraft such as Mariner 9 and MGS have been equipped with instruments capable of measuring the temperature structure of the atmosphere as they orbit the planet.

A major difficulty with trying to map weather systems from a single orbiting spacecraft, however, is that it can only observe a given part of the atmosphere for a

* Empirical Orthogonal Function (e.g., see Preisendorfer, 1988).

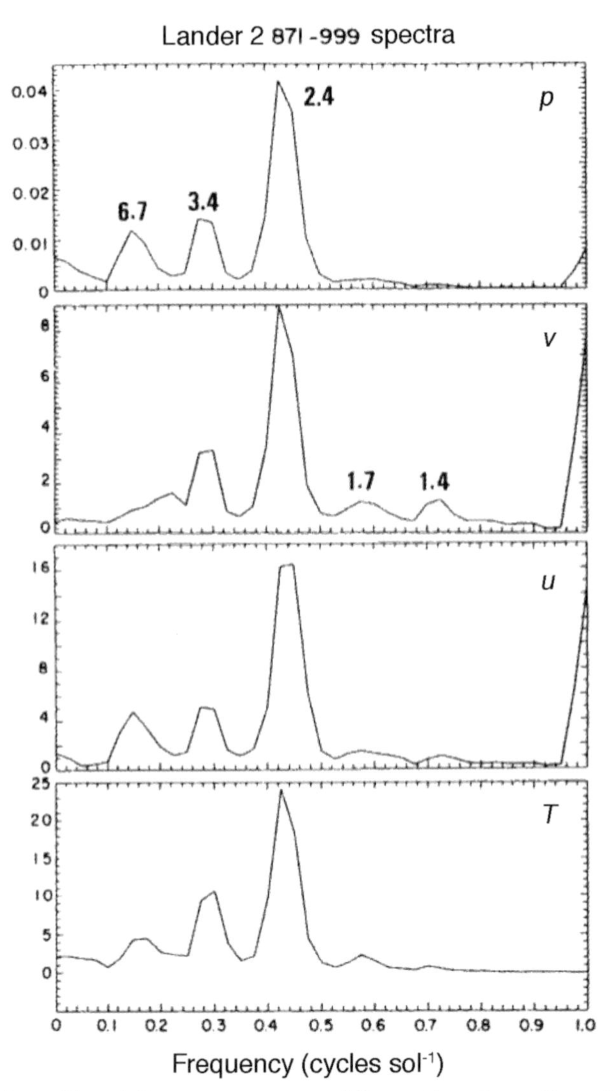

Figure 6.5. Bandpass-filtered frequency spectra of fluctuations in p, T, and wind from the Viking Lander 2 measurements for a ~128-sol period during northern winter on Mars. Peaks are labelled with periods in sols.

After Barnes 1981.

limited period during each orbit. Weather maps can then be assembled only by collecting together observations of different places obtained at different points within the orbit (or even from different orbits). As a result, it might take hours, or even days, to assemble enough observations to construct a complete map of the atmosphere. If the atmosphere changes during this time, then the map assembled from the observations may not represent a complete and self-consistent picture of the

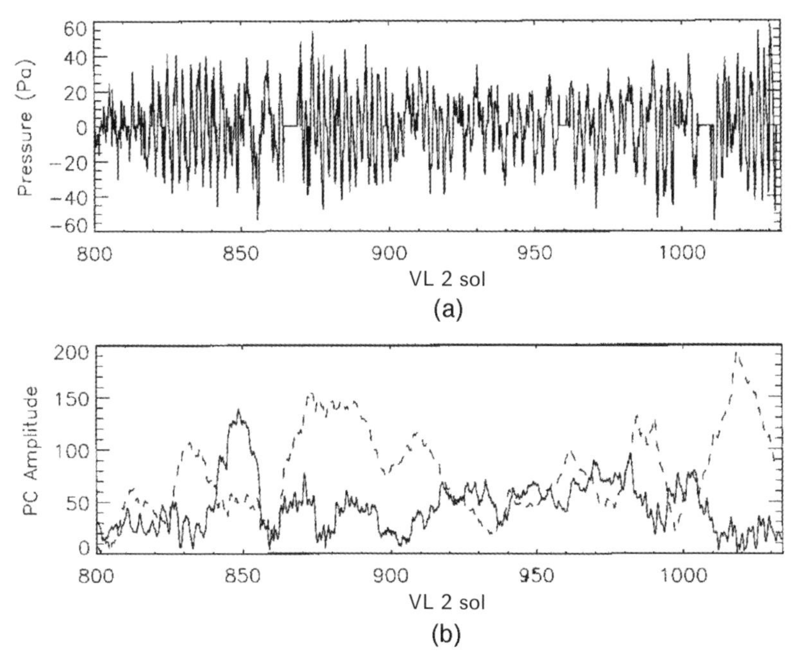

Figure 6.6. EOF analysis of VL 2 pressure time series, (a) shows the original detrended time series while (b) shows the corresponding amplitude of the two loading pairs of PCs.
From Collins *et al.* (1996).

atmosphere at any particular time, but rapidly evolving features (such as travelling weather systems) may be distorted by the interpolation of the map. In the case of a travelling wave pattern, this can lead to ambiguities in the identification of even the dominant wavenumber and the direction and speed of propagation.

This was a particular problem for the Mariner 9 spacecraft, which was placed into a highly elliptical orbit around Mars with an orbital period of around 45 hr. This meant that it would take several days to assemble enough observations, in the form of temperature profiles retrieved from nadir measurements using the Infrared Interferometer Spectrometer (IRIS) instrument, to draw up a map of the atmospheric structure. As a result, the maps obtained were somewhat ambiguous, and it was difficult to unscramble the identification of stationary from travelling wave structures. Conrath (1981) presented some clear evidence of planetary waves embedded in the northern hemisphere polar vortex from the IRIS observations, but he could not readily distinguish between the presence of travelling waves (which might be capable of accounting for the periodicities found in the VL time series) and a stationary $m = 2$ pattern (see Chapter 4).

The orbit of the MGS spacecraft is much more favourable for obtaining maps of atmospheric structure relatively quickly. With 12 near-circular, sun-synchronous orbits per sol, just 1 day's observations are sufficient to construct a reasonable map which might be capable of capturing structures provided they don't change too much. For the weather systems observed by the Viking landers, with periods

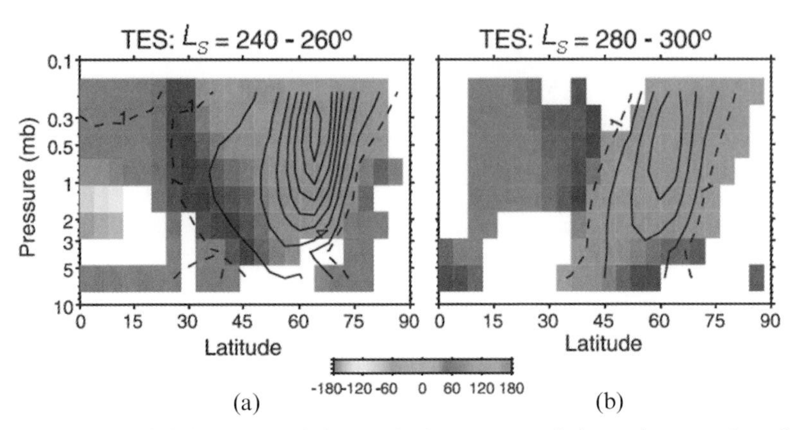

Figure 6.7. Latitude–height maps of the vertical structure of thermal waves found during northern hemisphere winter in the first Mars Global Surveyor (MGS) mapping year by Wilson *et al.* (2002). (a) $m = 1$, period 20 sol; (b) $m = 1$, period 10 sol. Contours represent temperature amplitude at intervals of 2 K (the 1 K contour is dashed) and the shading indicates the relative phase (in degrees longitude).

of 2–10 sol, the fastest moving systems are just at the margins of being confused in simple maps of MGS observations, though techniques such as Fast Fourier Asynoptic Mapping (e.g., Salby, 1982; Barnes, 2001) can be useful in this range. At the time of writing, analyses using these data were still under way. However, it has already proved possible to determine a number of features of the travelling thermal waves in the northern hemisphere. Wilson *et al.* (2002) have found evidence of a travelling $m = 1$ wave centred around 65°N specially during northern hemisphere early winter ($L_S = 220°–270°$) during the first year of the mapping phase of the mission. The strongest waves had a long period (~20 sol), though after $L_S = 270°$ they found $m = 1$ with a rather shorter period of around 10 sol.

Figure 6.7 shows latitude–height maps of the amplitudes of the two $m = 1$ waves found by Wilson *et al.* (2002), in which it is clear that the waves extend quite deeply in altitude, and are tilted towards the pole. This is in the same sense as the circumpolar jet is tilted during northern winter, which indicates a clear connection between the jet and the waves, either as a direct instability or as a neutral Rossby wave forced from elsewhere in the atmosphere. Based on GCM simulations, Wilson *et al.* (2002) interpret the shorter-period wave as a direct instability, but the longer-period feature appears to be due to forcing from the subtropics. The origins of these kinds of waves will be considered further later in this chapter.

To date, almost all of the available observations have referred to the northern hemisphere, in which it seems clear that travelling wave-like weather systems commonly occur between northern autumn and spring, often peaking in early winter and disappearing altogether during the summer season. Much less is known from observations about the southern hemisphere, however. This is partly because no lander spacecraft have yet been sent successfully to the southern hemisphere, and also because the available MGS observations, which might be able to distinguish travelling weather systems in the south, have yet to be properly analysed. However, some new observations from the radio-occultation experiment on MGS (Hinson and Wilson, 2002) have begun to reveal the possible existence of travelling waves during southern winter. Radio-occultation profiles suffer from a similar problem of intermittency of coverage to the Thermal Emission Spectrometer (TES) observations in providing vertical profiles of temperature at just one or two places per orbit. As a result, it takes at least a couple of days to build up enough information to begin to distinguish travelling and stationary structures.

A particular group of profiles were taken during midwinter in the south ($L_S \sim 134°$–$148°$) at around 68°S, and indicated the presence of a travelling wave with $m = 3$, an amplitude around 7 K at low levels (300 hPa), and apparent phase speed of 10–15 m s^{-1}. The amplitude also appeared to be strongly variable with longitude, and most activity concentrated into a possible 'storm zone' around 150°–330°E longitude. Such observations have provided but a tantalizing glimpse of the nature of weather systems in Mars' southern winter, but leave open many questions requiring further study.

6.3 BASIC THEORY OF BAROCLINIC AND BAROTROPIC WEATHER SYSTEMS

The close association observed between the travelling wave-like weather systems on Mars and the eastward circumpolar jetstreams in winter is almost certainly not an accident, but indicates a causal link between the two phenomena. As we mentioned at the start of this chapter, it is now widely recognized that the equivalent disturbances in the Earth's atmosphere arise from a fundamental instability ('baroclinic instability') of the mid-latitude zonal jetstream and its associated thermal structure. It is likely also that a similar kind of origin applies to weather systems on Mars, and we now explore this idea in a little more detail.

We have already given an outline of the physical mechanism behind one kind of instability – *barotropic instability* – in which a disturbance can transport momentum horizontally in the sense to reduce the basic horizontal shear of an initial zonal flow. Such a process can ultimately reduce the kinetic energy of the initial flow, releasing energy to feed the spontaneous growth of the momentum-transporting perturbation. It would, however, require a basic zonal flow with strong shears in latitude before anything much would happen. Now we have seen that the circumpolar jetstreams on Mars may indeed exhibit strong horizontal shears. As we have seen in Chapter 3, however, they may also have strong shears in the vertial direction too, and that may indicate the possibility of other sources of energy for growing perturbations and waves.

An alternative process which may also lead to the release of initial energy for growing perturbations, when the initial flow is stably-stratified, in geostrophic balance, and is produced by differential heating in latitude, is that of 'baroclinic instability' or 'sloping convection' (Hide and Mason, 1975). The differential heating may result in an equator–pole temperature difference which, in geostrophic

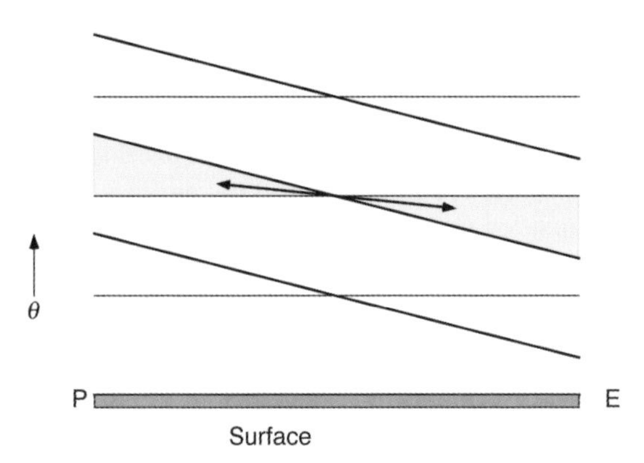

Figure 6.8. Schematic distribution of potential temperature with latitude in a rapidly rotating, differentially heated atmosphere like Mars'. Thick contours represent potential temperature and thin lines show the geopotential surfaces. Arrows indicate suitable trajectories for the notional exchange of fluid to release potential energy within the hatched 'wedge of instability' (see text).

balance, will lead to a zonal flow which has a strong vertical shear, with increasing westerly winds with height. The corresponding temperature distribution in the atmosphere would be in 'thermal wind balance' (see Chapter 3), and have potential temperature increasing with both increasing height and decreasing latitude (i.e., the potential temperature contours slope upwards from equator to pole since θ must increase with height in a stably-stratified flow), much as shown schematically in Figure 6.8.

As we showed in Chapter 3, if parcels of air are exchanged along sloping trajectories within the so-called 'wedge of instability' (Hide and Mason, 1975), now defined with respect to potential temperature θ (cf the trajectory indicated in Figure 6.8), then after the exchange the more buoyant parcel finds itself at a higher altitude and surrounded by less buoyant air, and vice versa for the less buoyant parcel. In this case, buoyancy forces will act in the same direction as the exchange itself, potential energy will be released and can enable the process to happen spontaneously without the need for an external source of energy. This latter process is what is sometimes referred to as 'sloping convection', since it works rather like conventional convection in releasing potential energy stored in the initial stratified fluid but, unlike conventional convection, it requires sloping trajectories to allow the energetically favourable transports to take place. It is also referred to as *baroclinic instability* since it must occur in a baroclinic fluid. The latter is defined as a fluid in which contours of pressure and density do not coincide – and for which contours of potential temperature (and density) must be inclined with respect to geopotentials.

6.3.1 'Classical' models of baroclinic and barotropic instability

The basic theory of these two most important processes for rapidly rotating atmospheres was worked out in detail mostly in the 1930s–1950s, mainly in the context of trying to account for large-scale wave motion and cyclogenesis in the Earth's atmosphere. These studies resulted in the development of some relatively simple, highly idealized, mathematical models of fluid motion, which have come to be regarded as the 'classical' illustrations of the basic processes. Despite the fact that these 'classical' models are highly idealized, subsequent experience has shown that their solutions contain the essence of much more complicated and realistic manifestations of the processes, and have provided valuable insights (both qualitative and quantitative) for interpreting the results of more complicated models.

The principles behind most of these models relate to the dynamics of wave-like, 'normal mode' perturbations to the basic flow whose stability is being investigated. The essential idea is to take the basic (typically 2-D) zonal flow and add a (3-D) perturbation of infinitesimal amplitude. The superposition of these two components (zonal flow and perturbation) are substituted into the equations of motion and heat transfer, and linearized (i.e., to neglect non-linear terms which involve products of two quantities of infinitesimal amplitude). The linearized equations may then (unlike their fully non-linear counterparts) be mathematically tractable, and solutions obtained, at least for some idealized cases of simple zonal flows. Instability of the basic flow may then be apparent if solutions for the perturbation can be found which retain their spatial structure but grow exponentially in amplitude with time.

Even without obtaining complete solutions, this approach can allow the formulation of some particular constraints on the form and structure of a basic flow which would permit the development of growing perturbations. Such constraints are generalizations of similar criteria derived, for example, by Rayleigh (1880), Kuo (1949), Charney and Stern (1962) (see also Holton, 1992 and Drazin and Reid, 1981), and focus attention on the structure of the distribution of quasigeostrophic potential vorticity in the basic flow. Thus, a necessary (though not sufficient) condition for growing perturbations is that the poleward gradient of zonal mean potential vorticity $\bar{q}_y$ should change sign somewhere within the domain, or at a boundary (in association with a poleward temperature gradient at the boundary). Such a criterion applies to both barotropic and baroclinic instabilities, the distinction being that for barotropic instability, $\bar{q}_y$ would change sign across a latitude circle more or less independently of altitude, whereas for baroclinic instability, $\bar{q}_y$ would change its sign in the vertical direction, or at a horizontal boundary. Thus, consideration of the structure of $\bar{q}_y$ in a basic zonal flow may enable us to rule out some types of instability without having to solve the problem in detail. The drawback of this approach is that the condition is only necessary and not sufficient – so just because it may be satisfied does not necessarily imply that instability will occur. In that case, it may be necessary to solve the problem more fully.

6.3.1.1 *The Eady model of baroclinic instability*

One of the neatest examples of this approach to the detailed solution of an instability problem is the well known Eady model of baroclinic instability (Eady,

1949). In this model, the objective is to devise a model of pure baroclinic instability (i.e., with no barotropic shear) which can be solved completely by conventional analytical mathematics, without the need to use a computer. Such a situation is extremely rare in fluid mechanics, but is particularly valuable in the present context. In common with other instability problems, the objective is to seek solutions to the linearized form of the equations of motion, most compactly represented as the quasigeostrophic potential vorticity equation. In general the linearized form of this equation (for constant N^2) is:

$$\left(\frac{\partial}{\partial t} + U\frac{\partial}{\partial x}\right)\left(\nabla_h^2\psi' + \frac{f_0^2}{N^2}\frac{\partial^2\psi'}{\partial z^2}\right) + \frac{\partial\bar{q}}{\partial y}\frac{\partial\psi'}{\partial x} = 0 \tag{6.1}$$

where ψ' is a horizontal streamfunction for the infinitesimal perturbation; U is the basic zonal flow (a function of both y and z in general); ∇_h^2 is the horizontal Laplacian; and

$$\frac{\partial\bar{q}}{\partial y} = \beta - \frac{\partial^2 U}{\partial y^2} + \frac{f_0^2}{N^2}\frac{\partial^2 U}{\partial z^2} \tag{6.2}$$

(again assuming N^2 = constant). In most circumstances, the presence of the last term on the LHS of Eq. (6.1) causes many mathematical difficulties which often make it very difficult to obtain a clear solution.

The genius of Eady's approach, however, was to choose a form of U designed to ensure $\bar{q}_y = 0$ everywhere inside the model domain so that the mathematically inconvenient term on the LHS disappears. In its simplest form, the Eady model basic state takes $U = \Lambda z$, where Λ = constant. The penalty for picking such a convenient zonal flow U is that the essence of the instability (and the way in which it satisfies the necessary condition on $\bar{q}_y$ for instability) depends on the boundary conditions. These boil down to setting the vertical velocity to zero at both the bottom, $z = -D/2$, and at the top of the model, $z = D/2$. In terms of the perturbation streamfunction ψ', this implies:

$$\left(\frac{\partial}{\partial t} + U\frac{\partial}{\partial x}\right)\frac{\partial\psi'}{\partial z} - \Lambda\frac{\partial\psi'}{\partial x} = 0 \tag{6.3}$$

at $z = \pm D/2$. By substituting a simple wave-like form for ψ' into Eqs (6.1) and (6.3) of the form:

$$\psi' = \psi_0(z)\exp\left[i(kx + ly - kct)\right] \tag{6.4}$$

the problem can be reduced to a dispersion relation for c (e.g., see Holton, 1992; James, 1994; Andrews, 2001), from which it can be shown that the phase speed is complex if the horizontal wavenumber $k \leq 2.4f_0/(ND)$. A complex phase speed implies the possibility of modes with exponentially growing amplitude, and hence instability. Furthermore, the model predicts that the horizontal wavelength of the mode with the maximum growth rate kc_i is $\lambda_{\max} \simeq 3.9ND/f_0$, at which the maximum growth rate is $kc_i \simeq 0.31(f_0/N)dU/dz$. Figure 6.9 shows a typical growth-rate curve for the Eady model, illustrating the variation of growth rate kc_i with non-dimensional wavenumber $K = f_0k/(ND)$. The real part of the phase speed also

Colour plates

igure 1.1 Panorama of the Martian surface from the Viking Lander 1 spacecraft, showing the dusty, boulder-
:rewn desert landscape with some of the Lander instruments in the foreground.

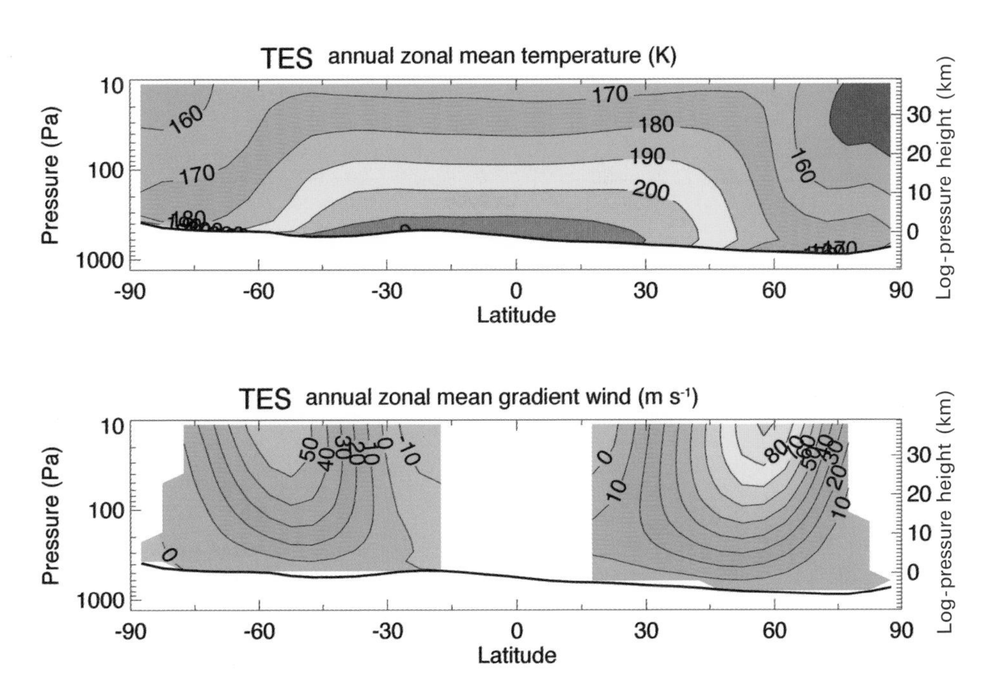

Figure 3.3 Annual and zonal mean temperature structure of the Martian atmosphere, as observed by the Thermal Emission Spectrometer (TES) instrument on board Mars Global Surveyor (MGS) during the first year of the mapping phase (late 1999–early 2001), together with the corresponding zonal wind in gradient wind balance. Contours show temperature in intervals of 5 K and zonal wind in intervals of $10\,m\,s^{-1}$.

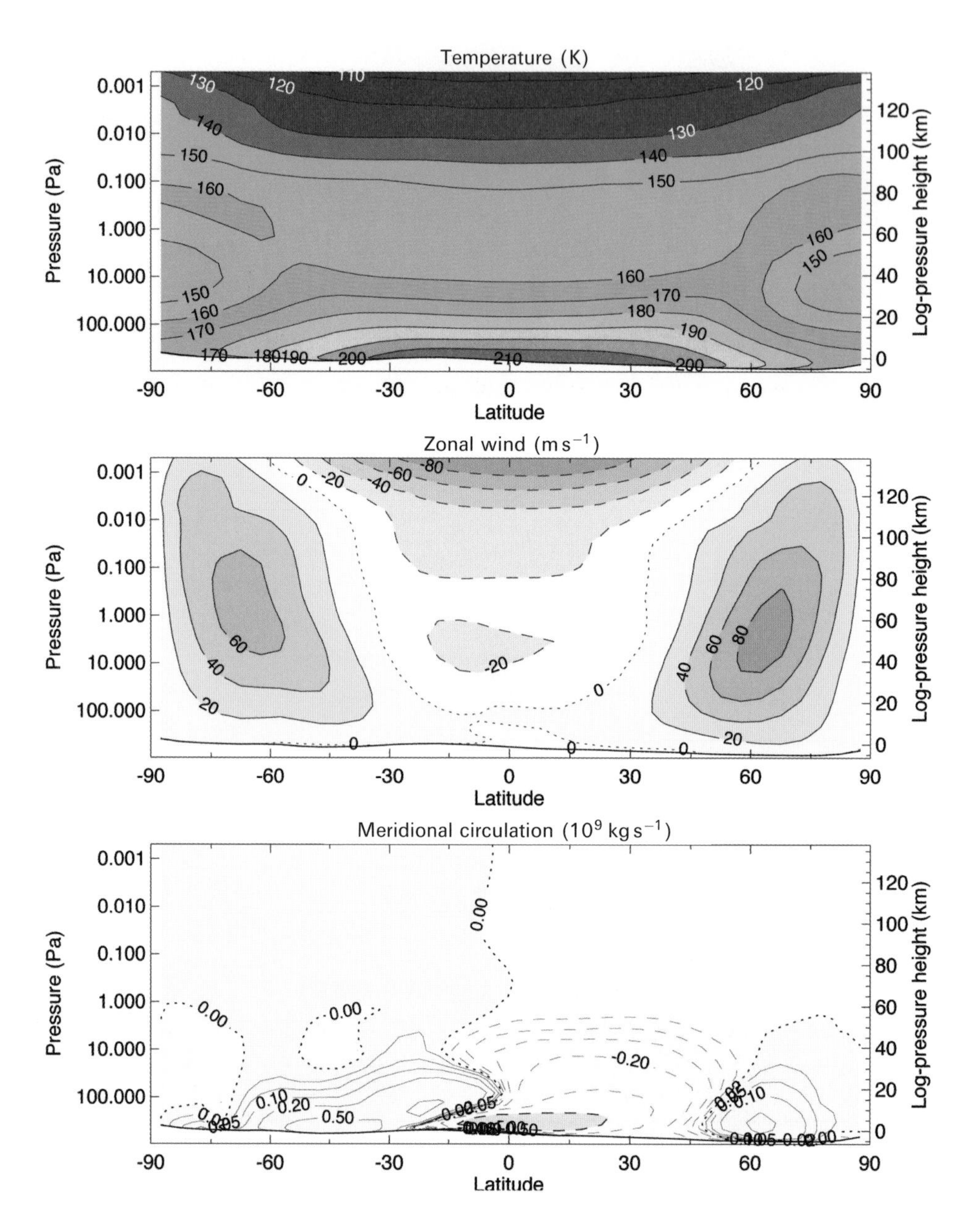

Figure 3.7 Latitude-height cross-sections of the annual and zonal mean flow in the Martian atmosphere, as simulated in the Oxford Mars GCM. The data were derived from a multiannual simulation using a distribution of dust to emulate conditions similar to those prevailing in the Viking era (see Lewis *et al.*, 1999). Frame (a) shows the zonally averaged temperature structure, (b) shows the corresponding pattern of zonal wind, and (c) the meridional mass streamfunction.

From Forget *et al.* (1999) based on data from Lewis *et al.* (1999).

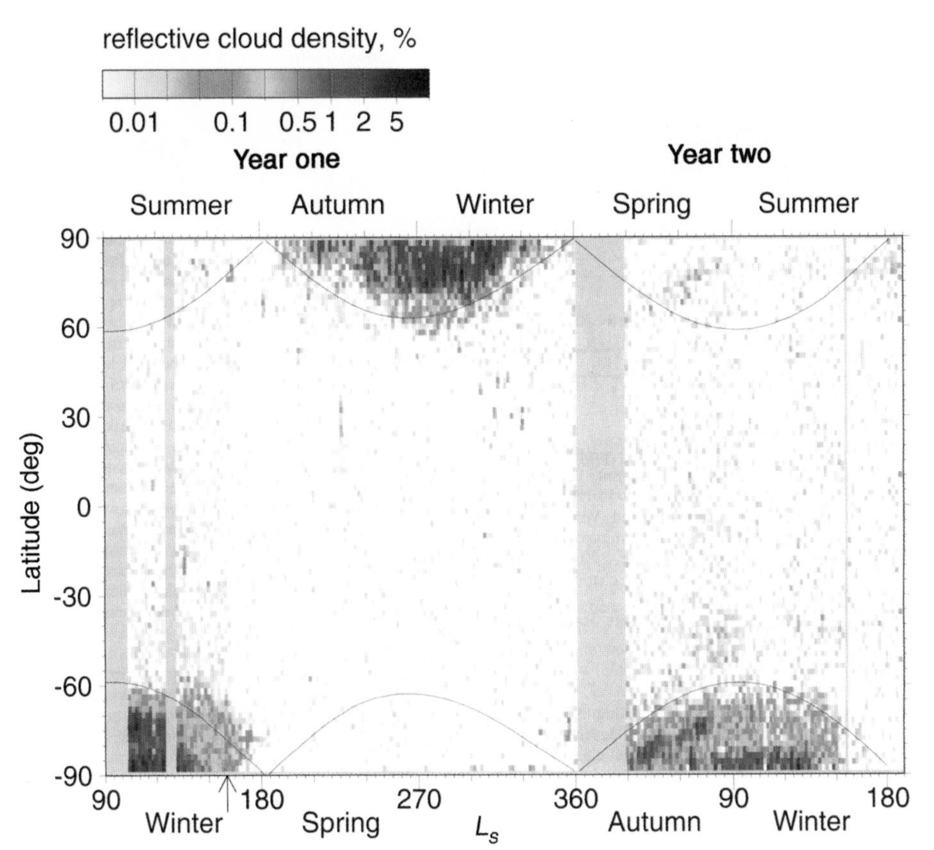

Figure 3.10 Variation of reflected returns from relatively highly reflective clouds in the Martian atmosphere, as measured by the Mars Orbiter Laser Altimeter (MOLA) instrument on Mars Global Surveyor (MGS). In this map, returns have been zonally-averaged and placed into 2° × 2° bins in latitude and L_S.

From Neumann *et al.* (2003).

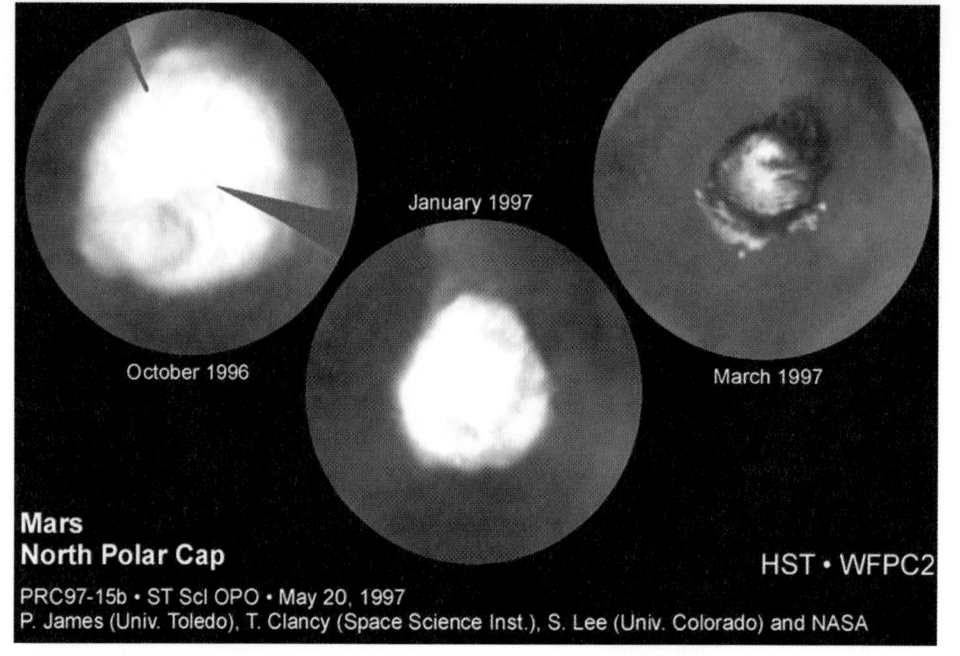

Figure 3.13 Sequence of Hubble Space Telescope (HST) images projected to show the North Pole of Mars, and illustrating the advance and retreat of the CO_2 ice cap.

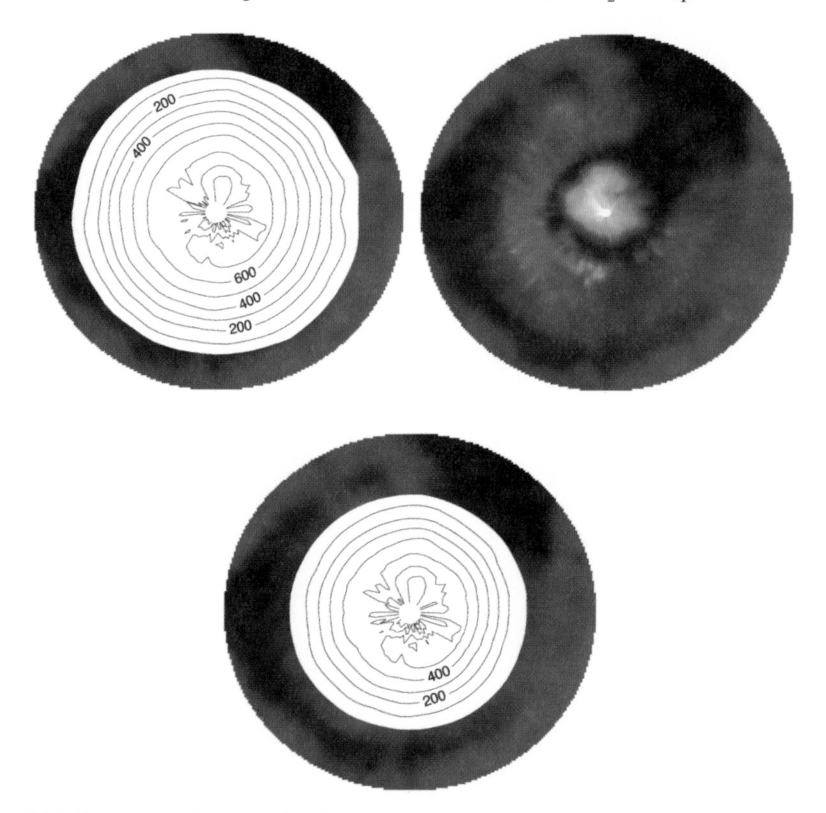

Figure 3.14 Sequence of maps of CO_2 ice cover over Mars' North Pole for seasons corresponding to the Hubble Space Telescope (HST) images in Figure 3.13 (see colour section), obtained during a simulation of Mars' seasonal variations in its atmospheric circulation in the Oxford and LMD Mars GCM under conditions close to those found during the Mars Global Surveyor (MGS) mission. Surface mass densities greater than $25\,\mathrm{kg\,m^{-2}}$ are shown in white, overlain by contours of mass density at intervals of $100\,\mathrm{kg\,m^{-2}}$. Note that all the CO_2 ice has evaporated by the time of the upper right frame.

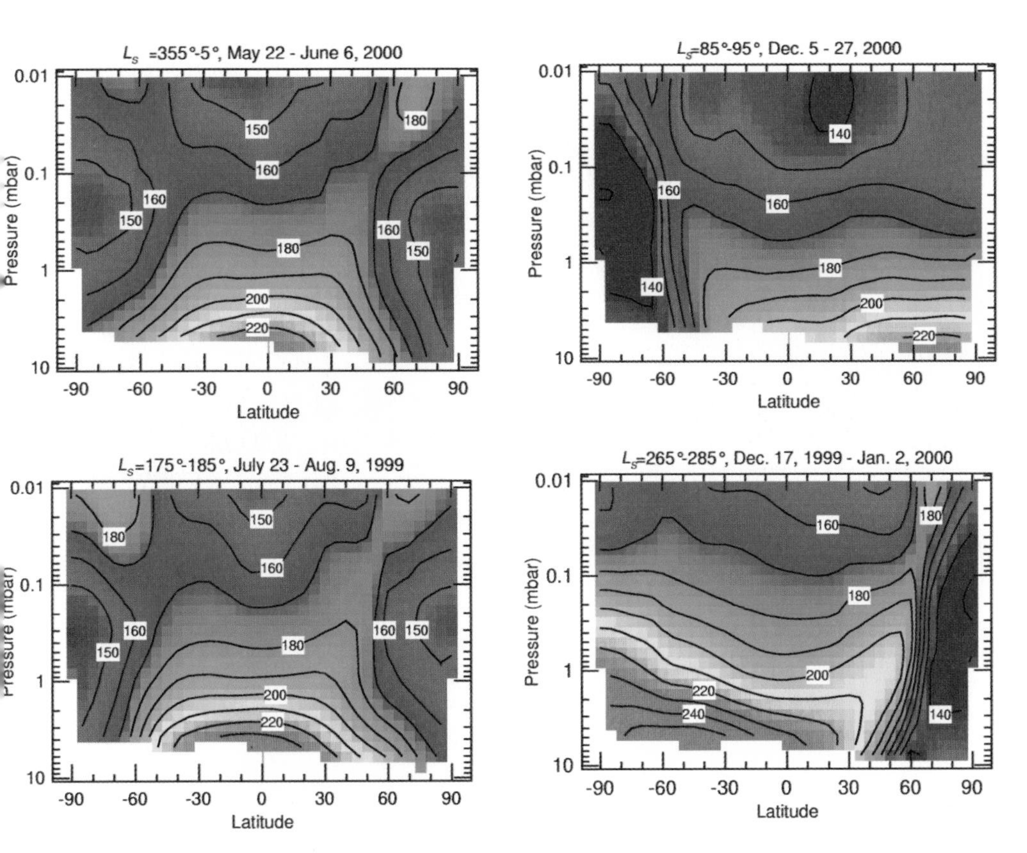

igure 3.18 Sequence of latitude–height maps of zonally averaged temperature, derived from Thermal mission Spectrometer (TES) observations from the Mars Global Surveyor (MGS) orbiter during the napping year 1999–2001. Maps are shown to represent each of the four 'cardinal' seasons, starting in the op left with northern spring ($L_S \sim 0°$), northern summer (lower left), northern autumn (upper right) and orthern winter (lower right).

dapted from Smith *et al.* (2003).

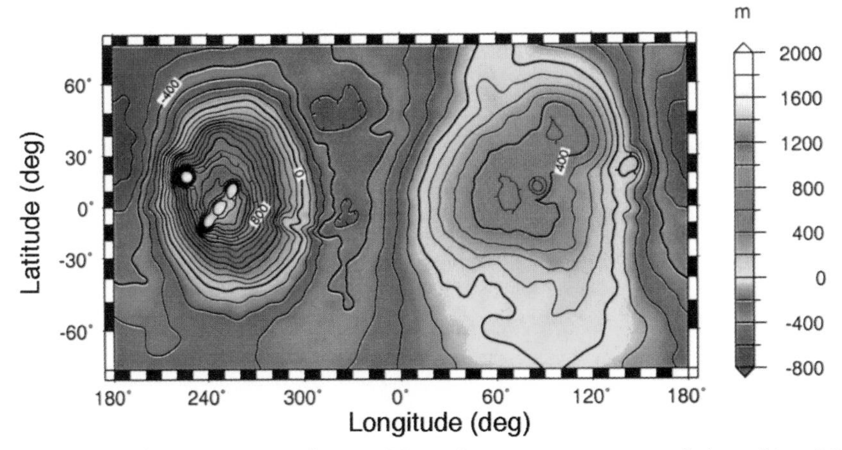

Figure 4.1 Map of gravity anomalies on Mars, from measurements of the orbit of Mars Global Surveyor (MGS). Contours show the departure of the areoid (effective mean 'sea level' on Mars) from a uniform oblate spheroid rotating at the same rate as Mars.

From Smith *et al.* (1999).

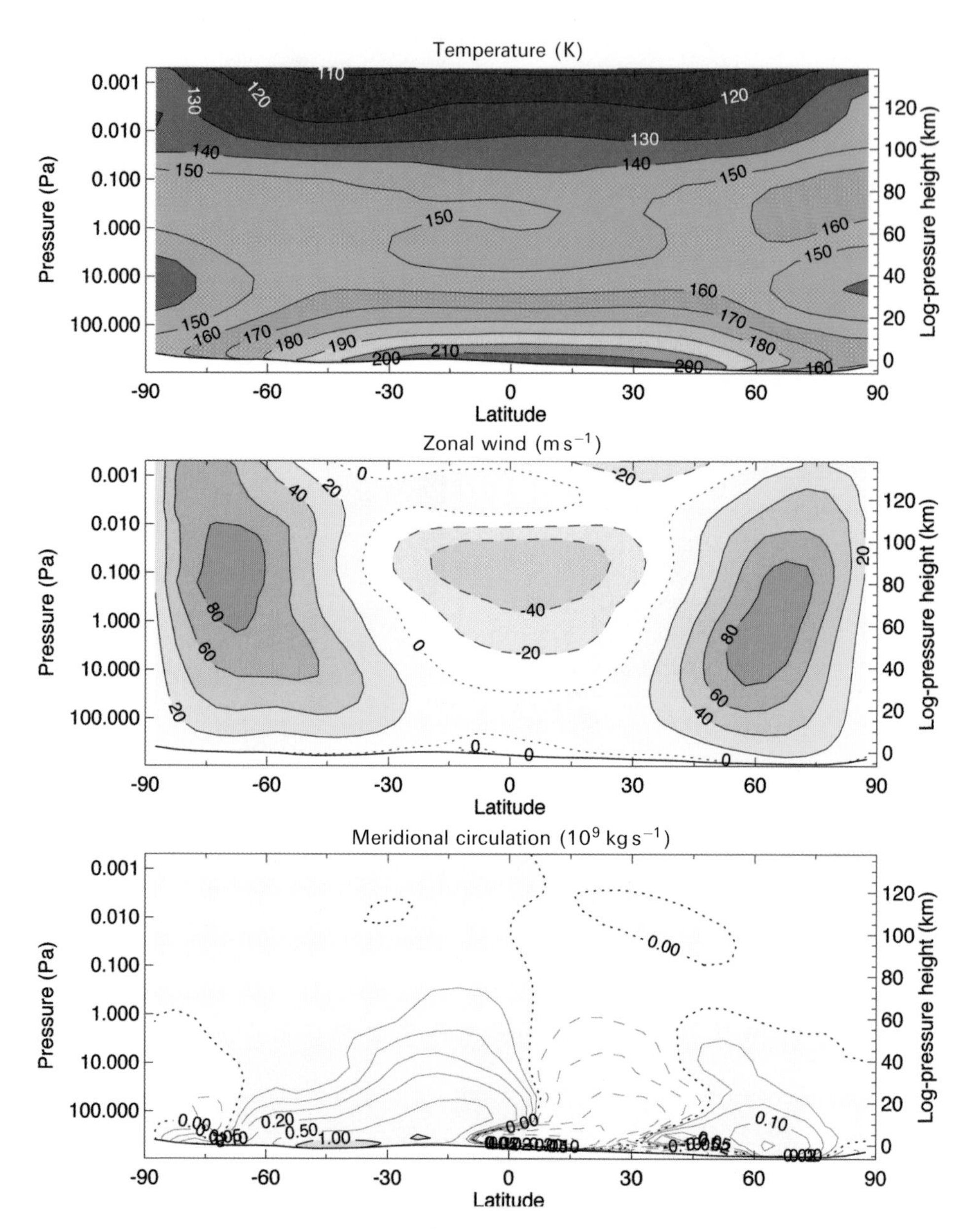

Figure 3.19 Mars GCM zonal-mean temperature, zonal wind and mean meridional circulation averaged over a 61-sol period following northern spring equinox, $L_S = 0°–30°$. For the mean meridional circulation, positive streamfunction values imply anticlockwise circulation in the plane of the paper. Note that the shading is linear, but there are pseudo-logarithmic contours around the zero value in order to show circulations in regions where they are weaker. Streamfunction contours are labelled in units of $10^9\ \mathrm{kg\,s^{-1}}$.

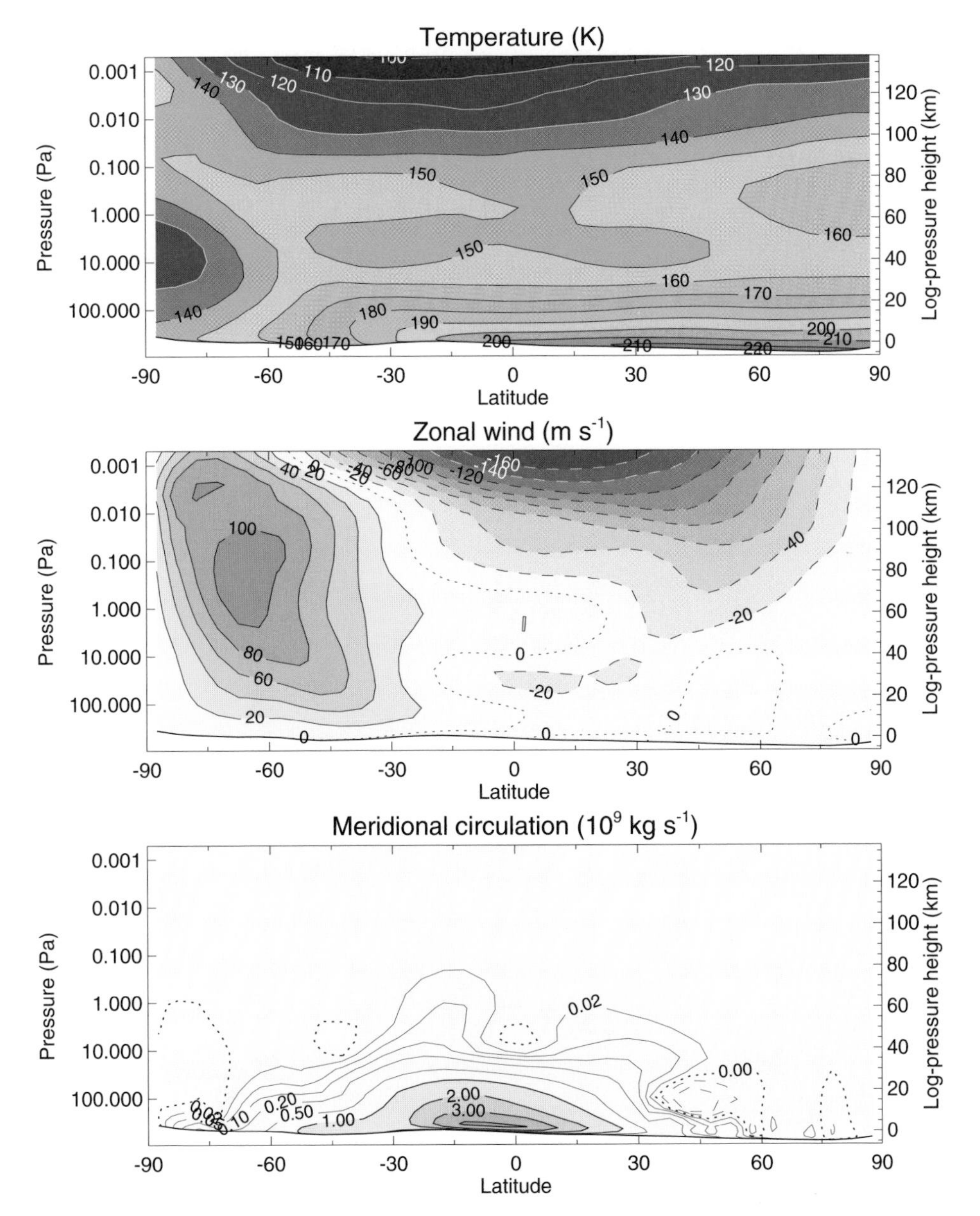

Figure 3.20 Mars GCM zonal-mean temperature, zonal wind and mean meridional circulation averaged over a 65-sol period following northern summer solstice, $L_S = 90^\circ$–120°.

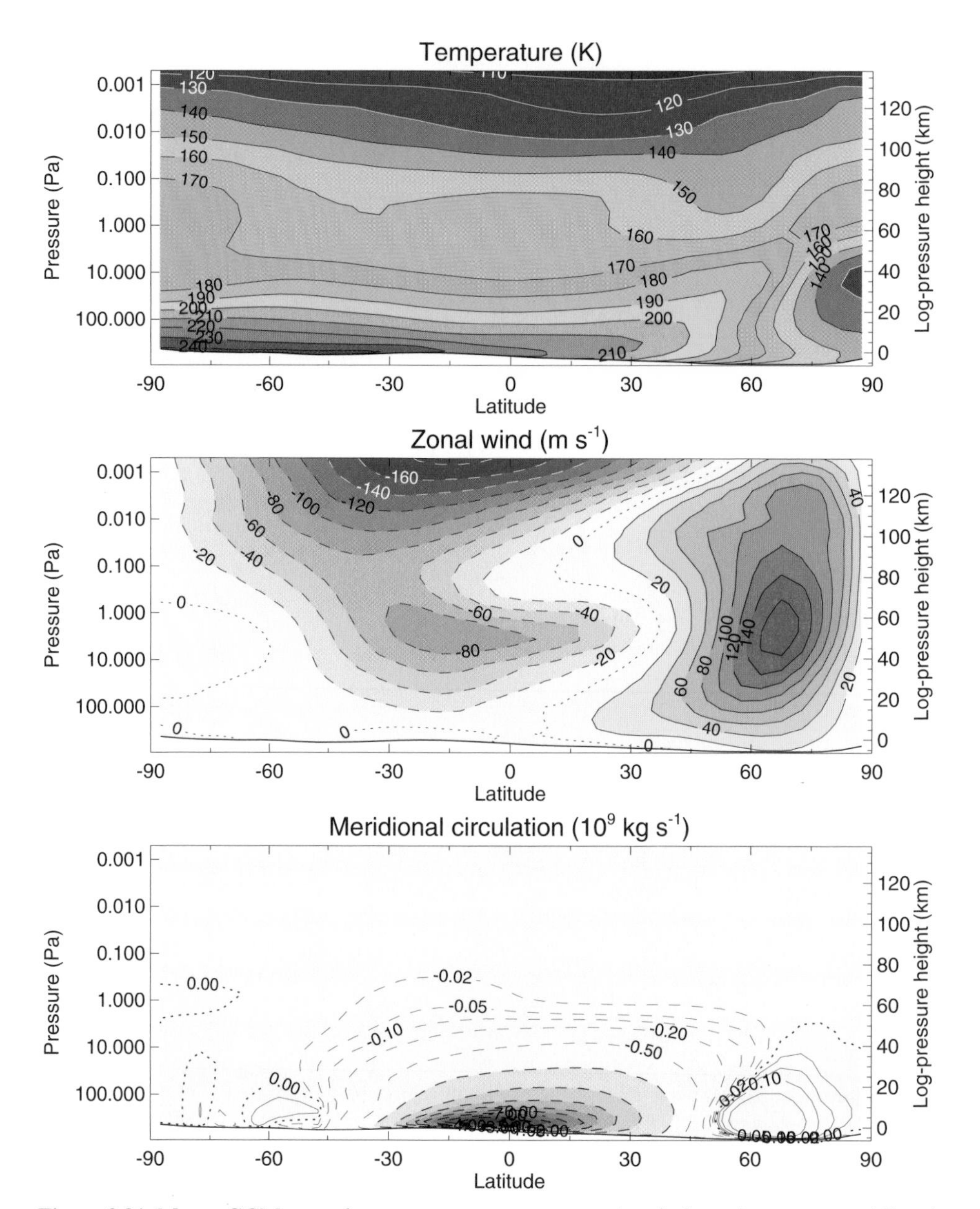

Figure 3.21 Mars GCM zonal-mean temperature, zonal wind and mean meridional circulation averaged over a 47-sol period following northern winter solstice, $L_S = 270°–300°$.

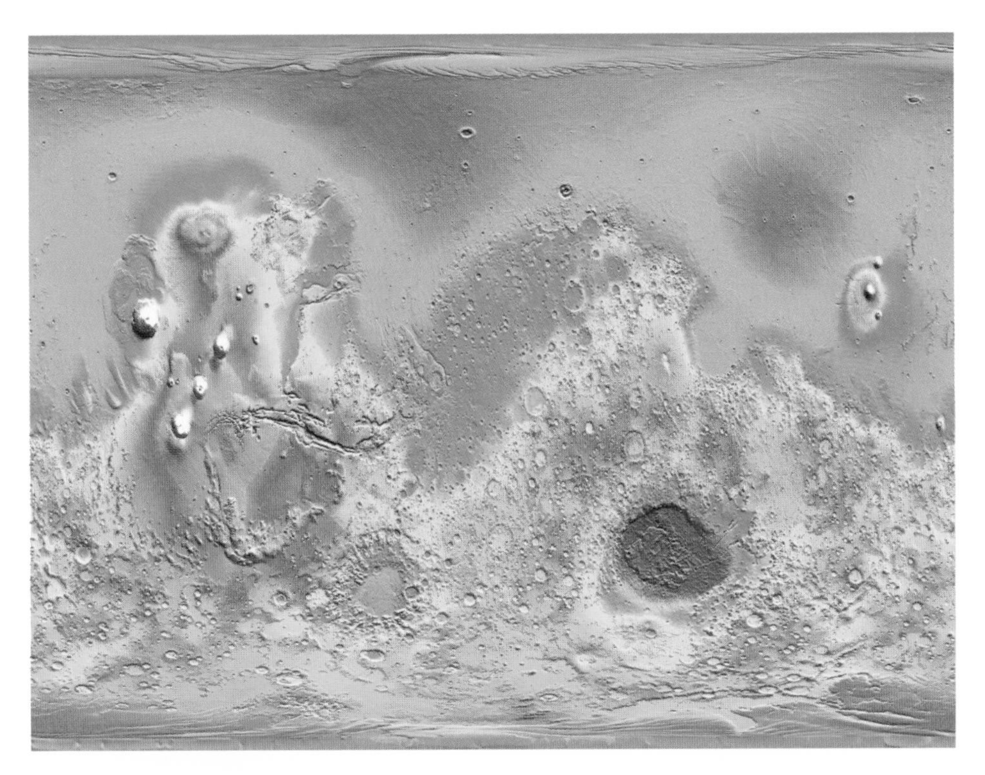

Figure 4.2 Map of Mars' surface topography, relative to an areopotential surface corresponding to a mean atmospheric surface pressure of 6.1 hPa, derived by the Mars Global Surveyor (MGS) Mars Orbiter Laser Altimeter (MOLA) team.

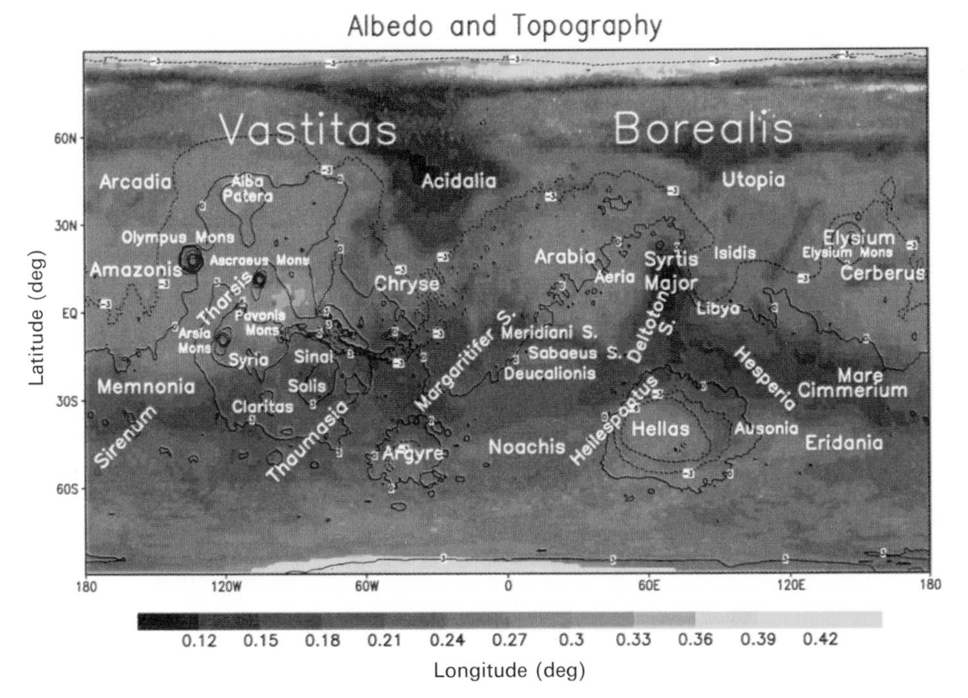

Figure 4.3 Labelled map of Mars, showing variations in surface albedo (colour) and contours of surface height every 3 km (derived from the Mars Orbiter Laser Altimeter (MOLA) digital terrain model of Mars).

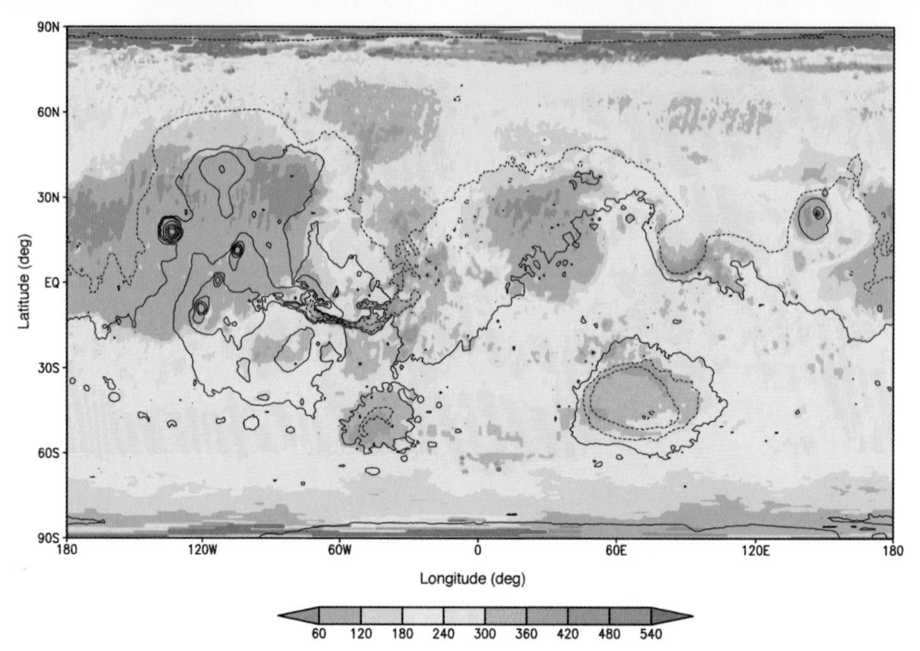

Figure 4.4 Map of the distribution of surface thermal inertia on Mars, measured in the infrared from variations in the surface response to the diurnal cycle in units of $\mathrm{J\,m^{-2}\,s^{-1/2}\,K^{-1}}$, superposed on contours of topographic height from the Mars Orbiter Laser Altimeter (MOLA) digital terrain model (at intervals of 2 km).

From Mellon *et al.* (2000), combined with adjusted polar fields from Paige *et al.* (1994) and Paige and Keegan (1994).

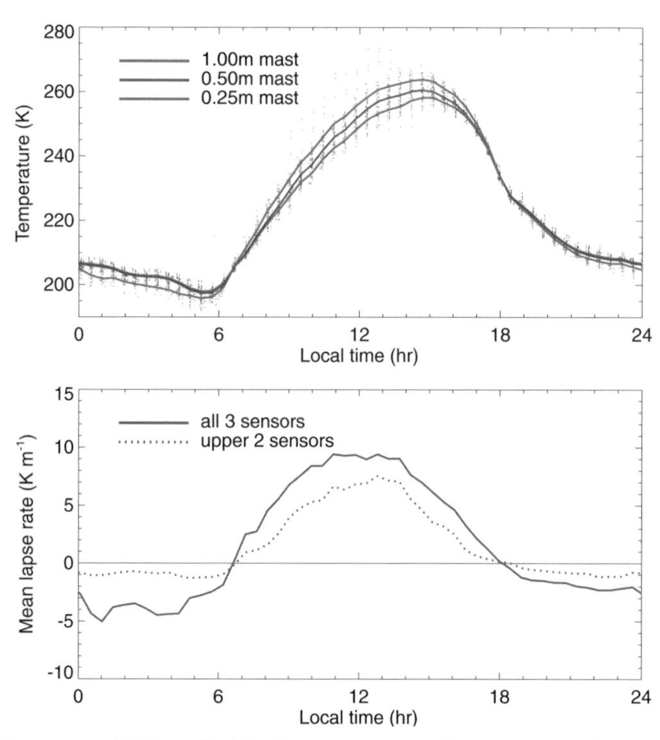

Figure 5.1 Upper panel: Mars Pathfinder temperature data, measured at 25 cm, 50 cm, and 1 m points on the meteorological mast over the first 30 sol of the mission against local time of day. Individual measurements are shown as small dots and a mean temperature cycle over all 30 sol for each point on the mast is shown as a line. Lower panel: A simple estimate of mean lapse rate as a function of time of day obtained by a linear fit to either all three sensors, or just to the upper two at 50 cm and 1 m.

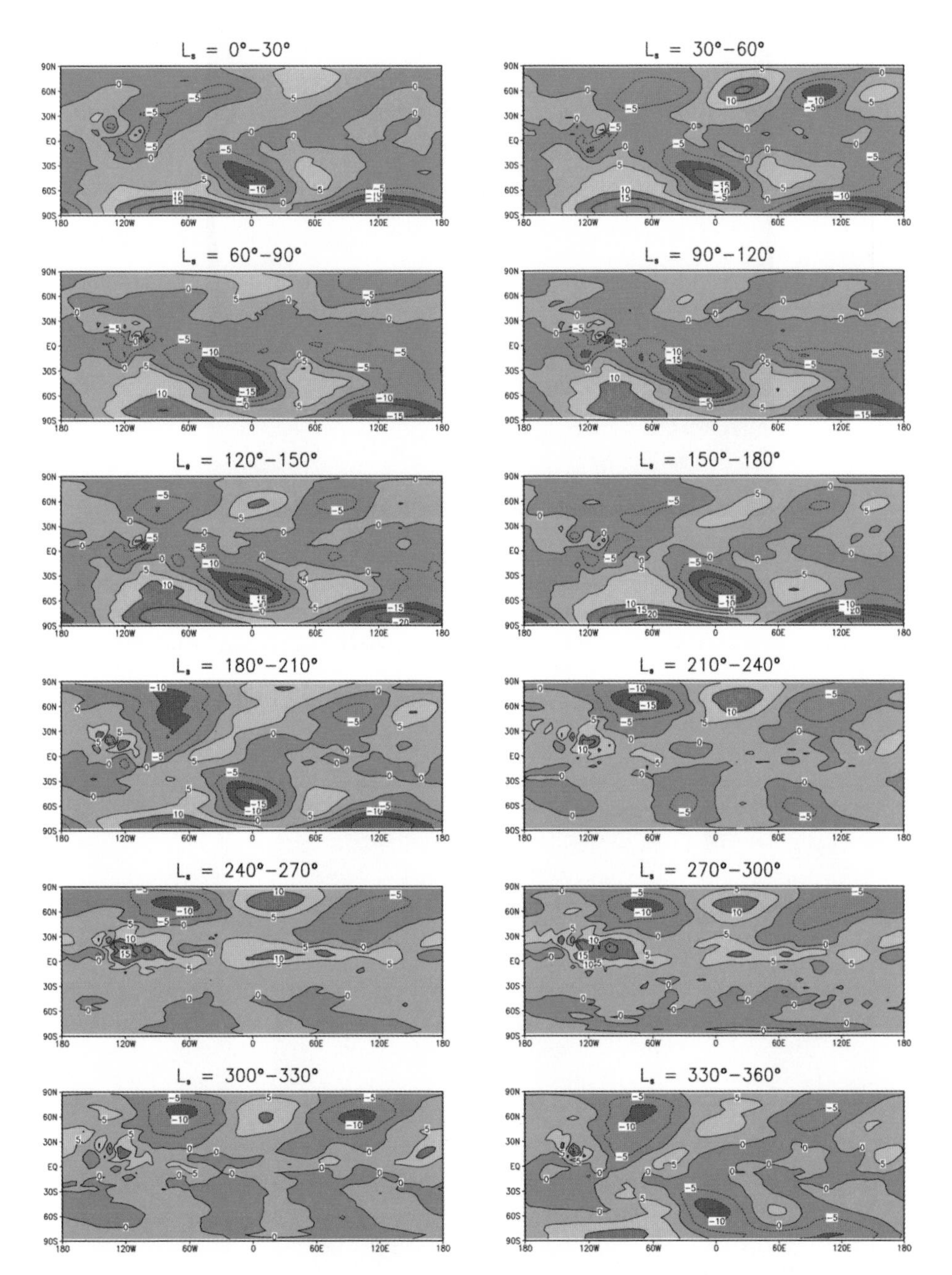

Figure 4.10 Sequence of maps of meridional (northward) wind ($\mathrm{m\,s^{-1}}$), averaged over 30° in L_S for each of the 12 'months' as defined by Lewis *et al.* (1999), taken from simulations using the Oxford version of the European Mars GCM. Maps are shown interpolated onto an altitude of approximately 21 km ($p \sim 1$ hPa) above the mean areoid.

From Forget *et al.* (1999) based on data of Lewis *et al.* (1999).

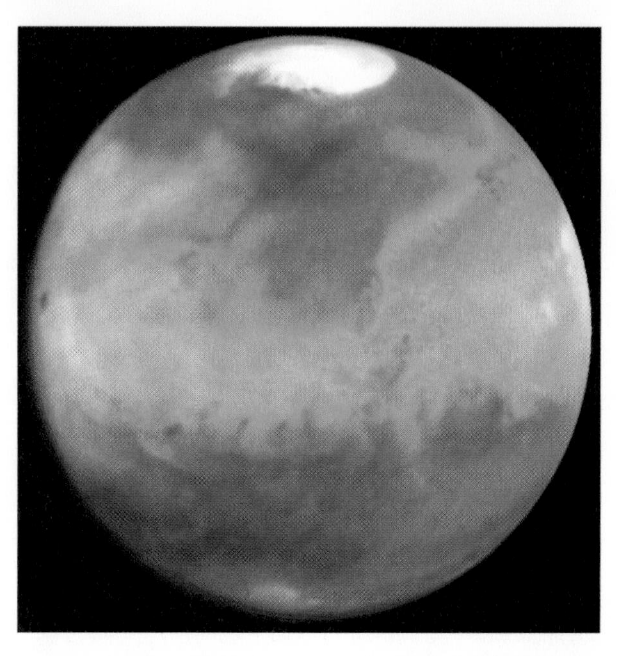

Figure 5.10 An image of Mars from the Hubble Space Telescope (HST) in northern hemisphere summer, viewed from almost the same angle as the Sun with local noon close to the centre of the image. The permanent ice cap on the North Pole can be seen in sunlight toward the top of the image. Toward the south, at the bottom of the image, there are signs of winter polar clouds. On the left side of the disc the white colour is mainly morning fog and frost over the Tharsis region, with the top of Olympus Mons still visible. On the right side, evening clouds are forming over Syrtis Major.

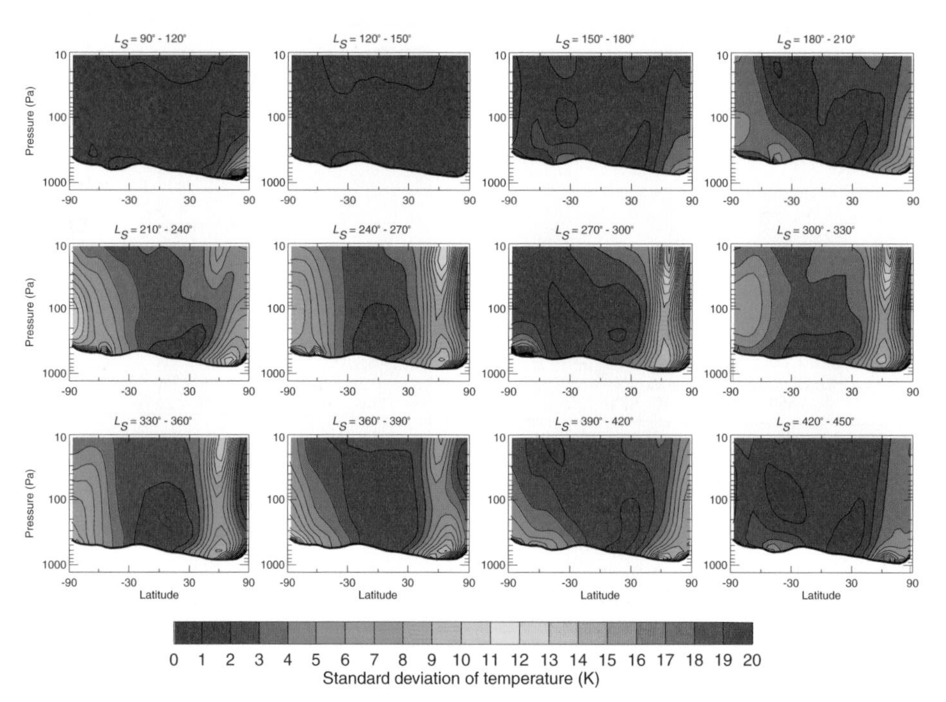

Figure 6.13 Series of latitude–height maps of transient rms eddy temperature perturbations, representing the (non-diurnal) eddy activity due mainly to mid-latitude weather systems at various martian 'seasons', as simulated by the Oxford Mars GCM.

From Forget *et al.* (1999) based on data of Lewis *et al.* (1999).

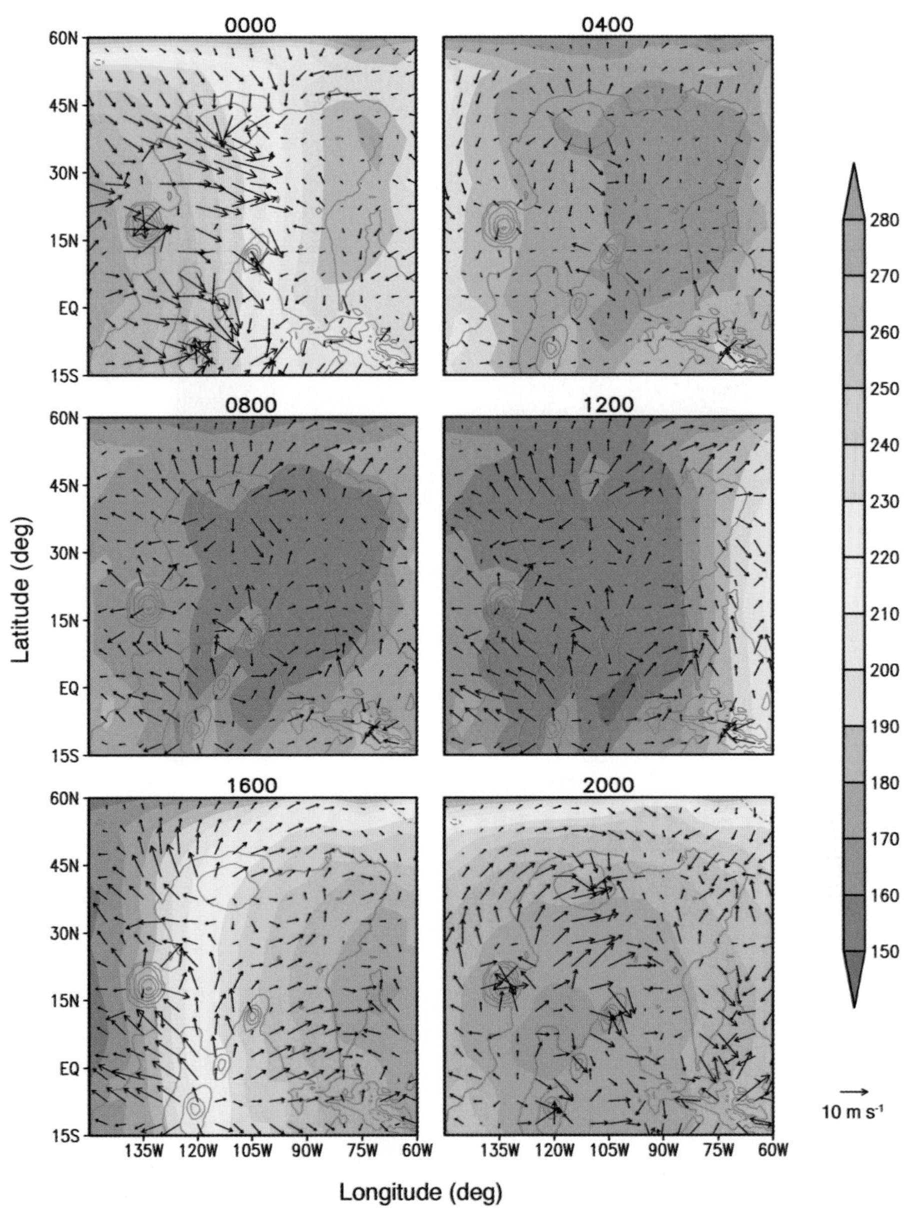

Figure 5.14 Surface temperature (colours), near-surface winds (vectors), and surface topography (grey contours) from a limited region of the GCM (15°S–60°N, 150°W–60°W), which covers the Tharsis Ridge, with Olympus Mons near the centre-left and Vallis Marineris near the lower-right. Plots are shown for six universal times, with the time at 0° longitude labelled at the top of each plot, at northern spring equinox, $L_S = 0°$. The local time on the plot is one hour earlier for every 15°W of longitude (i.e., ten hours earlier at the left of the plot, to four hours earlier at the right).

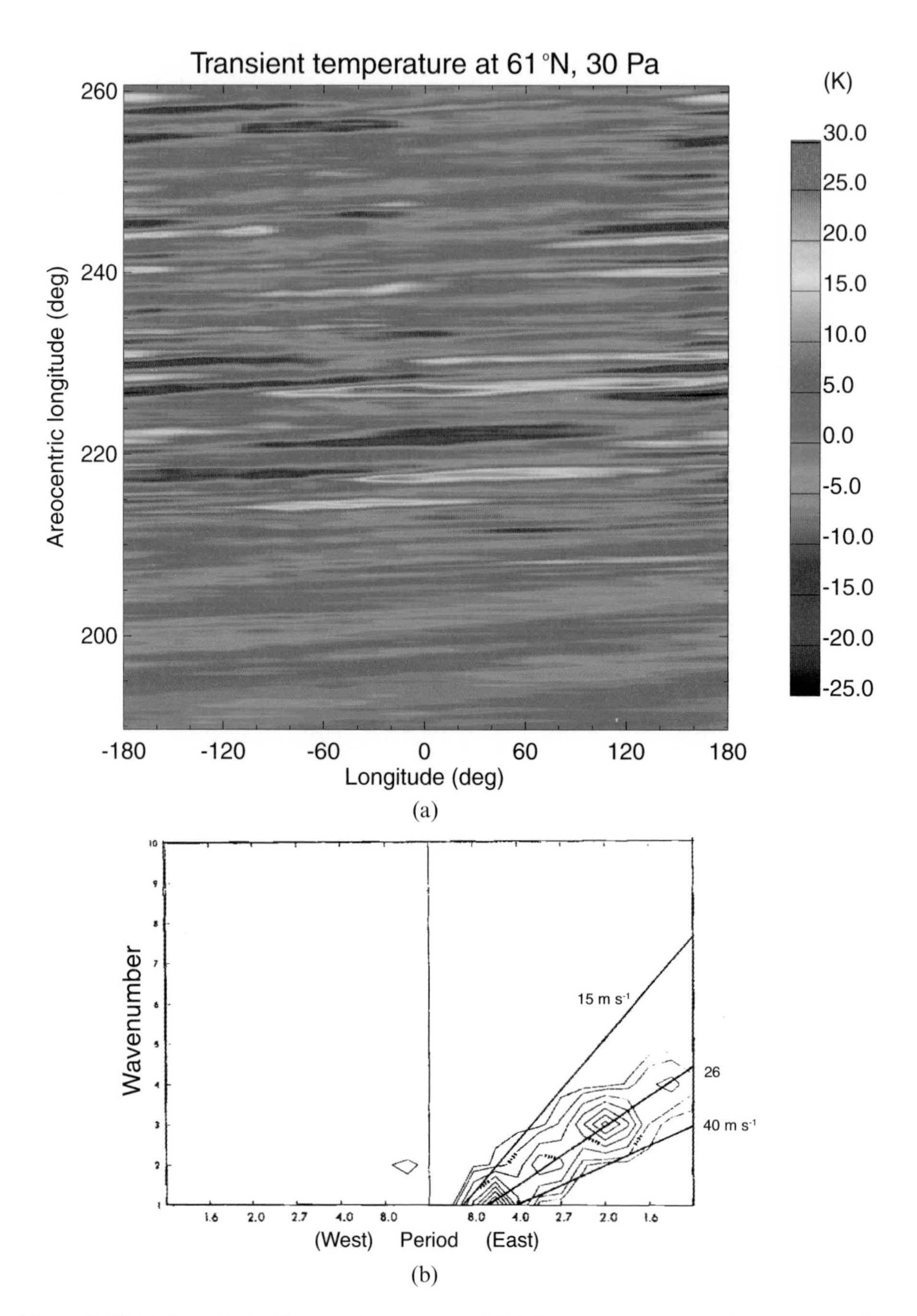

Figure 6.15 (a) Longitude–time contour map and (b) frequency–wavenumber spectrum from an analysis of the space–time structure in (a) temperature and (b) geopotential height, of travelling baroclinic weather systems. (a) is from a numerical simulation of the northern autumn circulation at 61°N using the Oxford Mars GCM, while (b) is a space–time spectrum. (b) From Barnes *et al.* (1993).

Figure 7.1 Image of the landscape around the Pathfinder Lander, taken by the camera on board the Pathfinder Lander spacecraft.

From NASA/JPL.

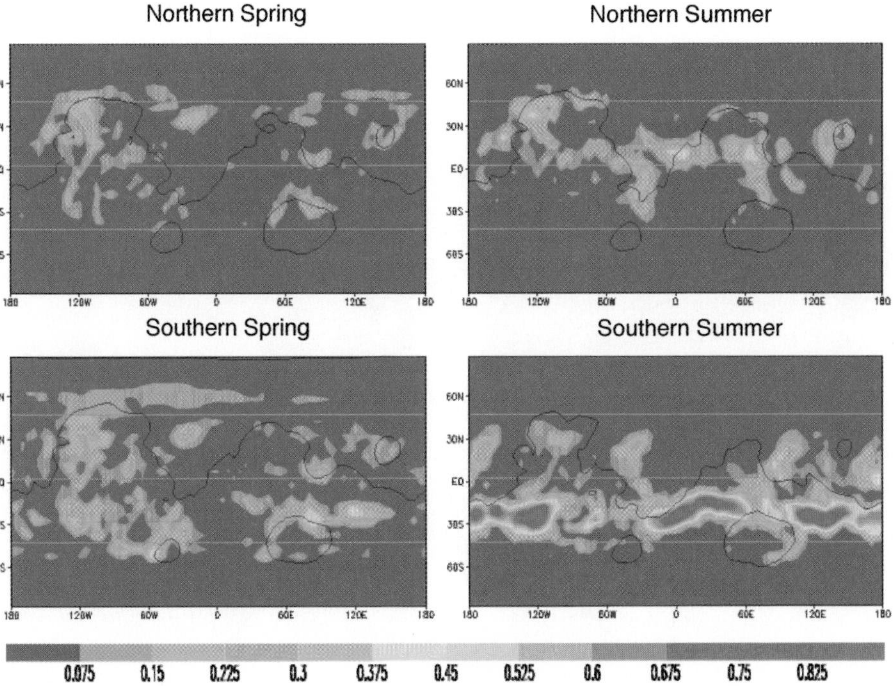

Figure 7.13 Shaded contour maps in longitude and latitude of the horizontal, seasonally-averaged distribution of dust devil activity index, Λ, for each of the four main seasons during the Martian year.

After Newman (2001).

Figure 7.14 Photograph of a haboob dust storm on Earth, in the High Plains of eastern Colorado.

From David O. Blanchard (www.dblanchard.net/script/main.html), with permission.

Figure 7.15 Viking image of a local dust storm, which occurred towards the end of the 1977 great dust storm during early afternoon. This image (VO image 248857) is about 700 km wide, and the storm cloud is about 200 km across.

From NASA/JPL.

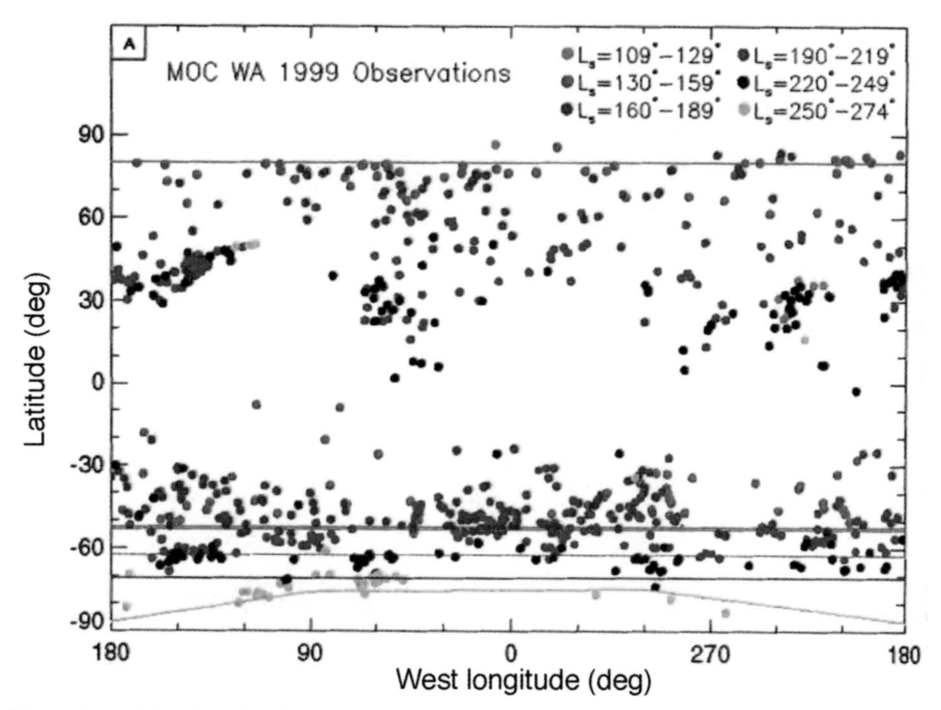

Figure 7.16 Map showing the location of local dust storms observed throughout 1999 (a) as a function of latitude and longitude, and (b) as a function of latitude and season (L_S).
From Cantor *et al.* (2001).

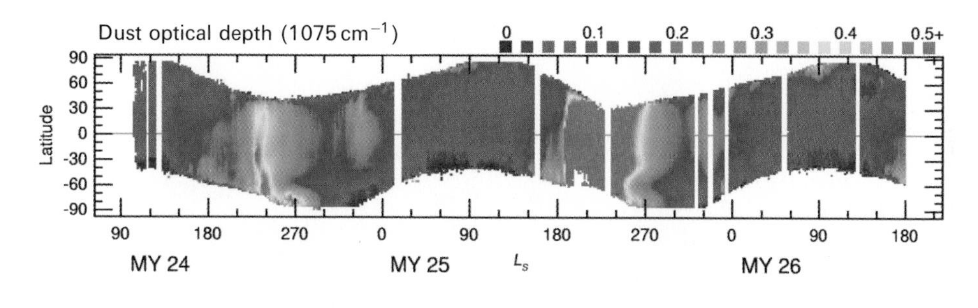

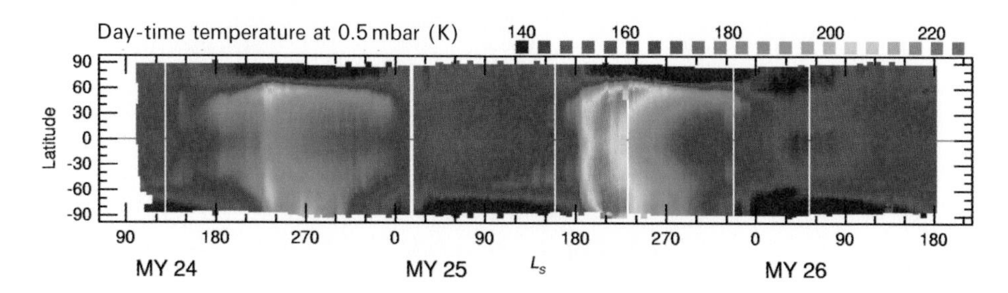

Figure 7.18 Variations in (a) the zonal mean dust optical depth and (b) the zonal mean temperature at 0.5 hPa, derived from Thermal Emission Spectrometer (TES) observations during the Mars Global Surveyor (MGS) mapping period.
From Smith (2003).

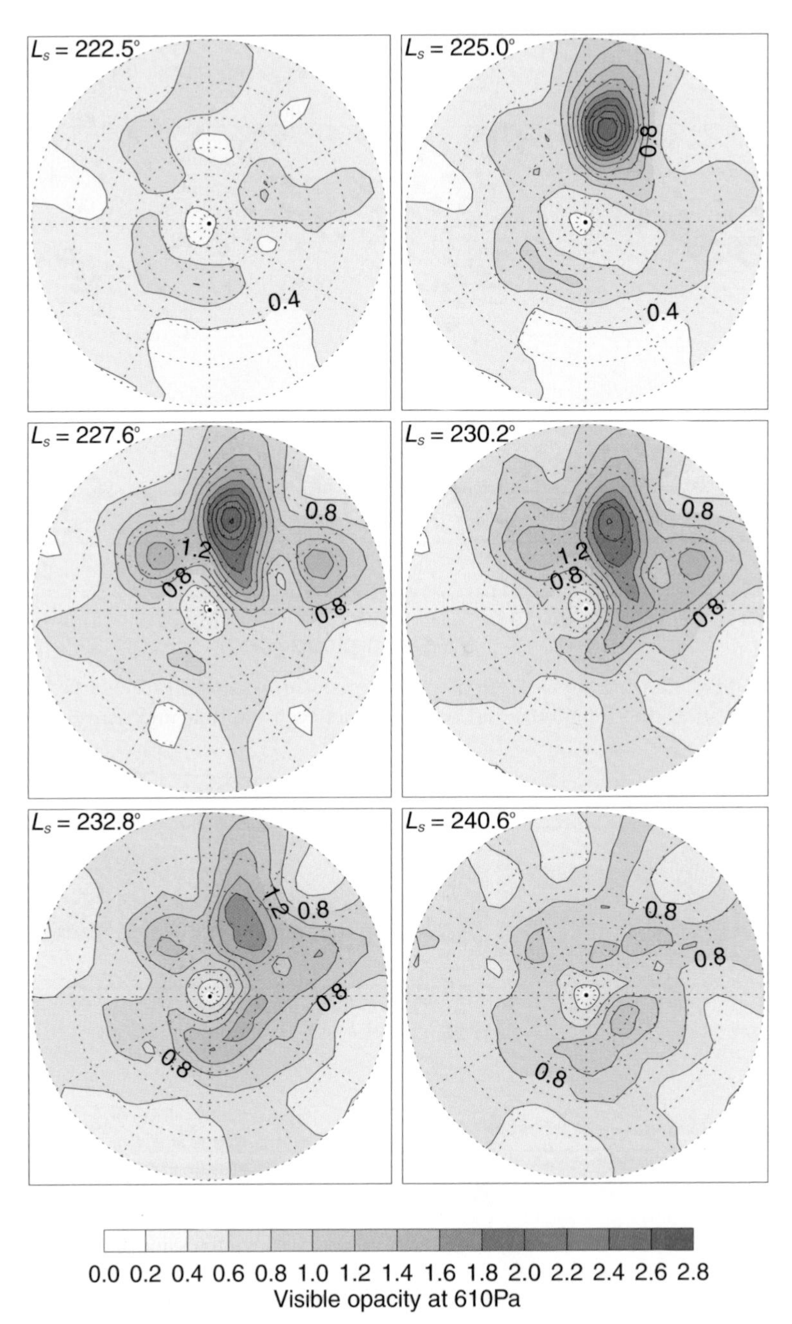

Figure 7.19 Sequence of maps of dust optical depth, showing the evolution of the Noachis dust storm of 1999, from a combination of Thermal Emission Spectrometer (TES) measurements of temperature and dust optical depth assimilated into a Mars GCM simulation.

From Lewis *et al.* (2003).

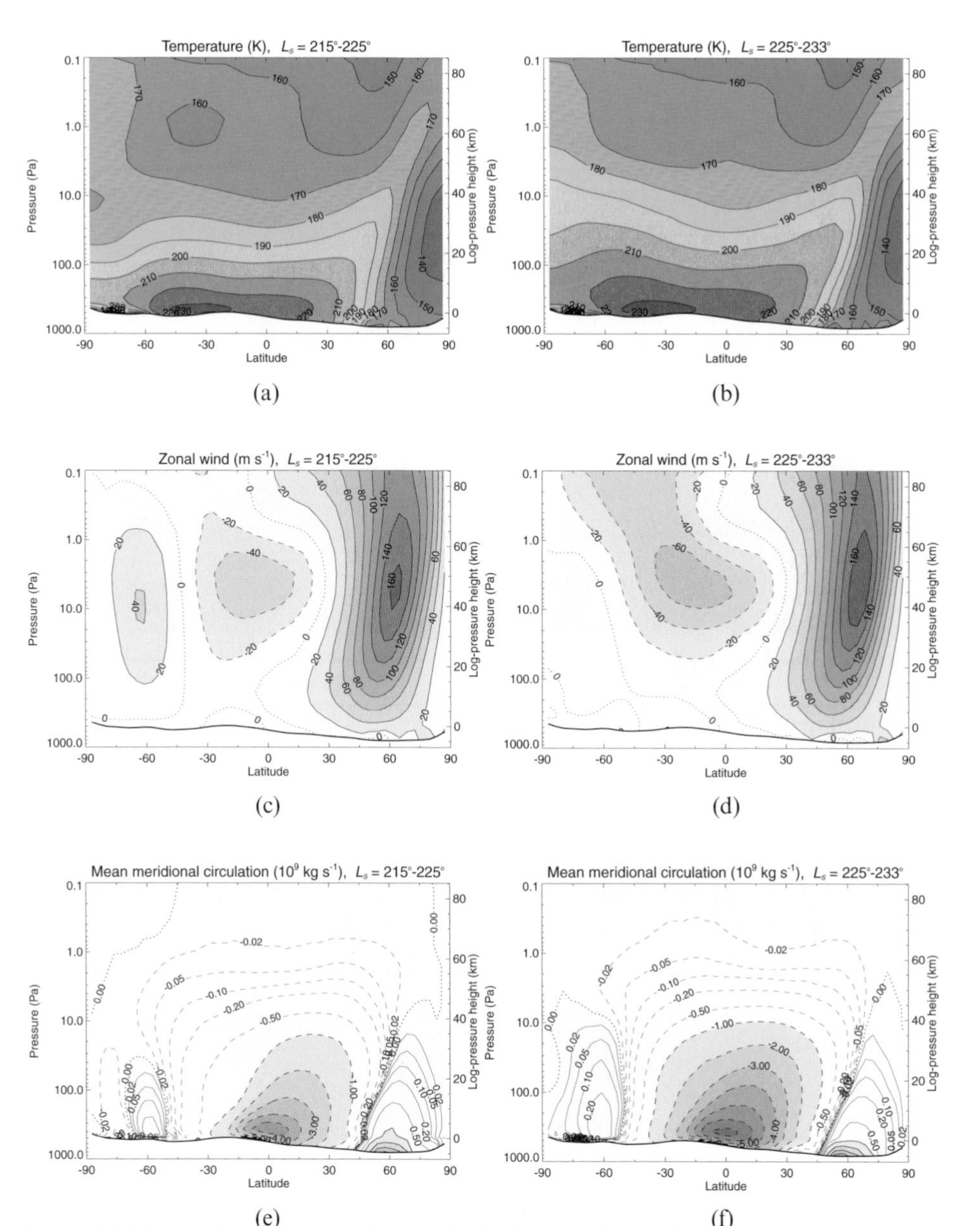

Figure 7.20 The influence on the large-scale Hadley circulation during the evolution of the Noachis dust storm of 1999, from a combination of Thermal Emission Spectrometer (TES) measurements of temperature and dust optical depth assimilated into a Mars GCM simulation. Frames show [(a) and (b)] the zonal mean temperature, [(c) and (d)] zonal mean zonal wind, and [(e) and (f)] mean meridional streamfunction for the periods [(a), (c), and (e)] just before the Noachis regional storm and [(b), (d), and (f)] during the storm.

From Lewis *et al.* (2003)

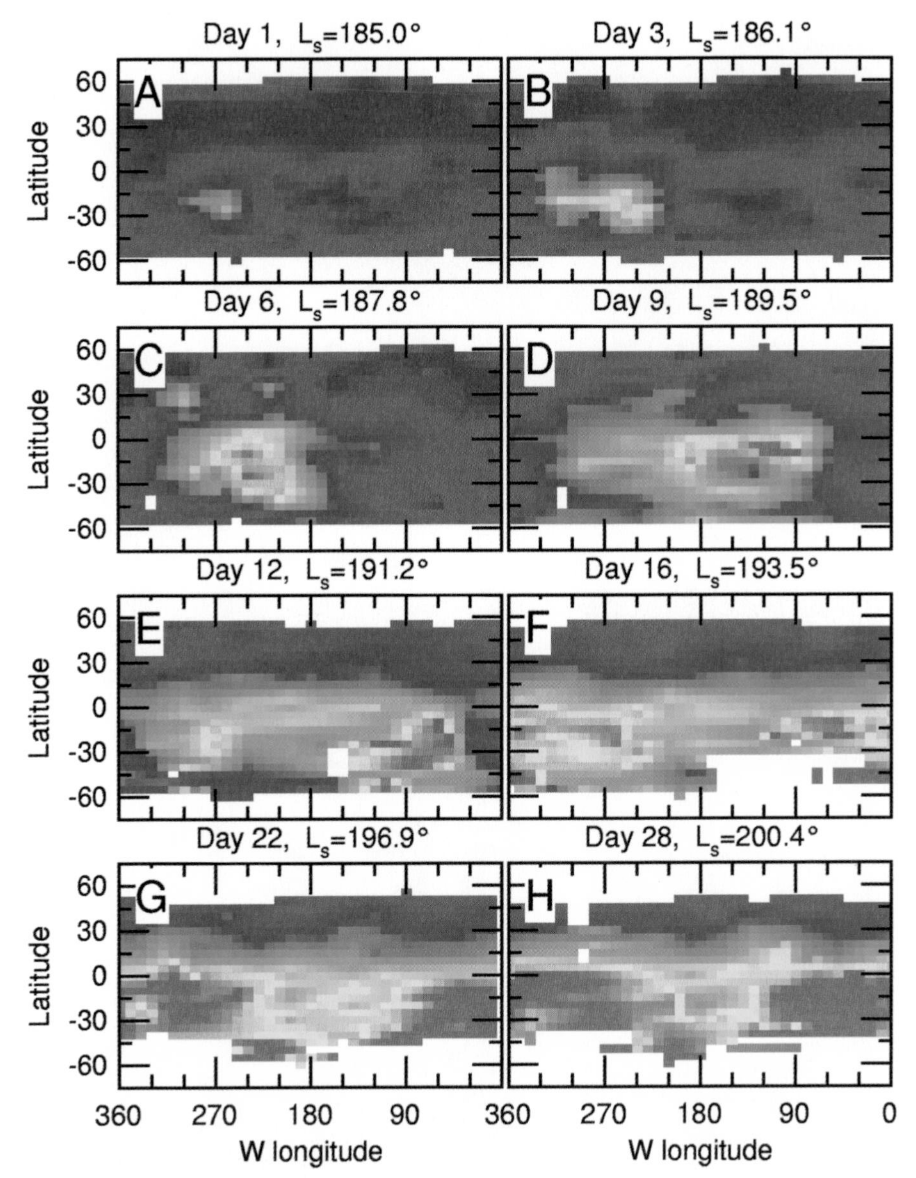

Figure 7.23 Thermal Emission Spectrometer (TES) maps of dust optical depth during the development of the global dust storm of 2001.

From Smith *et al.* (2002).

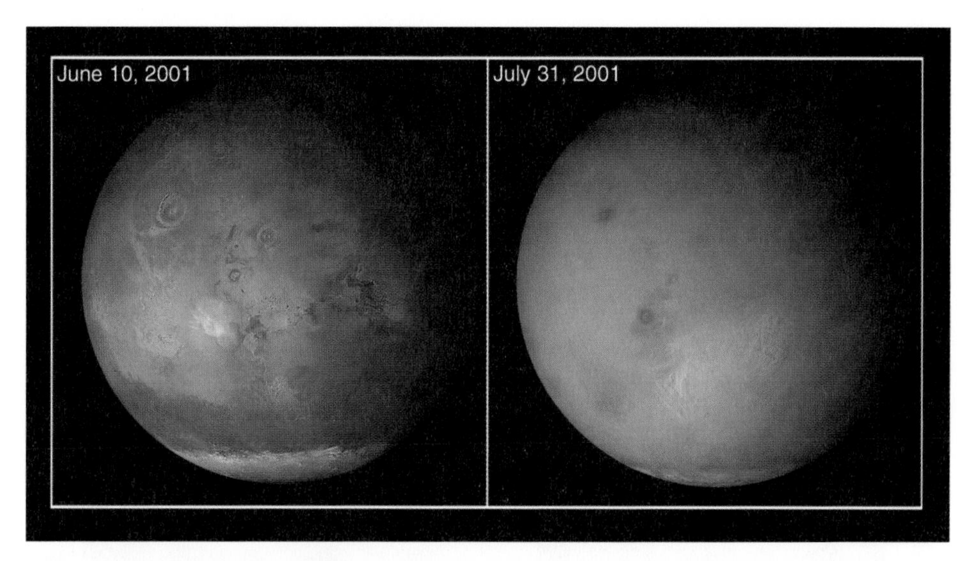

Figure 7.24 Mars Orbiter Camera (MOC) wide-angle images of Mars shortly before, and during, the global dust storm of 2001.

From NASA/JPL and Malin Space Science Systems.

Figure 7.25 Hubble Space Telescope (HST) wide-angle images of Mars shortly before, and during, the global dust storm of 2001.

From NASA/Space Telescope Science Institute.

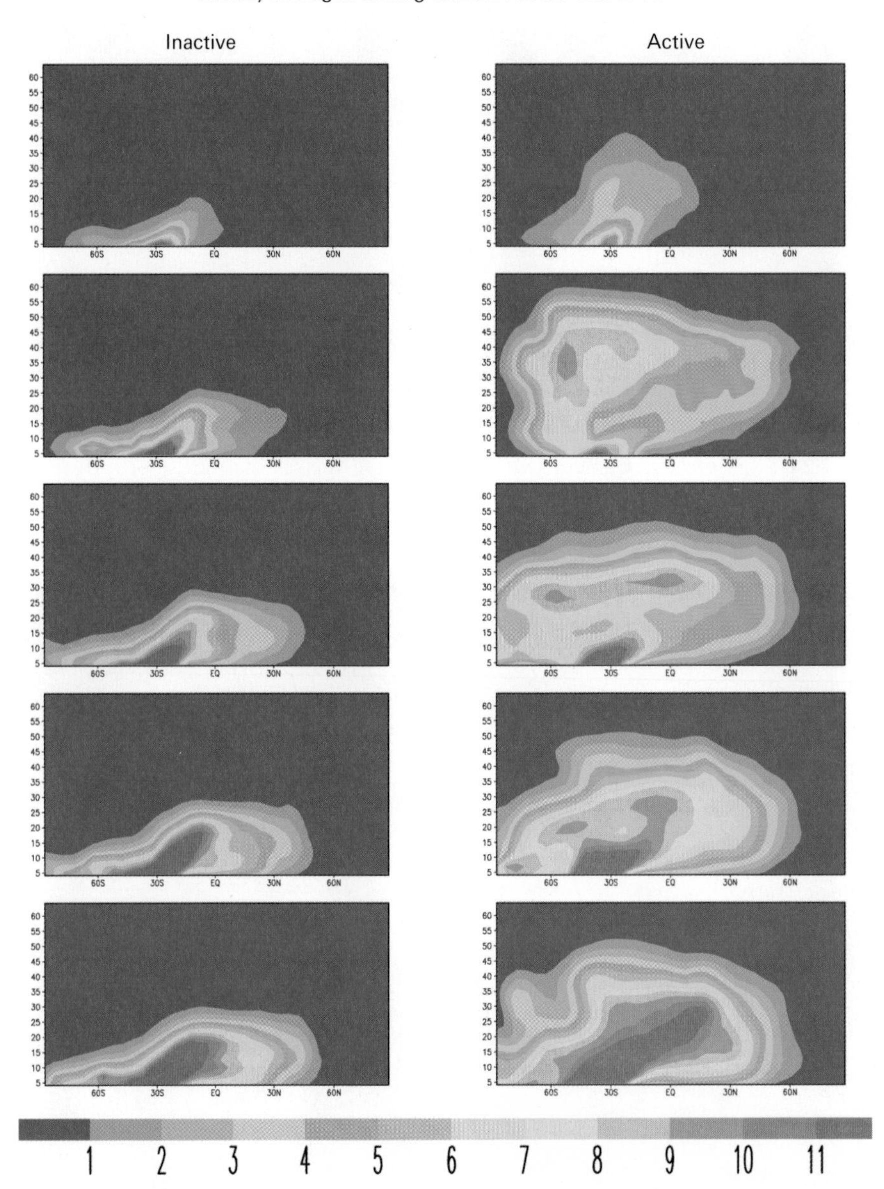

Figure 7.26 Latitude–height maps of zonal mean dust concentration every 2 sols (starting at the top of each column), comparing the transports produced when dust is radiatively active (right-hand column) with those of a passive dust tracer (left-hand column).

Simulations were carried out using the Oxford Mars GCM by Newman (2001).

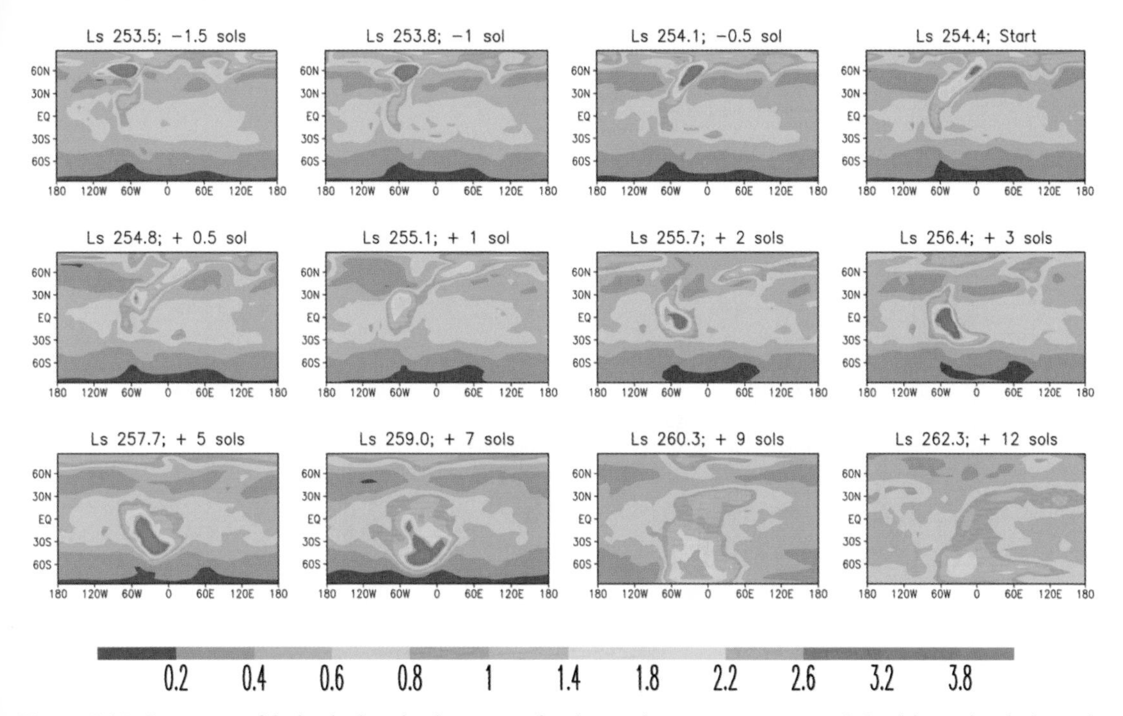

Figure 7.27 Sequence of latitude–longitude maps of column dust amounts every 0.5 sol in a simulation of a regional dust storm around $L_S \sim 255°$.

The sequence was computed by Newman *et al.* (2002b) using the Oxford Mars GCM, with both near-surface wind stress and dust devil lifting parameterizations active.

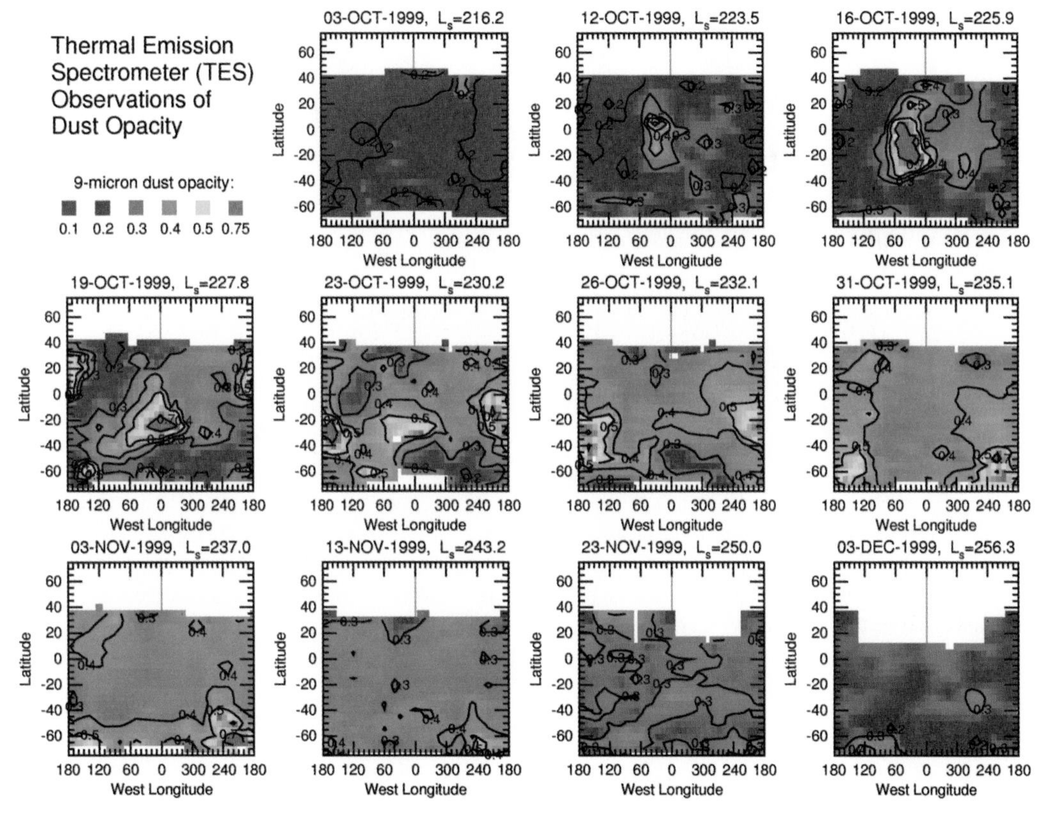

Figure 7.28 Sequence of latitude–longitude maps of dust infrared optical depth during the development of the Chryse dust storm, observed by the Thermal Emission Spectrometer (TES) instrument on board Mars Global Surveyor (MGS) in 1999.

From Smith *et al.* (2000).

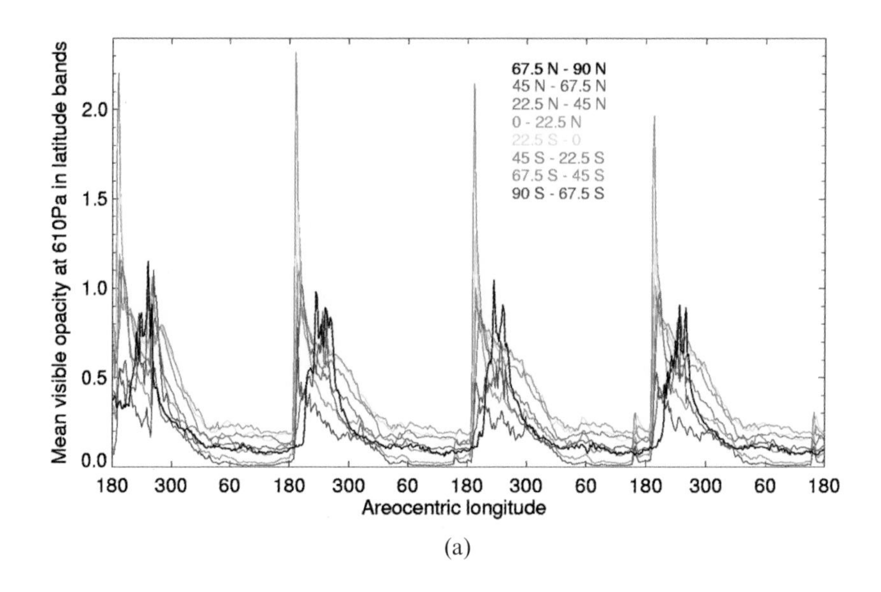

(a)

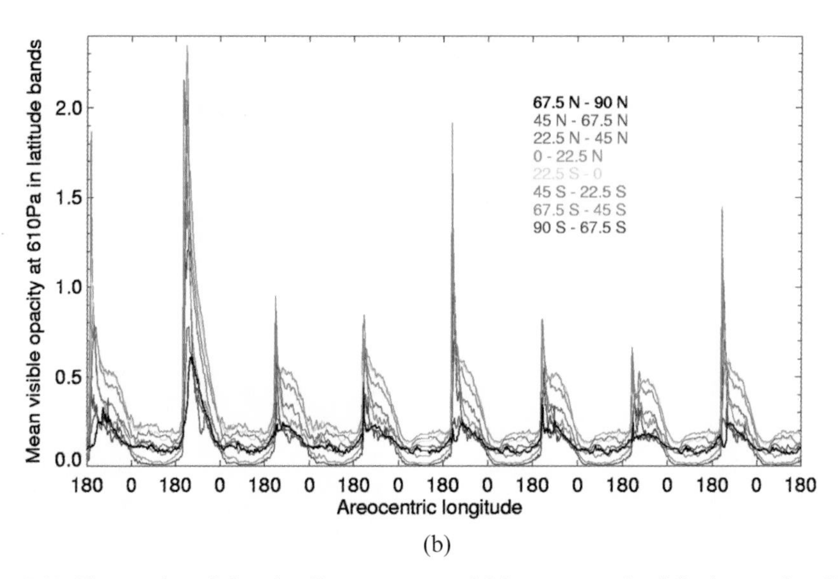

(b)

Figure 7.31 Time series of dust loading over several Mars years of a fully-interactive GCM simulation of the dust cycle: (a) a nearly periodic case, and (b) a more chaotic example.
From Newman (2002b).

(a)

(b)

Figure 8.8 Illustrations of the concept of the great northern ocean or 'Oceanus Borealis' on Mars, which some suggest may have occurred during the late Noachian/early Hesperian period of Mars' history. (a) Artist's impression of what Mars might have looked like during the existence of Oceanus Borealis, including the presence of clouds (painting by Michael Carroll), (b) computer simulation of a northern ocean, visualizing the flooding of all regions below a particular areopotential height on the Mars Orbiter Laser Altimeter (MOLA) topographic data set, in this case showing the region between Olympus Mons and Elysium.

From NASA/Greg Shirah (b) and Michael Cannott (`www.spacedinoart.com`) (a).

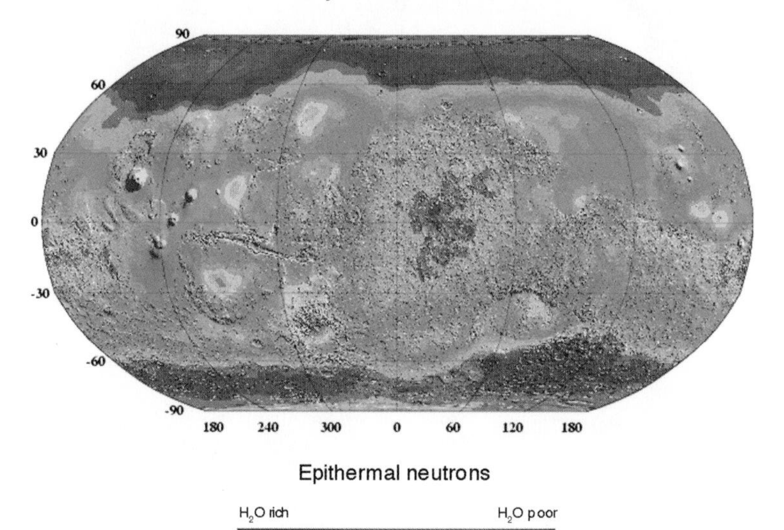

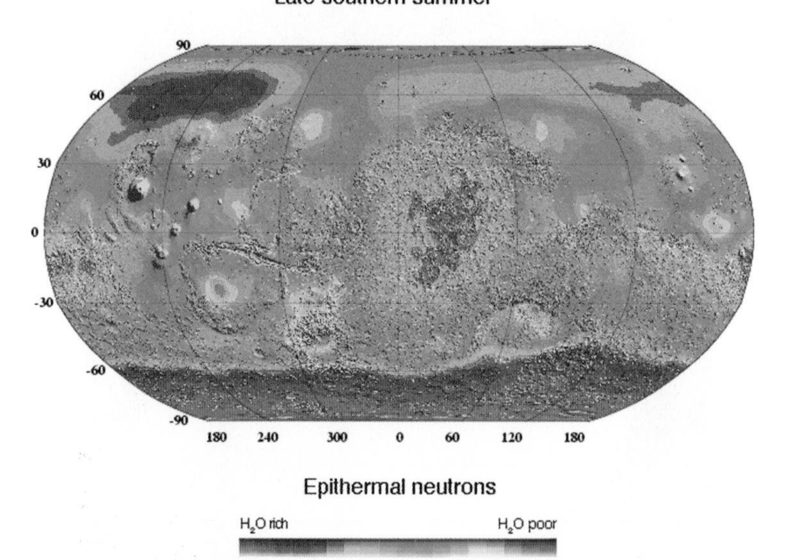

Figure 8.11 Colour-shaded maps of subsurface water concentration in the uppermost 1–2 m of the Martian surface, as determined from the epithermal neutron flux measured by the Gamma Ray Spectrometer on board the Mars Odyssey spacecraft in 2002 (Boynton *et al.*, 2002). The colour shading indicates relative water concentration, with blue indicating high concentrations and red low concentration. The map is superposed onto a shaded relief representation of the Mars Orbiter Laser Altimeter (MOLA) topography. The two maps show changing exposure of subsurface ice between northern and southern summer.

From *Science* magazine, with permission.

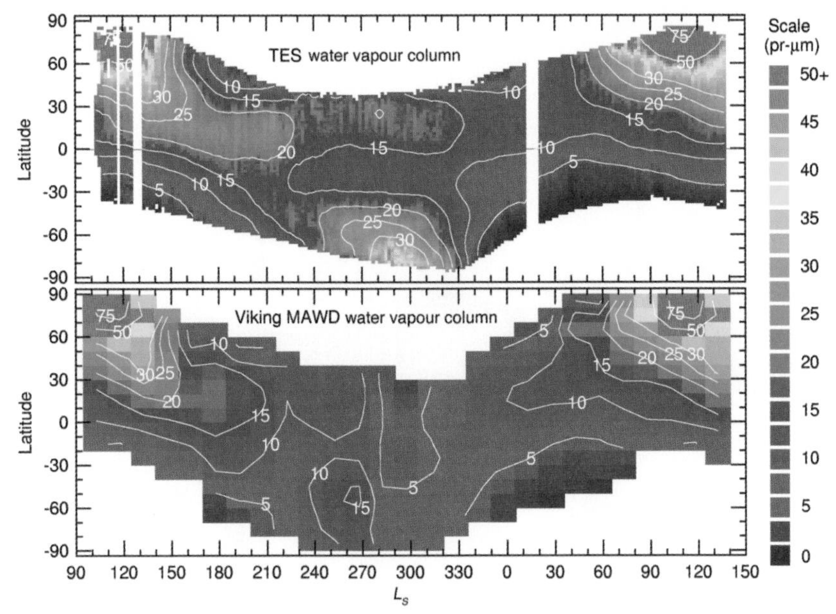

Figure 8.12 Maps of the zonally-averaged column abundance of water vapour in the atmosphere of Mars (in units of precipitable μm of liquid water) as a function of solar longitude L_S, as measured by (a) the Mars Global Surveyor (MGS) Thermal Emission Spectrometer (TES) instrument and (b) the Viking Mars Atmospheric Water Detector (MAWD) instrument.

(a) From Smith (2002). (b) From Jakosky and Farmer (1982).

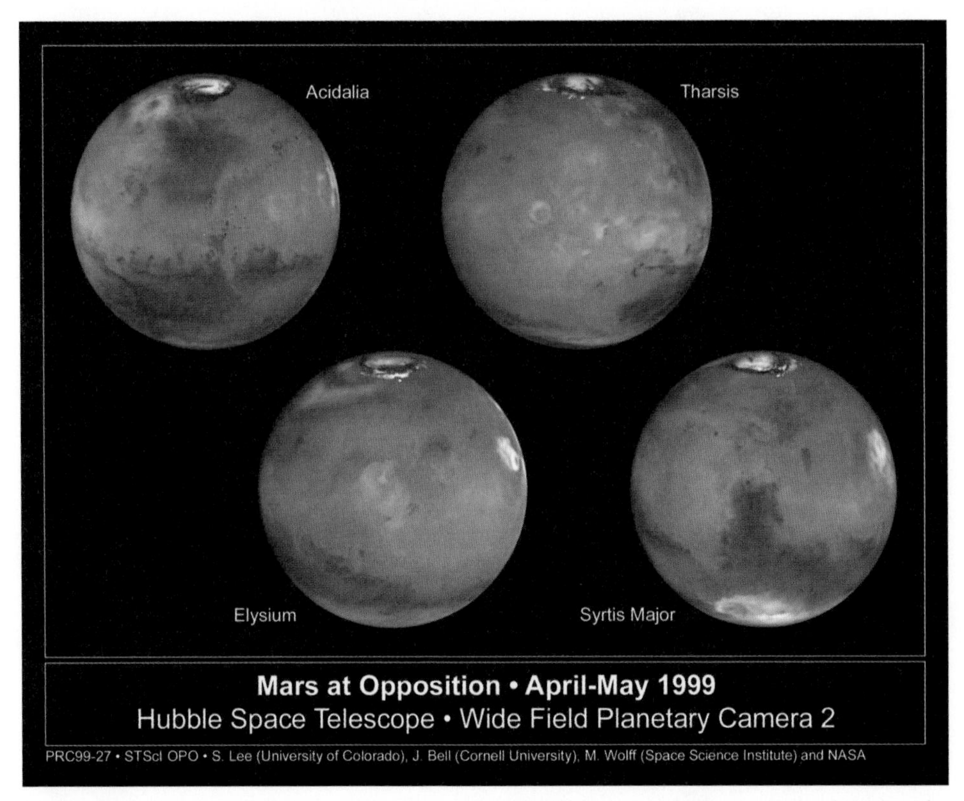

Figure 8.14 A series of Hubble Space Telescope (HST) images of Mars, obtained during the opposition in April and May of 1999, showing different aspects of the planet with a number of tenuous water ice clouds.

From NASA/Space Telescope Science Institute.

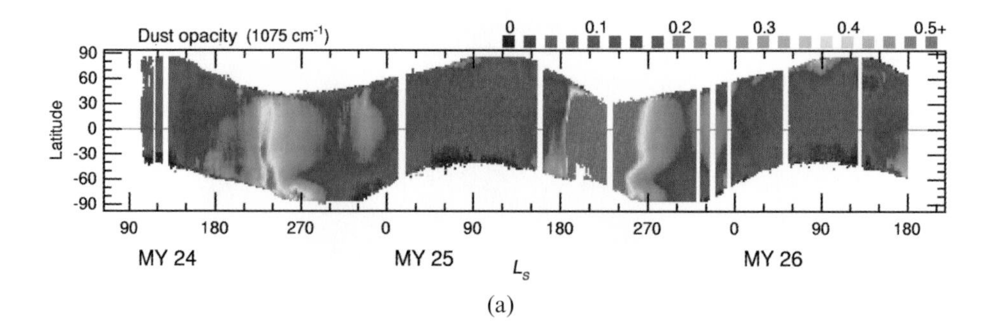

(a)

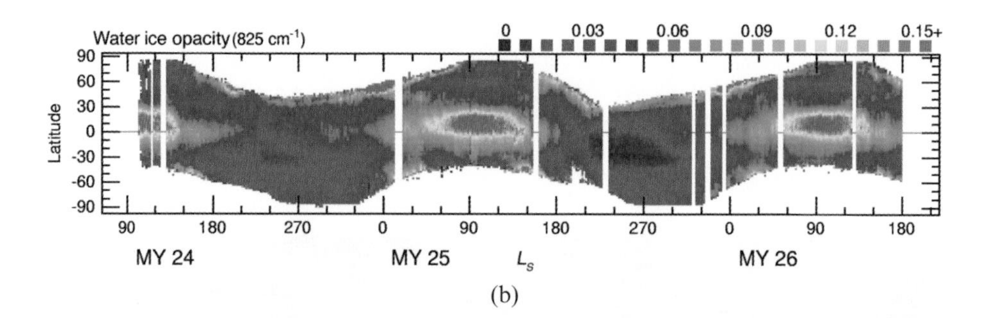

(b)

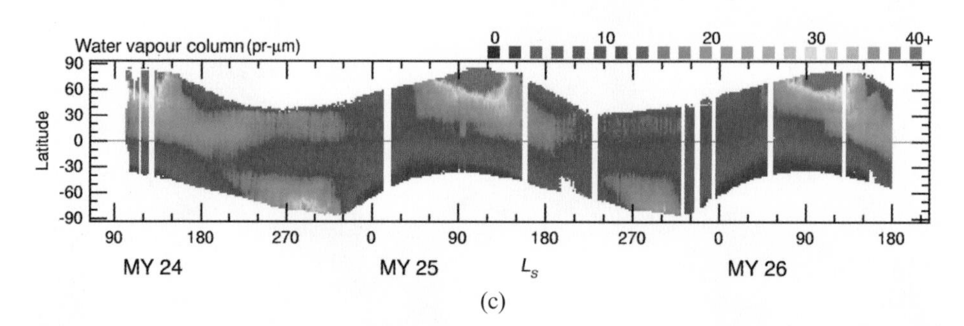

(c)

Figure 8.13 Maps of (a) the zonally-averaged column abundance of dust in the atmosphere of Mars, (b) total column amounts of cloud water ice, and (c) total column amounts of water vapour (in units of precipitable μm of liquid water) as a function of solar longitude L_S, as measured by the Mars Global Surveyor (MGS) Thermal Emission Spectrometer (TES) instrument for two complete Mars years.

Adapted from Smith (2003).

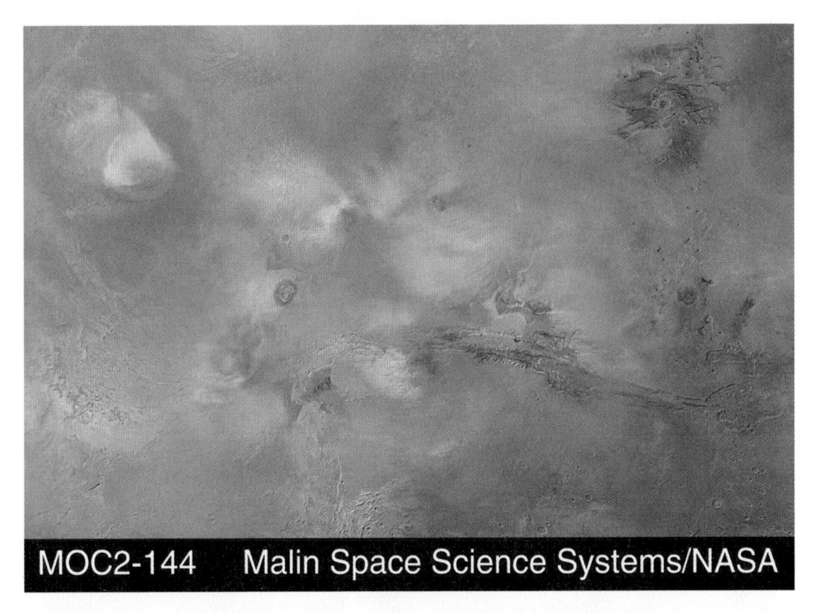

Figure 8.15 A map-projected mosaic of wide-angle Mars Orbiter Camera (MOC) images of Mars, showing a number of tenuous water ice clouds associated with features on the Tharsis Plateau.

Figure 8.17 Two images taken by the Pathfinder Lander spacecraft, illustrating different types of cloud formation observed from the surface of Mars.
From NASA/JPL.

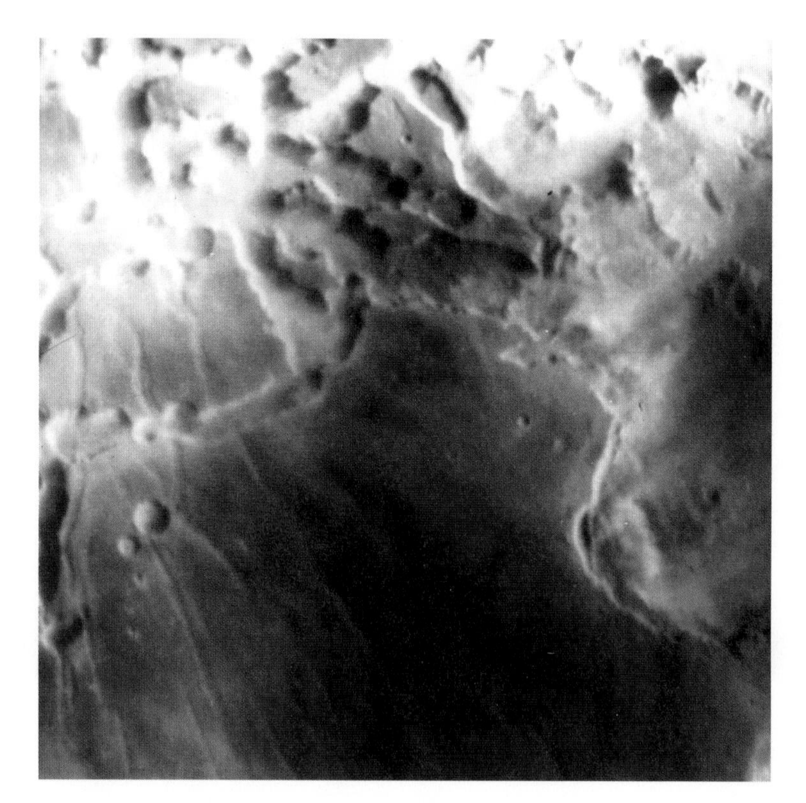

Figure 8.19 Viking Orbiter image of near-surface fogs lingering just after dawn in the Noctis Labyrinthus region of Mars, at the western end of Valles Marineris.
From NASA/JPL.

Figure 8.20 Viking Lander 1 image of surface frost formation just after dawn during late autumn.
From NASA/JPL.

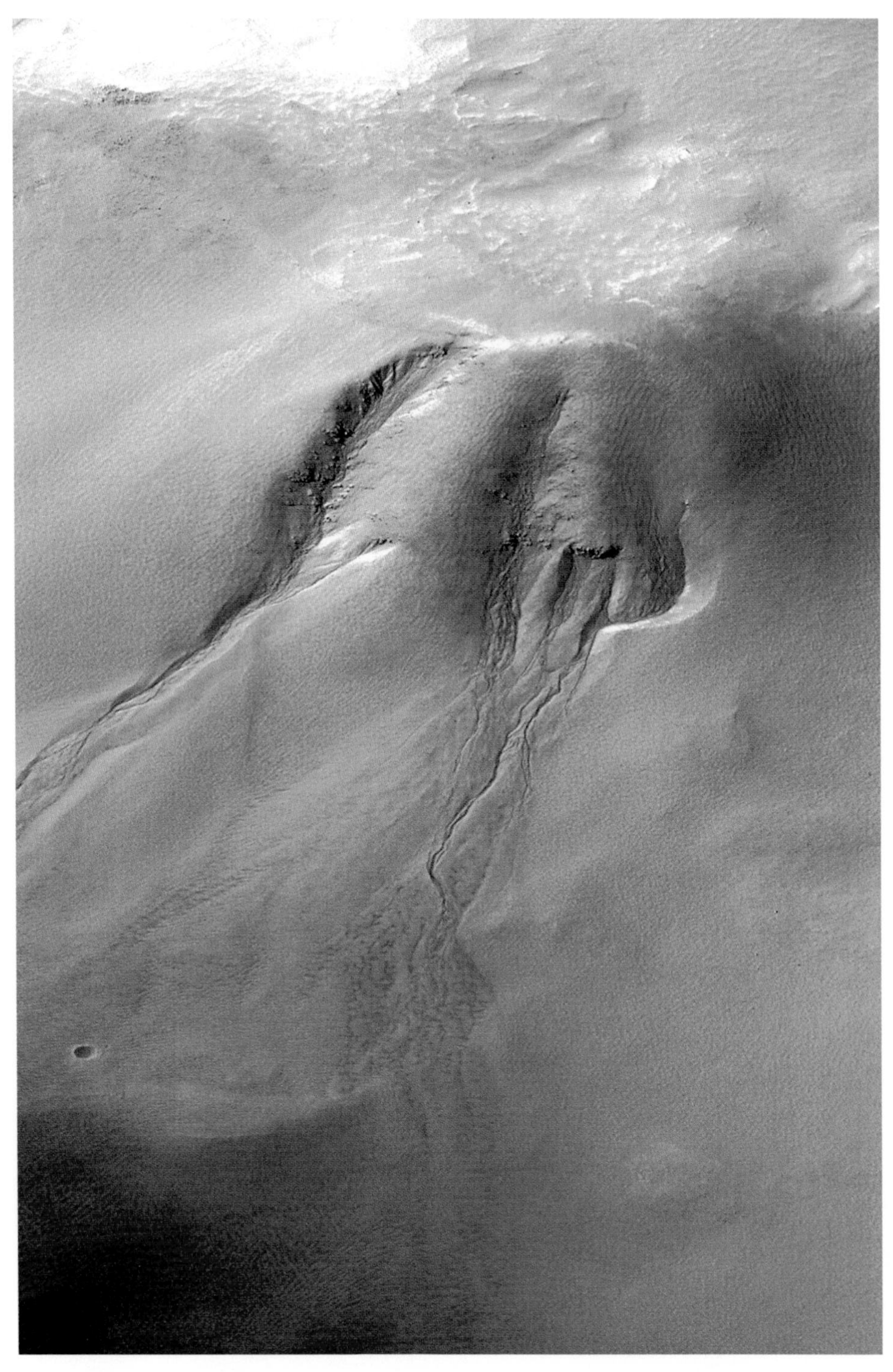

Figure 8.22 Mars Orbiter Camera (MOC) image of some recent gullies found in the southward-facing slopes of a crater in the southern region of Noachis Planitia.

From Malin Space Science Systems and NASA.

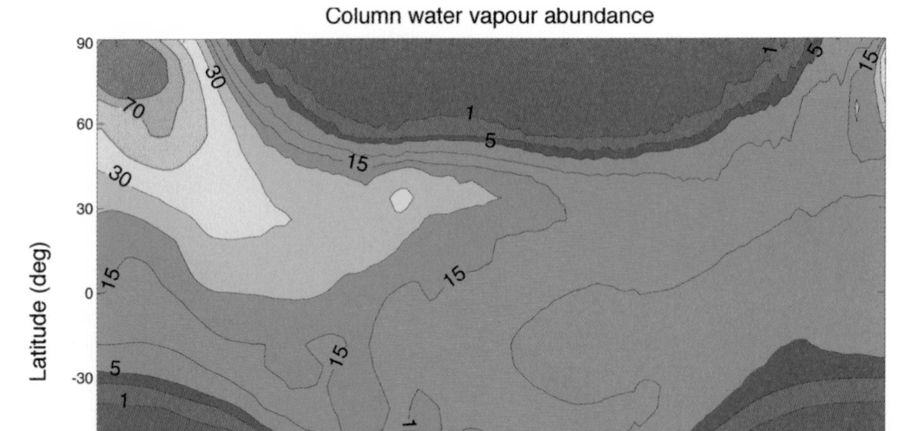

(a)

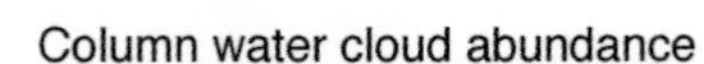

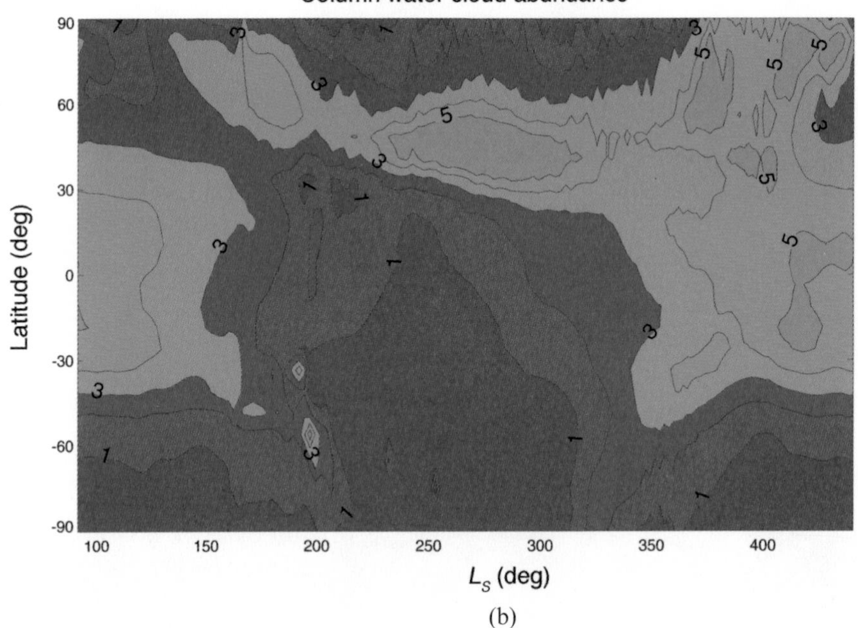

(b)

Figure 8.24 Variations of the zonal mean atmospheric column water vapour (a) and cloud ice (b) as a function of latitude and season (L_S) in a simulation of the hydrological cycle on Mars using the Oxford Mars GCM.

From Böttger (2003).

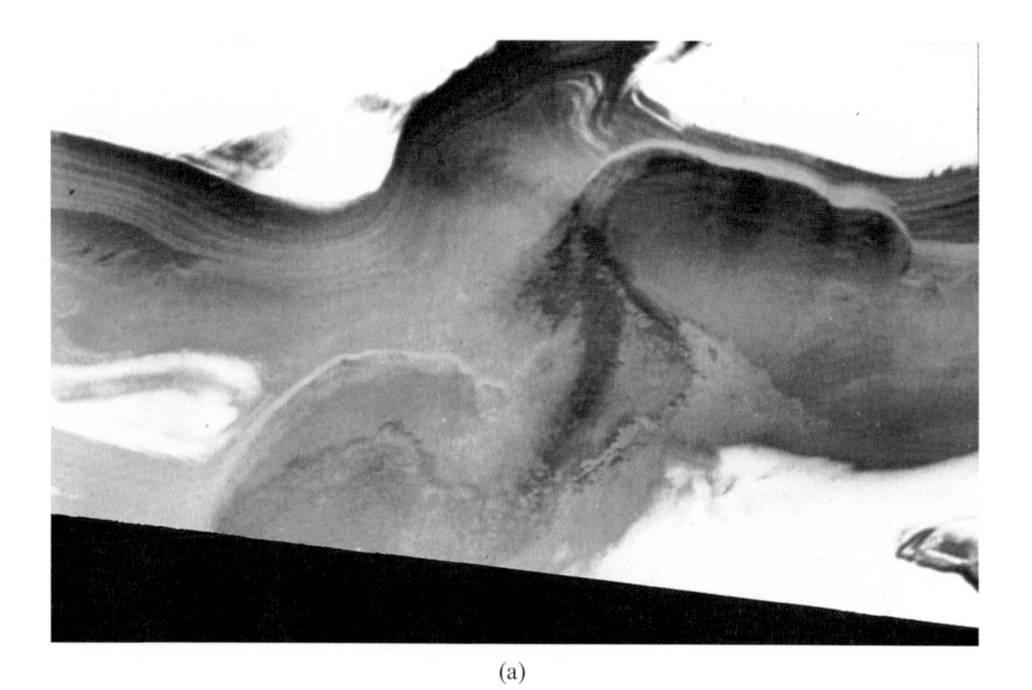

(a)

(b)

Figure 9.1 Two views of the polar layered terrains on Mars: (a) Viking image of layered terrains in the northern polar regions of Mars at the fringes of the summer residual cap, and (b) a narrow-angle image from the Mars Orbiter Camera (MOC) on Mars Global Surveyor (MGS) of a small region of the northern polar layered terrain (PLT) in early spring. This picture covers an area about 3 km wide near 85.2°N, 4.4°W.

From NASA/JPL and Malin Space Science Systems.

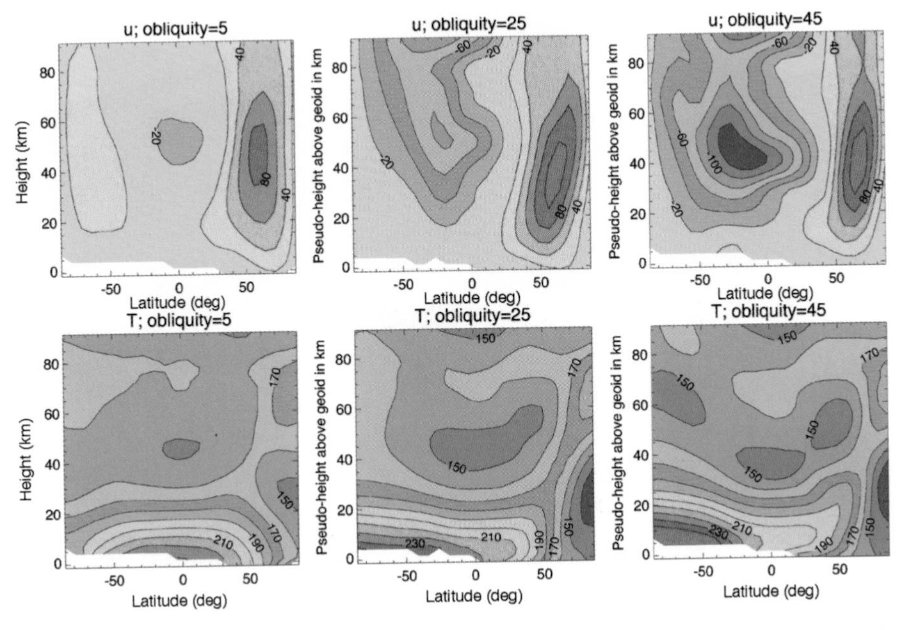

Figure 9.8 Variations in the zonal mean wind pattern (upper row) and temperature structure (lower row) as a function of obliquity for northern winter solstice, as simulated by the Oxford Mars GCM.

From Newman *et al.* (2003).

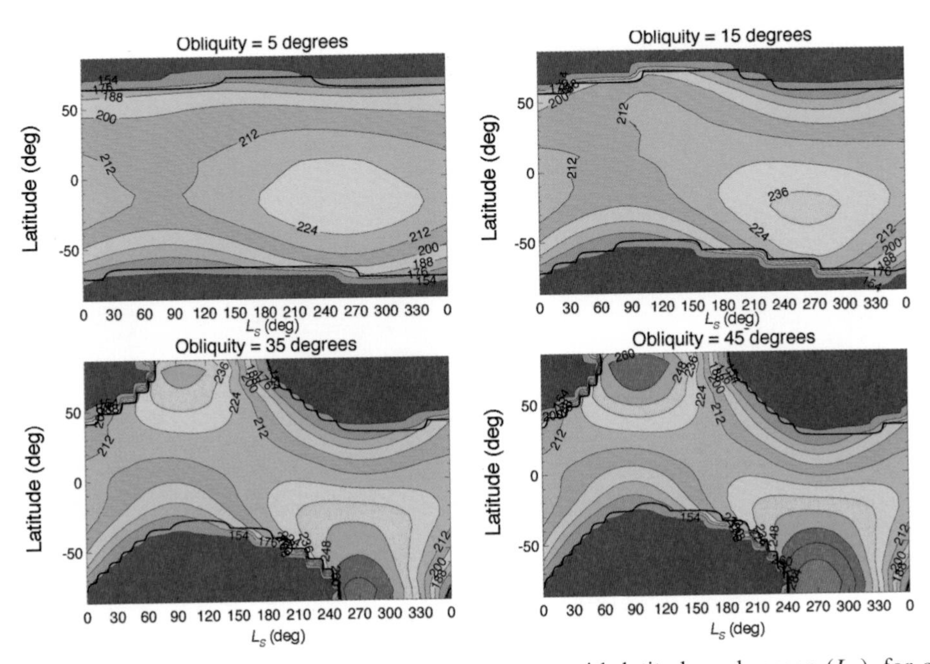

Figure 9.9 Variations in the zonal surface temperature with latitude and season (L_S), for a range of obliquities, as simulated by the Oxford Mars GCM.

From Newman *et al.* (2003).

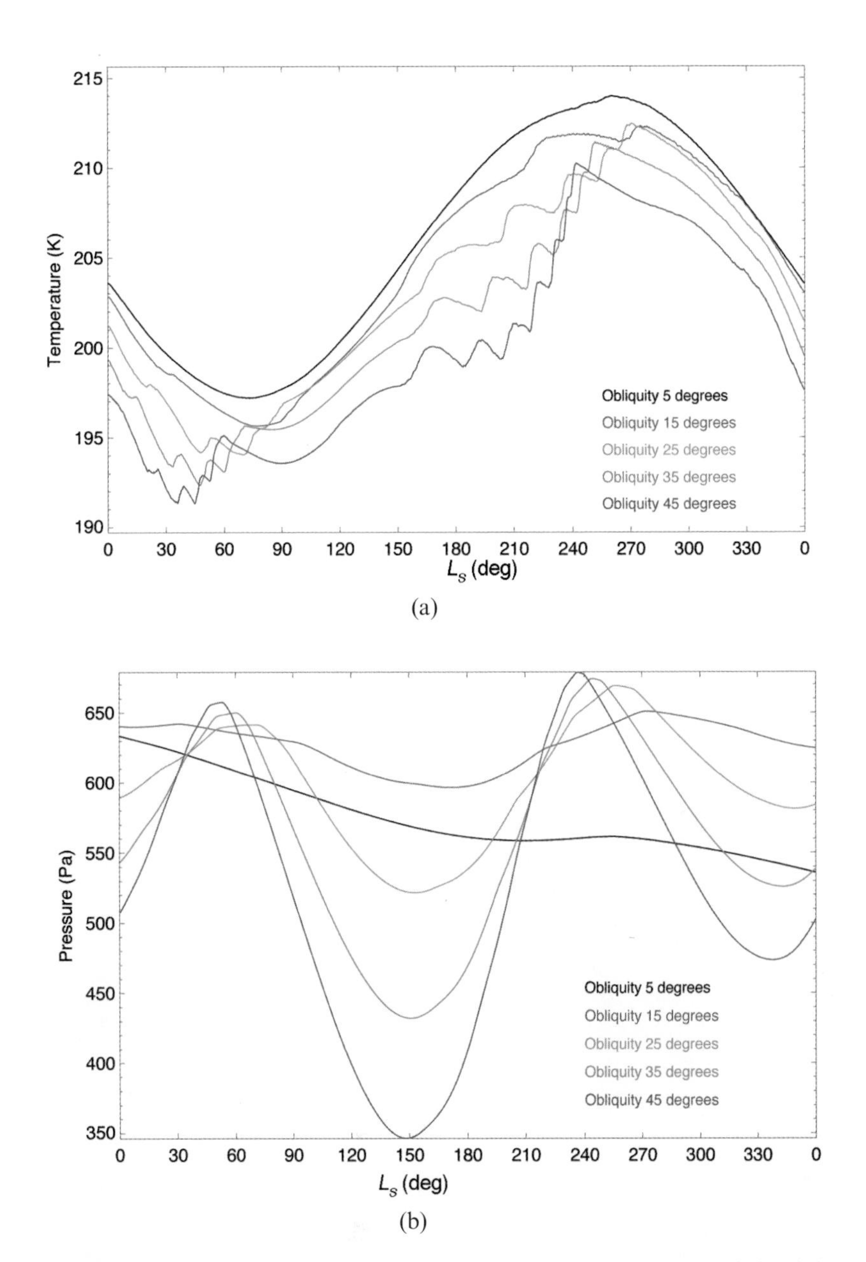

Figure 9.10 Seasonal variations in (a) global mean surface temperature and (b) global mean surface pressure for various values of obliquity, as simulated in the Oxford Mars GCM with fixed boundary conditions and atmospheric mass.

From Newman *et al.* (2003).

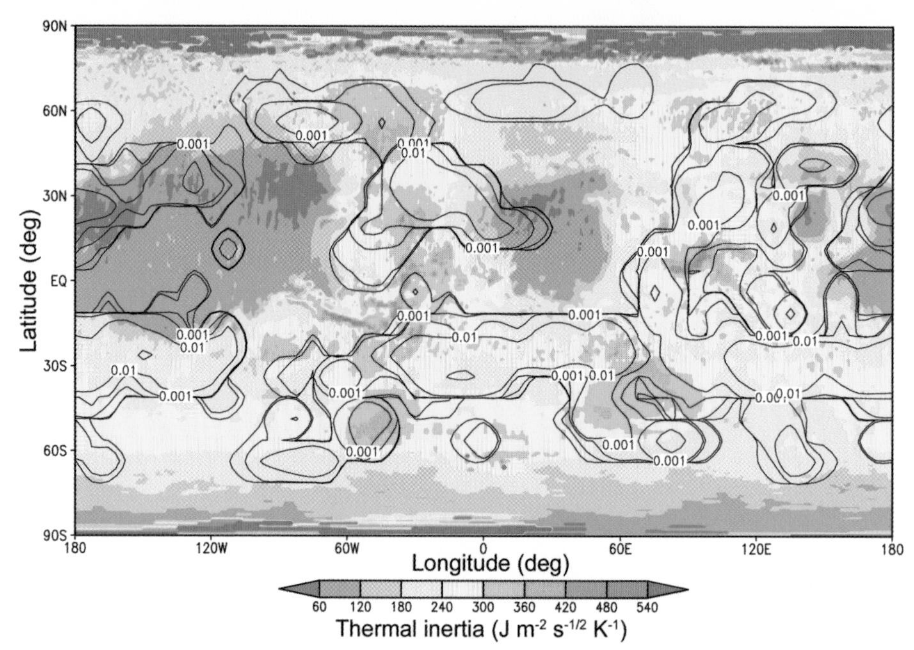

Figure 9.11 Map of annual 'deflation potential' (representing the mean local aeolian erosion rate due to near-surface wind stress) at high obliquity (45°), as simulated by the Oxford MGCM. Contours show deflation potential in $cm\,yr^{-1}$, with a logarithmic interval of one contour per decade, superimposed on a shaded map of surface thermal inertia.

Figure 10.3 Artist's impression of the human exploration of Mars, assuming that it is possible to modify the environment sufficiently to enable humans to walk freely on the surface with a simple breathing apparatus.

From a painting by Michael Carroll (www.spacedinoart.com).

(a)

(b)

Figure 10.2 Artist's impressions of the human exploration of Mars, assuming that it is possible to place teams of astronauts and equipment on the surface of Mars, showing (a) establishing a base immediately after landing on Mars, and (b) an expedition using mobile vehicles into Mars' North Polar Region.

From paintings by David Hardy (www.astroart.org).

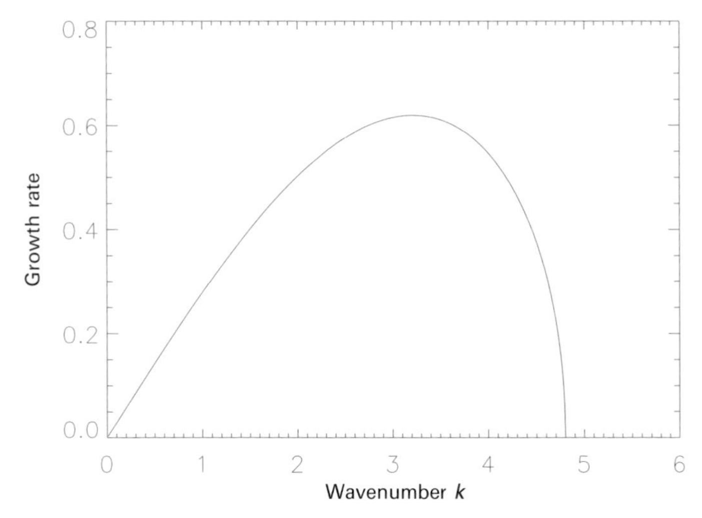

Figure 6.9. Growth-rate curve for the Eady model of baroclinic instability, assuming the lateral wavenumber $l = 0$.

indicates the propagation speed of the developing baroclinic wave. For the Eady model, it is found that the most rapidly growing mode moves precisely at the speed of the zonal flow at mid-height ($z = 0$), known as the 'steering level' for this mode.

An extraordinary feature of this model is that, if appropriate values for N, D, ΔU, and f_0 are taken, for example, for the Earth's mid-latitude troposphere ($N \sim 10^{-2}\,\mathrm{s}^{-1}$, $D \sim 10\,\mathrm{km}$, $\Delta U \sim 50\,\mathrm{m\,s}^{-1}$, and $f_0 \sim 10^{-4}\,\mathrm{s}^{-1}$), the model predicts $\lambda_{max} \sim 4{,}000\,\mathrm{km}$ and a growth timescale ($1/kc_i$) of $\sim$0.75 day. Such numbers are remarkably close to the scales for developing baroclinic weather systems on Earth, and the success of this model was an important historic factor in the interpretation of such weather systems as arising from baroclinic instability gaining widespread acceptance in the 1950s and 1960s. It is also interesting to note that the distinguished early pioneer of planetary meteorology, Seymour Hess, made some rather similar estimates of λ_{max} and growth rates in 1950 (Hess, 1950), based on some ingenious ways of estimating N and other parameters before they could be directly measured, to make a very early assessment that baroclinic instability might also be important for Martian meteorology. For example, taking $N \sim 10^{-2}\,\mathrm{s}^{-1}$, $D \sim 10\,\mathrm{km}$, $\Delta U \sim 50\,\mathrm{m\,s}^{-1}$, and $f_0 \sim 10^{-4}\,\mathrm{s}^{-1}$ leads to the expectation that the wavelength of maximum growth rate, and the growth rate itself, should be quite similar to the Earth, anticipating a favoured wavenumber for baroclinic waves on Mars of around $m \sim 2$–3.

6.3.1.2 The Charney model

Although the Eady model has many advantages as an elementary illustrative model of baroclinic instability, it would seem to have many drawbacks and limitations as a 'realistic' model for a planetary atmosphere. In its classical form, for example, the Eady model requires a rigid top boundary and eliminates the planetary vorticity gradient β. It also ignores the compressibility of the atmosphere (though the latter two features can be taken into account within a 'generalized Eady model', e.g., see Read, 1988). If all of these effects are included in a more realistic model, then this leads to significant mathematical complications, although the resulting model (due to Charney, 1947) in fact predates that of Eady in the history of the subject.

In practice, however, despite these additional complications (which necessitate solving some aspects of the model on a computer), the overall conclusions from the Charney model are not very different from that of the Eady model, at least for conditions prevailing on Earth. The Charney model still predicts $\lambda_{\max}$ around a few thousand km and a corresponding minimum growth timescale ~ 0.75 day. A major difference, however, lies in the structure of the most unstable modes. In particular, the most rapidly growing modes typically have their maximum amplitude close to the ground and decay monotonically towards zero at high altitude. This is also reflected in the distribution of horizontal heat and momentum fluxes, which are also confined to within $z \leq H$ of the surface, where $H \sim H_p$, the pressure scale height (e.g., see Pedlosky, 1987 for some computed examples of solutions to the Charney problem). This is in part a consequence of the way the Charney model satisfies the necessary condition for instability. $\bar{q}_y$ is positive in the free atmosphere (because of $\beta > 0$), and can only change sign in association with a horizontal temperature gradient at the bottom boundary.

A further difference from the Eady model lies in the expected dependence of the wavelength of maximum growth rate on the parameters of the problem. As mentioned above, the Eady model predicts $\lambda_{\max} \sim ND/f_0$, where D is the depth of the model domain. In fact this dependence applies even if $D > H_p$ (Read, 1988), though this is not widely recognized in the literature. In the Charney model, however, the depth of the model domain is infinite. In that case the model solutions pick their own vertical scale, depending on the parameters of the problem. In practice, it is found from a scaling analysis (Held, 1978; Branscombe, 1983) that the characteristic vertical scale of the most rapidly growing Charney mode typically depends on the parameter Γ, given by:

$$\Gamma = \frac{\beta N^2 H_p}{f_0^2 dU/dz} \tag{6.5}$$

If the basic shear is relatively strong, so that $\Gamma \ll 1$, then the wavelength of maximum growth rate is found to be $\lambda_{\max} \sim NH_p/f_0$ to within a factor O(1). In the other limit where $\Gamma \gg 1$ and vertical shear is relatively weak, however, it is found that $\lambda_{\max} \sim NH/f_0$, where $H = H_p/\Gamma$. In that case, the most rapidly growing mode is relatively shallow ($H \ll H_p$) and trapped close to the surface, and is of correspondingly short horizontal wavelength. Such a situation may be relevant, for example, to northern summer conditions near the polar cap on Mars, when relatively small and shallow spiral eddies are occasionally found close to the residual polar cap edge (see Section 6.2.1).

The Charney model is actually quite a bit more complicated than summarized here (e.g., see Pedlosky, 1987 for more detailed discussion). In particular, the simple 'Charney mode' outlined above, with a single maximum in activity close to the ground and uniform decay with height, is not the only kind of unstable mode. Longer-wavelength instabilities are also found, including classes of mode with one or more nodes in the vertical direction (i.e., where the amplitude actually goes locally to zero, by analogy, for example, with vibrations on a guitar string), and the zonal phase of the wave flips by 180°. Such modes are often referred to as 'Green' modes (after the English meteorologist John Green, who discovered them in the late 1950s (Green, 1960)), and may have very deep structures extending much further than H_p upwards from the ground. Even the longer wavelength Charney modes (without nodes in the vertical) may also penetrate more deeply than H_p, though, in common with the so-called 'Green modes', their growth rates are typically quite weak compared with the most rapidly growing Charney modes.

6.3.1.3 Barotropic instability

As briefly outlined above, barotropic instability arises from a horizontally-sheared zonal flow for which $\bar{q}_y$ changes sign in latitude. Historically, however, there are no widely cited 'classical' illustrative models of this instability, unlike baroclinic instability. This is partly because of the difficulty in finding particularly simple zonal flows for which analytical solutions can be found (because the mathematics is too complicated), and also because the properties of the resulting solutions turn out to be highly dependent on the characteristics of the basic zonal flow. In particular, there is no obvious scale for the most unstable mode which emerges independent of the properties of the zonal flow. Moreover, some friction may also be needed to result in a preferred scale in some circumstances.

The same kind of linearized perturbation approach as was used for the baroclinic cases above can also be applied to pure barotropic instability and, like baroclinic instability, this involves applying Eq. (6.1) to a zonal flow which is vertically uniform (so $dU/dz = 0$). In this case:

$$\bar{q}_y = \beta - \frac{d^2U}{dy^2} \tag{6.6}$$

and the necessary condition for instability reduces to the so-called Rayleigh–Kuo condition that $\beta - \bar{u}_{yy}$ change sign within the model domain. For specific examples, however, it is necessary to keep the last term on the LHS of Eq. (6.1) non-zero, otherwise the solutions are trivial and no instability is found.

Michelangeli *et al.* (1987), for example, investigated the stability of various barotropic jets with shapes and intensities which represented the observed upper-level zonal jets on Mars, Venus, and the Earth. Their study obtained growing waves for the circumpolar northern winter Martian jet, located close to the latitude of the peak of the jet, and with maximum growth rates for planetary wavenumbers $m = 1$ and 2 around 0.5 sol^{-1}. However, the predicted drift periods for such growing waves were only around 1 sol, which is much shorter than the observed waves on Mars near the surface. Since the actual circumpolar jet varies in strength considerably between the surface and its peak around 40–50-km altitude, it is likely that purely barotropic instabilities would be concentrated at high altitudes, and be relatively weak near the ground. Thus, it is likely that the main source of variability observed by the VL spacecraft was not due to simple barotropic instability, but to some more complicated mixture of barotropic and baroclinic effects.

6.4 MORE REALISTIC LINEARIZED MODELS

Although simple, idealized models, such as those discussed in the previous section, can provide helpful insight and order of magnitude estimates of properties such as the scales of the most rapidly growing instabilities, their application to specific problems often leave unresolved questions such as whether a particular kind of instability is actually occurring in the system under investigation. In the present situation, for example, the circumpolar jet in Mars' atmosphere is clearly seen to have both horizontal and vertical shears, suggesting that either baroclinic or barotropic instablility might occur (or some mixture of the two)? In this situation, it may be desirable to investigate the instability of a more realistic kind of zonal flow, with a structure closer to observations. In practice, this generally entails solving the linearized equations of motion on a computer, which is less convenient than an analytical model such as the Eady problem, but enables the detailed properties of the unstable modes to be determined fairly accurately.

6.4.1 Mixed baroclinic–barotropic instability

Early attempts to investigate the linear instability of more realistic baroclinic zonal flows made use of relatively simple zonal flows, and were based closely upon either two-level, two-layer, Eady or Charney models (Mintz, 1961; Leovy, 1969; Blumsack and Gierasch, 1972; Gadian, 1978). As well as using zonal flows of plausible shape and structure (though only for the northern hemisphere), some of these early studies also included the effects of zonally-symmetric topography and radiative damping. Their results generally came to rather similar conclusions, in favouring zonal wavenumbers $m = 2$–5 as the most unstable modes. Radiative effects were found to have rather little effect on the results, while the topography of the northern polar regions was found to decrease the most unstable wavelength and the corresponding growth rates.

The most detailed linear instability studies to date have been carried out by Barnes (1984), using the linearized quasigeostrophic equations of motion, and by Tanaka and Arai (1999) using linearized forms of the full meteorological primitive equations. Both these studies have concentrated on investigating the instability of the northern winter jet on Mars. The structure of the basic jet flow was taken from a GCM simulation of mid-latitude flow on Mars (Barnes, 1984; Barnes and Haberle, 1996), which is illustrated in Figure 6.10. This shows a typical jet associated with the strong thermal contrast at the edge of the developing winter polar cap, and with a

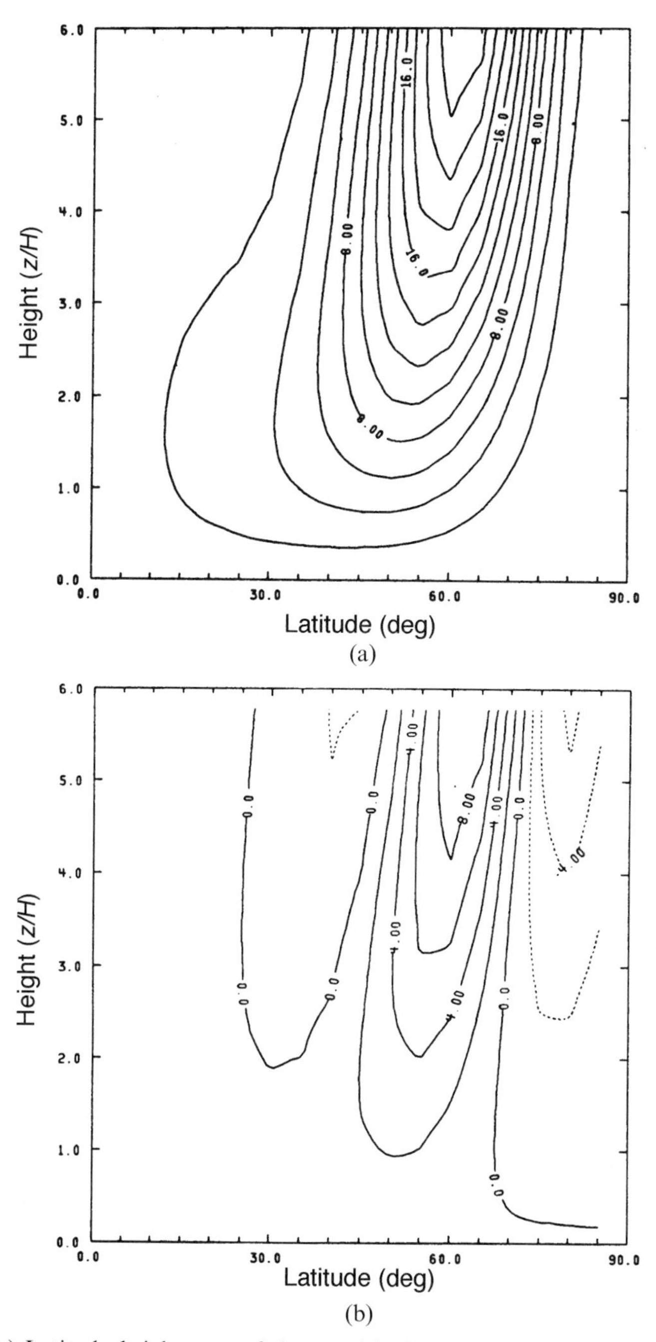

Figure 6.10. (a) Latitude–height map of the zonal jet investigated by Barnes (1984) and (in slightly modified form) by Tanaka and Arai (1999); (b) corresponding horizontal quasi-geostrophic potential vorticity gradient across the jet.

From Barnes (1984).

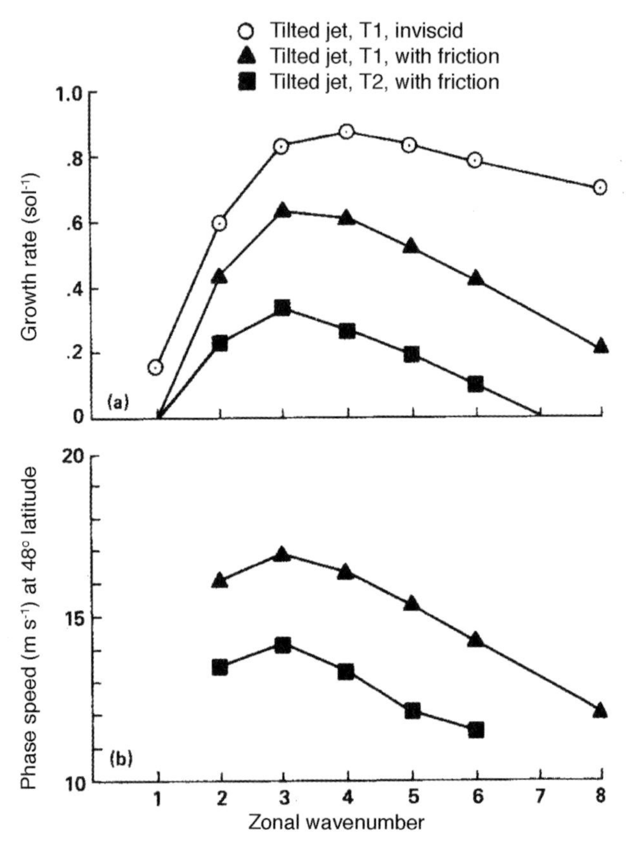

Figure 6.11. Growth rates (a) and phase speeds (b) computed for three cases of the instability of a baroclinic zonal jet on Mars by Barnes (1984) as a function of zonal wavenumber.

characteristic poleward tilt in the axis of the jet with height (cf. Figures 3.3, 3.7, 3.20 and 3.21).

The potential vorticity gradient in this jet is also illustrated in Figure 6.10(b), which clearly changes sign in latitude around 40°N and 65°N. Thermal gradients along the lower boundary may also lead to a change of sign of $\bar{q}_y$ in the vertical too, much as in the Charney model of baroclinic instability, indicating that the flow may satisfy necessary conditions for both barotropic and baroclinic instabilities. Barnes (1984) also investigated several different assumed profiles of static stability (including a case representative of a major dust storm), as well as effects due to zonally-symmetric topography and Ekman and radiative damping.

Both Barnes (1984) and Tanaka and Arai (1999) used numerical eigenvalue techniques to determine the phase speeds, growth rates, and structures of the unstable modes under various conditions, and obtained growth rate curves such as the results illustrated in Figure 6.11. Barnes' results clearly indicate that the basic zonal flow is unstable to wavenumbers $1 \leq m \leq 5$–9, depending upon the stratification used, and with a wavenumber of maximum growth rate around $m = 3$–4. Growth rates themselves were around 0.2–0.6 sol^{-1}, and with phase speeds around 15 m s^{-1}. Tanaka and Arai (1999) found broadly similar results, though used a somewhat different assumed static stability profile. They tended to find wavenumbers of maximum growth rate around $m = 5$ and rather higher growth rates overall than Barnes. The dust storm conditions, however, had a stronger stable stratification and this was found to lead to significantly reduced growth rates in Barnes' model, although the wavenumber of maximum growth rate was roughly the same as before.

These kinds of linear instability calculation not only allow computation of the dispersion relation (including growth rates and phase speeds) for the unstable modes, but also the actual structures of the modes themselves. Figure 6.12, for example,

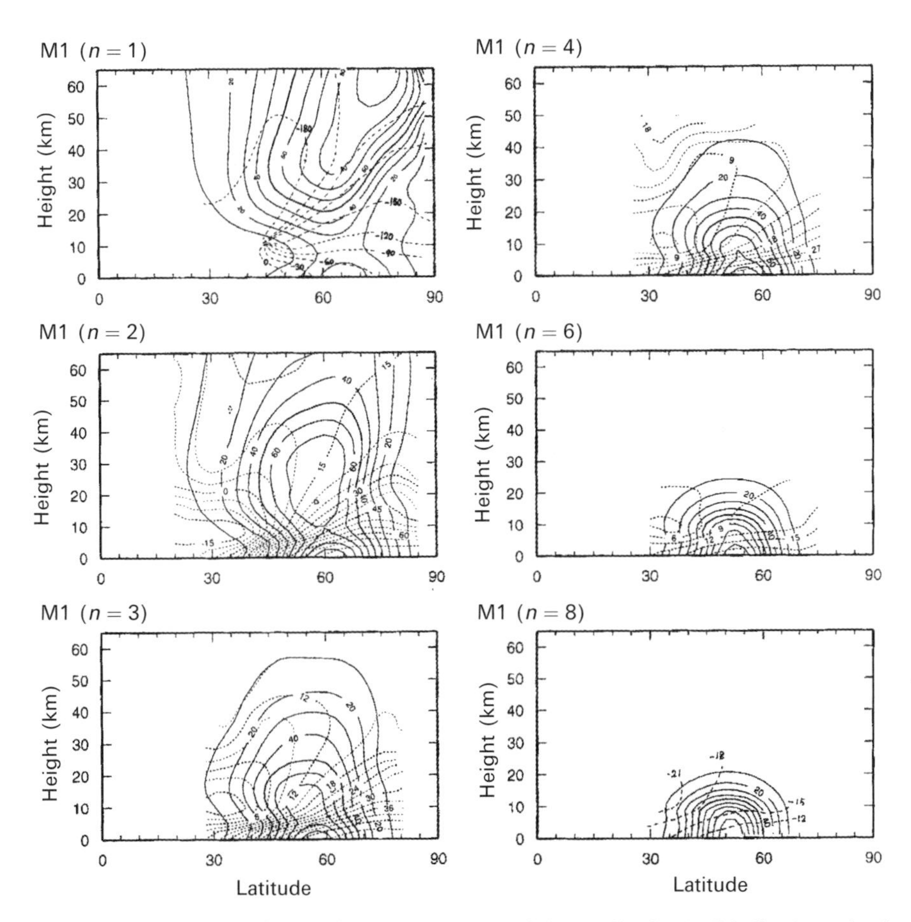

Figure 6.12. Latitude–height sections of geopotential amplitude (solid lines) and phase (dotted lines) of the fastest growing modes for a range of wavenumbers m. The wavenumber of maximum growth rate for this case is around $m = 5$–6.

From Tanaka and Arai (1999).

shows results from a computation of a range of unstable modes by Tanaka and Arai (1999) for a similar form of circumpolar jet to the one computed by Barnes (1984). Like Barnes' calculations, the instability problem was computed in full spherical geometry, but used the more accurate primitive equations (although this probably makes only minor differences to the final result in this case). This set of results clearly shows some important trends in the structures of the most unstable modes in this situation. All of the modes have a peak in amplitude close to the ground, much like the classical Charney modes, though the longer-wavelength modes ($m \leq 3$) also have a secondary maximum at higher altitude and extend upwards much more deeply than the shorter-wavelength modes. The amplitude also peaks in latitude around 50–65°N, close to the polar cap edge and the latitude of strongest zonal mean thermal gradients. The shortest wavelength modes (e.g., $m = 8$ in Figure 6.12) are evidently very shallow, extending only to around 15 km above the ground. For the deeper modes, also, the zonal phase changes rapidly at low levels and relatively little at high altitudes, giving the modes an 'equivalent barotropic' character at upper levels and strongly baroclinic character near the ground.

This mixture of baroclinic and barotropic character in the unstable modes is also reflected in the energy exchanges produced by the growing modes. The most rapidly growing modes have strong upward and poleward heat fluxes, which enable them to gain energy at the expense of the potential energy of the basic zonal flow pattern. They also produce significant poleward momentum fluxes, especially at fairly high altitudes (e.g., Barnes, 1984), in association with strong horizontal shears in the circumpolar jet. This also enables the mode to extract kinetic energy directly from the circumpolar jet, though the energy input from the baroclinic heat fluxes is substantially greater than the barotropic energy conversion. Like the equivalent instabilities in the Earth's atmosphere, therefore, these unstable modes have a mixed, though predominantly baroclinic character. Thus, in many respects the properties of the most unstable modes of these linear models (length scales, latitude of maximum activity, growth rates, etc.) seem to be closely consistent with the observed characteristics of travelling weather systems on Mars.

6.5 TRANSIENT WEATHER SYSTEMS IN GCM SIMULATIONS

Although the linear instability calculations described above can produce rapidly growing solutions which seem to have properties which resemble the observed weather systems on Mars, we must remember that any linearized model makes some pretty restrictive assumptions about the dynamics of the eddies it represents. In particular, the amplitude of the perturbations investigated in all these studies is assumed to be infinitesimal – in other words so small that the eddies they represent would be unobservable. The moment the eddies grow to an amplitude which could be detected, then the assumptions upon which the model was formulated become invalid and the model (at least formally) breaks down. This is because at finite amplitude we can no longer neglect the quadratic products of perturbation quantities in the equations of motion, and the problem becomes *non-linear*. This also means that the problem almost always becomes mathematically intractable and the only feasible way of solving it is to resort to a computer.

In practice, however, it has often been found that the fully non-linear solutions to problems in fluid mechanics still retain recognizable characteristics from the corresponding linearized problem. In the Earth's mid-latitudes, for example, even a model as simple as the baroclinically-unstable Eady wave retains many features qualitatively in common with the fully developed baroclinic weather systems in the atmosphere, including the dominant length scales and even the characteristic phase tilts with height of the pressure, wind and temperature fields. The dominant length scale of the fully developed flow is not always exactly that of the most rapidly growing instability, however, because in growing to finite amplitude the instability may significantly modify the structure of the zonal flow and change its own stability in favour of another mode. The dominant wavenumber of equilibrated, fully developed baroclinic eddies is therefore often found to be lower than that of the most rapidly growing linearized instability, and their detailed structure is likely to differ in detail from the linearized perturbation theory. However, there is usually enough in common for insights gained (more readily) from the linearized system to be useful in interpreting features of the (more complicated and difficult) non-linear system.

6.5.1 Seasonal variations in baroclinic transients

With the advent of comprehensive atmospheric models, it has become possible to perform detailed simulations of the continuously evolving atmosphere of Mars over periods of several Mars years. We have already seen in Chapters 2–5 the power of such simulations in capturing features such as the changing zonal mean structure of the atmosphere during the seasonal changes of the Martian year, and the development of features associated with the Martian surface topography. Such simulations also resolve the detailed time-varying components of the circulation, down to time scales of a few tens of minutes, and so are capable of representing the evolution of large-scale weather systems too.

The first thing to note is that the distribution and intensity of time-varying weather activity is highly dependent on the time of year and state of the seasons on Mars. Figure 6.13 (see colour section) presents a series of latitude–height sections, showing the distribution of transient kinetic energy (i.e., with the zonal and time mean flow, and the regular diurnal variations, removed) averaged over a Martian 'month' (where a 'month' is defined as 1/12 of the Martian year). The peaks of activity represent the typical location and intensity of weather activity predicted by the model.

From these statistics, it is clear that baroclinic weather systems occur mainly during the autumn, winter, and spring seasons in the relevant hemisphere, with most activity disappearing during the summer season. Furthermore, the model simulations would seem to suggest that eddy activity in the northern winter season is generally somewhat more intense than in the corresponding season in the south. This behaviour in the northern hemisphere is at least consistent with the observations from the VL time series discussed above, in that they also found weather variability to disappear during northern summer.

The disappearance of such weather activity during summer is quite consistent with an interpretation as the outcome of baroclinic instability since, as we have seen in Chapter 3, during summer the equator–pole temperature gradient at high latitudes reverses from that found during winter, such that the summer pole actually becomes warmer than the equator. This reversal of the temperature gradient at the ground has important implications for the horizontal gradient of potential vorticity. Recall that Charney-style baroclinic instability relies on a change of sign of the potential vorticity gradient at the ground from the northward gradient due to the spherical curvature of the planet in order to satisfy the necessary (Charney–Stern) condition for instability. This requires an equatorward thermal gradient at the ground, unless $\bar{q}_y$ changes sign somewhere else in the atmosphere. If the thermal gradient at the ground becomes poleward (so that the polar regions are warmer than at lower latitudes), then $\bar{q}_y$ will no longer change sign anywhere with height and the flow cannot be baroclinically unstable. Such a situation is further enhanced on Mars because summer zonal winds are also weaker than during winter, and shears are correspondingly less significant. Hence, summer in both hemispheres on Mars must be baroclinically stable. This is quite unlike the corresponding situation on Earth, where the effect of the oceans is to moderate the seasonal change of temperature at high latitudes so that the average thermal gradient at the ground is virtually always equatorwards. Hence baroclinic weather systems on Earth persist throughout the summer, although they may become somewhat weaker than in winter.

6.5.2 Wave structures and frequencies

Where the travelling weather systems do occur in the model simulations, they are found to be concentrated close to where the horizontal zonal mean temperature gradient is large, near the polar cap edge at the surface and leaning polewards at higher altitudes. The dominant wavenumbers are typically $m = 1$–3, and penetrate deeply into the high atmosphere from the surface (e.g., see Figure 6.14(a)). Moreover, their relative zonal phase increases with height (Figure 6.14(b)), indicating that the waves incline westwards with height, though with the steepest slopes near the ground. In this respect, they are very similar to the linear baroclinic instabilities found by the 'realistic' stability models of the previous section.

Such similarities also extend to the propagation speeds of the waves and their corresponding frequencies. Figure 6.15 (see colour section) shows an example of a frequency–wavenumber spectrum, taken from a GCM simulation during northern winter, in which contours of spectral power are shown as a function of zonal wavenumber and oscillation period. From this, it is clear that the spectrum is dominated by a small number of discrete wave components with $m = 1$, 2, 3, and 4 with uniformly increasing frequencies. Lines in Figure 6.15(b) indicate the directions of constant phase speed, so that points lying along such a line correspond to different wavenumbers all moving at the same speed relative to the ground. Thus, it would appear that the various wavenumbers $m = 1$–4 shown in Figure 6.15(b) are all

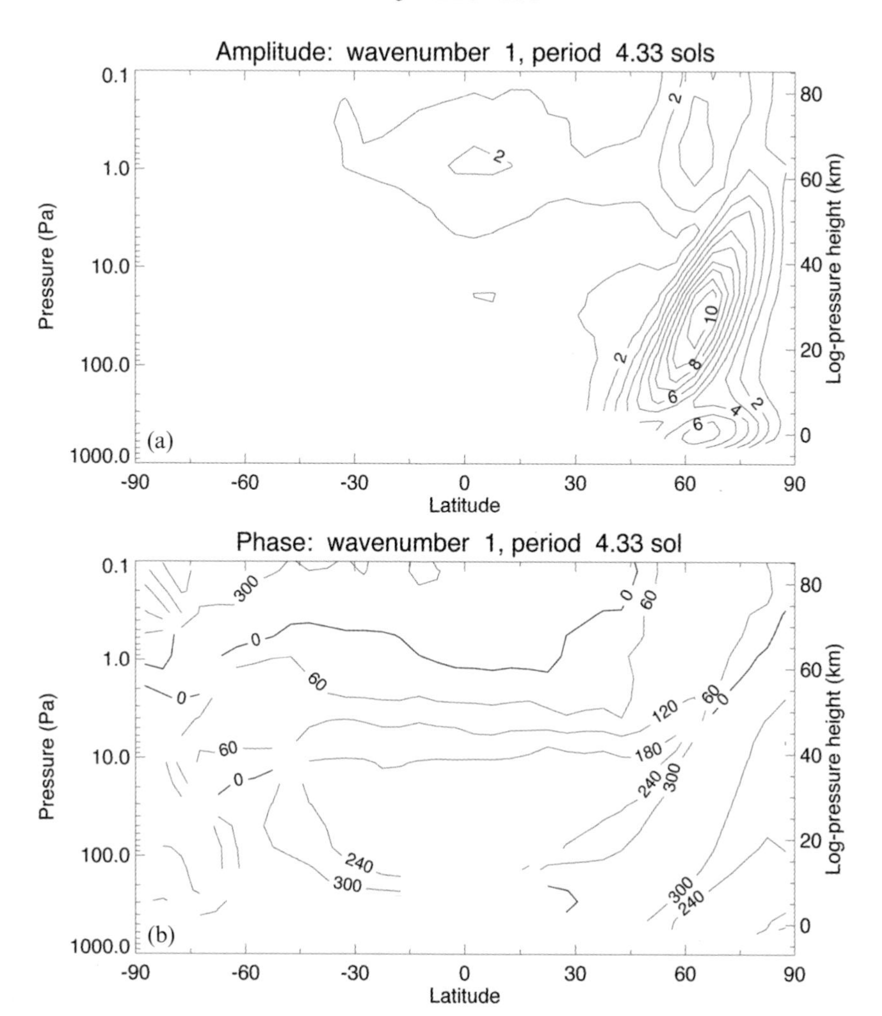

Figure 6.14. Latitude–height maps of (a) amplitude and (b) zonal phase of the meridional velocity in transient travelling wave disturbances in an assimilated analysis of TES observations of northern hemisphere winter circulation during 1999 using the Oxford Mars GCM. From Lewis *et al.* (2003).

moving together at the same phase speed as a coherent wave train. Furthermore, the spectrum is very clean and sharp, indicating that the waves are fairly steady in amplitude and moving at a roughly constant speed throughout the interval being analysed (112 sol).

This again is consistent with the analyses of the VL time series, in which it was found that the travelling weather systems during winter could persist as regular, almost periodic disturbances, for up to 60–80 sols before breaking down into a different pattern. Such a periodic behaviour is quite different to what is found on Earth, and suggests that weather forecasting on Mars should be much easier(!).

Weather systems in the southern hemisphere have been much less extensively observed, though Mars GCM simulations (and emerging observations from MGS – e.g., see Hinson and Wilson, 2002) suggest they should occur during the corresponding seasons either side of winter, much as in the north. The preferred wavenumbers are again on the planetary scale ($m = 1$–3), though they appear to be somewhat less regular in behaviour than their counterparts in the north. We return to the nature of these regular waves in more detail below.

6.5.3 Topography and 'storm zones'

Although the wavenumber–frequency spectra of the previous section suggest that transient weather systems have a nearly regular, periodic structure on a planetary scale, GCM simulations show that this is upset to some extent by zonally-varying topography. Figure 6.16(a) shows a time-mean map of transient eddy kinetic energy ($(u'^2 + v'^2)/2$) around northern winter solstice, from a simulation by Hollingsworth *et al.* (1996).

This clearly shows the transient eddy activity to be concentrated into preferred regions of longitude or 'storm zones', mainly centred in the extensive lowland plains of Arcadia, Acidalia, and Utopia. This is in clear contrast to what would occur (according to the GCM) if there were no topography present – a kind of 'thought experiment' which can be done easily in a model, though not in practice. Figure 6.16(b) shows results from a simulation under the same conditions as Figure 6.16(a), but with a flat underlying surface. In that case, the eddy kinetic energy is more or less uniform with longitude, showing that the travelling waves persist all the way around the planet at approximately uniform mean amplitude. Clearly, therefore, the topography is acting in the case shown in Figure 6.16(a) to modulate the transient waves and to break them up into 'storm zones'.

The reason for this may be understood from the theory of baroclinic waves, in which it is well established that where topographic slopes become comparable with the slope angle of the underlying topography, this acts to modify the sloping trajectories of moving air parcels and makes it harder for them to stay within the 'wedge of instability' mentioned above. This has the effect of making the flow less unstable (and may also modify the propagation speed of the waves), leading to a reduction in amplitude. This was confirmed in calculations of the ratio of the topographic and isentropic slopes in the model simulations of Hollingsworth *et al.* (1996), who showed that this ratio became close to 1 in the highland regions where the eddy kinetic energies (and corresponding heat fluxes) were reduced.

This effect is also well known in the Earth's atmosphere, especially in the northern hemisphere, where the ocean–continent contrasts modulate the activity of baroclinic cyclones and anticyclones and concentrate them into 'storm tracks' extending from the west coast of the major continents across the ocean and then ending over the eastern side of the land masses (e.g., see James, 1994). In that case,

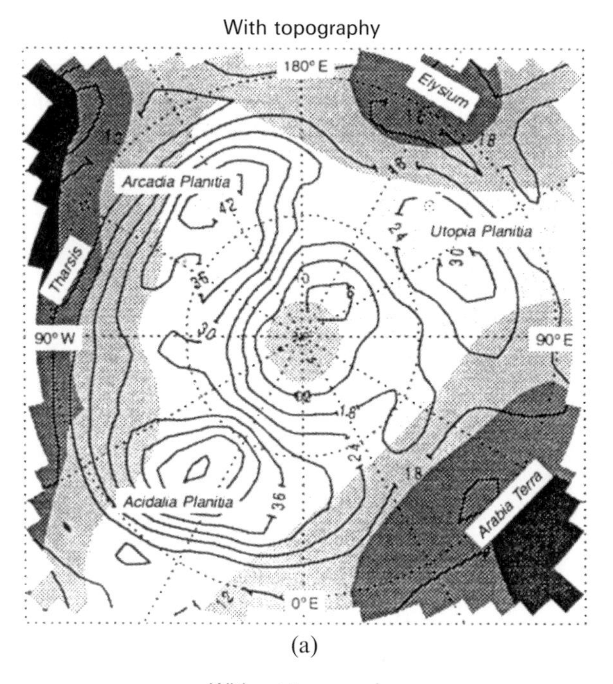

(a)

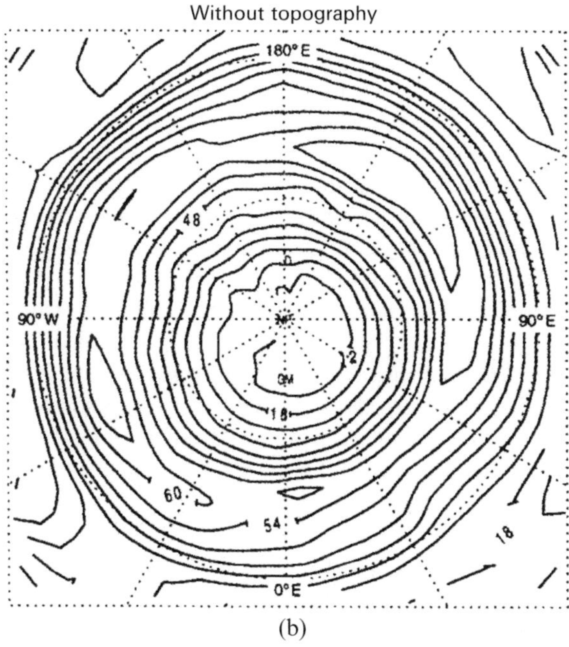

(b)

Figure 6.16. North polar projection maps of transient eddy kinetic energy at 5 hPa from northern winter solstice conditions on Mars in a GCM simulation (a) with full Martian topography, and (b) without topography.

From Hollingsworth *et al.* (1996).

however, it is the contrast in thermal properties, rather than the topographic slopes, which primarily brings about the concentration into 'storm tracks'.

A similar effect seems to be important in the southern hemisphere of Mars. Certainly both GCM simulations and observations (Hinson and Wilson 2002) suggest enhanced transient wave activity in the longitude region 150–330°E, with much weaker amplitudes especially near 90°E. A continuing puzzle, however, is what may be causing this modulation, since, unlike in the north, these longitudes do not obviously correlate with topographic features near the South Pole.

6.5.4 Transient eddy heat and momentum transports

We have discussed above how baroclinic and barotropic instabilities develop by transporting heat and momentum within the general circulation of the atmosphere in such a way as to release kinetic and potential energy stored in the zonal mean flow (itself driven by differential solar heating and other thermodynamic processes). In this respect, they are somewhat like a car engine, which takes in fuel and air (at the front of the vehicle) and converts them into exhaust gases which are then expelled at the back. In converting the fuel and air, (chemical) heat energy is released which can be used to keep the engine turning over and, as a by-product, causing the rest of the car to move along.

In the case of the atmosphere, the 'by-product' of baroclinic and barotropic instabilities may be to transport significant amounts of heat and momentum from one part of the atmosphere to another. To this end, the complicated structure of transient weather systems is important in enabling the eddies to achieve the transports they need to sustain themselves against the damping effects of friction (mainly at the ground) and radiative cooling. Transports of heat and momentum occur whenever there are significant coherent correlations between eddy velocity fields and other quantities such as temperature perturbations. Given that GCMs compute the complete time evolution of all significant meteorological variables, it is reasonably straightforward to calculate the average correlations of velocity, heat, and momentum in order to compute the eddy fluxes and integrated transports in the circulation.

Figure 6.17 shows some typical results from a GCM simulation, in which the poleward and vertical eddy correlations (fluxes) are computed and averaged in both the zonal direction and in time over a Martian 'month' (roughly 40–60 sol) close to the northern winter solstice. From this it is clear that transient eddies are producing a strong poleward horizontal heat flux over a deep region close to 60°N in association with the transient baroclinic eddies in the model (Figure 6.17(a)). This is also associated with strong upward heat transports (Figure 6.17(b)), which are necessary to allow the eddies to release potential energy in the mean zonal flow. The eddies also evidently transport zonal momentum as well as heat, and Figure 6.17(c) shows the transports to be mainly poleward north of around 40°N and equatorwards south of this.

Detailed calculations of the energy conversions resulting from these fluxes show that the eddies primarily gain their energy from the heat fluxes acting

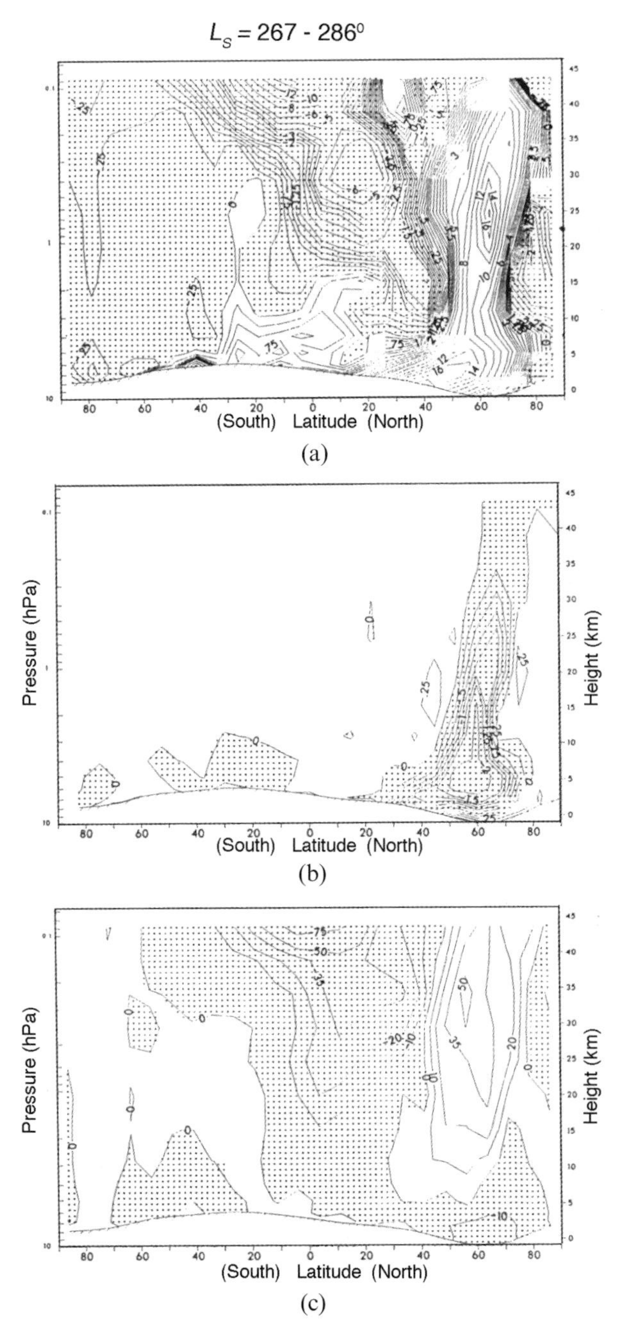

Figure 6.17. Zonally- and time-averaged transient eddy fluxes of heat and momentum for a typical northern hemisphere winter period on Mars, computed from a GCM simulation by Barnes *et al.* (1993): (a) meridional heat flux $(\overline{v'T'})$, (b) vertical heat flux $(\overline{\omega'T'})$, and (c) meridional momentum flux $(\overline{u'v'})$.

down the horizontal gradient of zonal mean temperature, and upwards, which confirms the character of the transient eddies as primarily baroclinic waves. The momentum fluxes are quite substantial, however, and suggest that the eddies gain almost a third of their energy from barotropic energy conversions. This clearly shows that transient weather systems on Mars have a very mixed baroclinic–barotropic character which is quite different in detail from the corresponding eddies on the Earth. On Earth, cyclones and anticyclones typically gain almost all of their energy by baroclinic conversions from the zonal flow, and actually lose energy via barotropic conversions back into the mean zonal flow (e.g., Holopainen, 1983).

The meridional eddy heat fluxes are important on a global scale for Mars, and play an important role in the heat balance of the polar regions. Calculations indicate (Pollack *et al.*, 1990; Barnes *et al.*, 1993) that the heat carried by the integrated horizontal eddy fluxes around 60°N may be $\sim 5\,\mathrm{W\,m^{-2}}$, which is at least as large as the heat transports due to the mean meridional (Hadley) circulation, and corresponds to about 20% of the heat exchanges resulting from CO_2 condensation during winter. Substantially greater fluxes may also occur at times, indicating that these eddies play a very important role in the heat balance of the entire atmosphere.

6.6 PREDICTABILITY AND CHAOS?

The highly coherent nature of Martian baroclinic eddies during winter is one of the most startling and remarkable features of Martian meteorology. We have already seen from analyses of the time series from the VLs that transient weather systems were found to track past each lander as if they were a simple, near-sinusoidal wave train moving along with a fixed amplitude and speed for intervals of several tens of days. The pattern would then switch for no apparent reason to a different periodicity, suggestive of a different zonal wavenumber, and then perhaps switch back and forth sporadically during the storm season until activity died away in the summer. In this way, weather patterns on Mars would seem to be much less chaotic and unpredictable than on Earth. Even so, the behaviour of Mars' weather systems is not entirely regular and periodic, with intermittent switching between periodic states which seems to occur at irregular intervals and for no apparent cause.

Similarly near-periodic behaviour is also apparent in GCM simulations of Martian winter conditions, as we have seen in Section 6.5.2, and so we may gain a lot of insight from studying the dynamics of these waves in model simulations. In fact, as we have already seen, GCMs offer another advantage in allowing us to perform some 'thought experiments' in which we can change the conditions on our model planet to try out an idea or hypothesis using the full dynamics of the model atmosphere.

6.6.1 A GCM 'thought experiment': turning night into day

Just such an experiment was carried out almost inadvertently during a time when the Oxford Mars GCM was being developed in the early 1990s. By this stage, most of the components of the model were in place and ready to be implemented, including the full dynamics, surface boundary conditions, and topography, and most of the radiative transfer parameterization schemes. As part of a test, some simulations were carried out for intervals of a few Mars years with an accurate representation of the varying solar heating during the annual cycle, but in which the diurnal variations of solar heating were not included. This meant that the incoming solar radiation remained the same at all longitudes during the day and night, preventing the excitation of thermal tides and other diurnal phenomena. When the results of this simulation were examined, a remarkable phenomenon was seen, in which baroclinic eddies developed during northern autumn but equilibrated to an almost perfectly regular state with a constant amplitude.

Figure 6.18 (taken from the paper by Collins *et al.*, 1996) shows time series of surface pressure, recorded at the location of VL 2 during a GCM simulation without a diurnal cycle, lasting up to 8 Mars years. On a time scale of several hundreds of days the pressure varies substantially due to the annual cycle of the

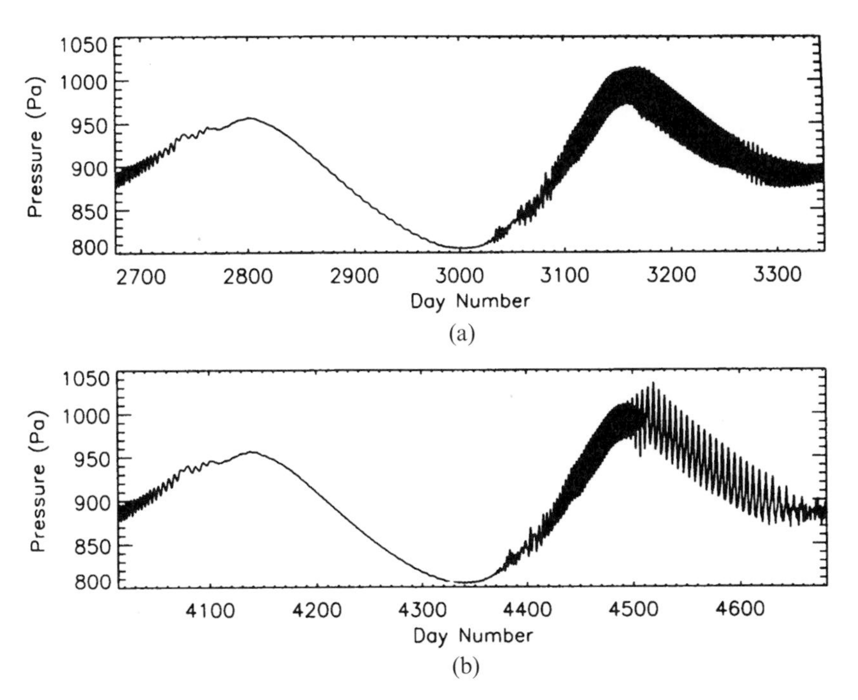

Figure 6.18. Simulated Viking Lander 2 pressure time series from a simulation using the Oxford Mars GCM lasting up to 7 Mars years without a diurnal cycle (Collins *et al.*, 1996). Extracts are shown (a) from year 5 and (b) from year 7 of the run.

condensing CO_2 polar caps, and this is clearly reproduced almost identically in each year. Note that the diurnal tidal signal is completely absent, even during summer, in these runs without a diurnal cycle. During the northern winter period, however, a highly periodic signal could be seen setting in around day number 3,100 in year 5 and day 4,400 in year 7. The initial frequency of the periodic signal is roughly the same in both years, with a period of 2.2 sol. In year 7, however, the 2.2-sol wave persisted only until around midwinter, at which point the flow switched to a lower frequency wave with a period of 5.5 sol. Out of the 8 years of the simulation, just two of the years kept the 2.2-sol oscillation throughout the winter seasons, while the rest underwent a transition to the 5.5-sol oscillation around midwinter. However, the precise timing of this transition varied from year to year, leading to some irregular interannual variability.

On closer examination of the circulation during the times of the 2.2- and 5.5-sol oscillations, it was clear that the higher frequency corresponded to an eastward propagating $m = 2$ wave while the 5.5-sol oscillation corresponded to $m = 1$. Figure 6.19 shows a series of maps of transient surface pressure at regular (1/4 period) intervals, showing the two cases corresponding to $m = 2$ and $m = 1$. This clearly shows that the two frequencies observed at the VL 2 site relate to highly regular wave patterns in each case, though some modulation in amplitude as the waves propagate eastwards around the pole can be seen due to the topographic 'storm zone' effect discussed in the previous section.

6.6.2 Intransitivity and hysteresis

Since the conditions (in terms of dust amounts and distribution, boundary conditions, heating, and cooling etc.) were exactly the same in each year of the simulation (unlike on the real Mars, where regional variations in dust amounts occur differently every year), the fact that different wavenumber states could be found in different years of the simulation without a diurnal cycle suggests that Mars (at least in the model) could 'chose' to keep either $m = 1$ or $m = 2$ during this time. Such a situation, where a system can 'chose' to exhibit two or more possible states under the same parameters is now well known to be common in many non-linear systems, and is sometimes called 'intransitivity'. The factors which lead the system to make its 'choice' between the two states might include the precise initial conditions or the immediate past history of the system, and is also related to a phenomenon called 'hysteresis'.

Hysteresis is found whenever the critical point in parameter space at which a system can be made to switch between two possible states depends on whether the controlling parameter is being increased or decreased (see Figure 6.20), and often arises in non-linear systems in association with subcritical bifurcations (e.g., Thompson and Stewart, 1986; Drazin, 1992). Hysteresis and intransitivity have not, however, featured strongly in discussions of the Earth's meteorology, which is generally considered to be too chaotic and complicated to manifest such a clear signature of coherent non-linear behaviour, though they have been discussed in

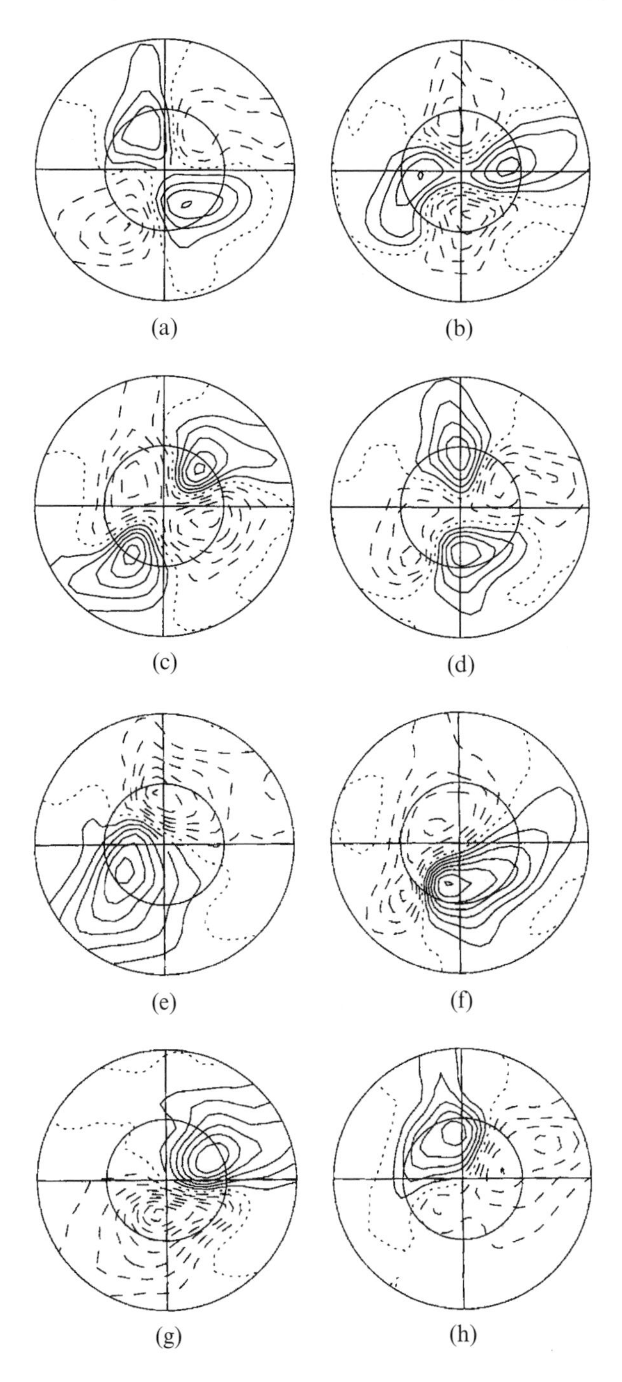

Figure 6.19. Snapshots of transient surface pressure over the life cycles of the high frequency (2.2 sol: (a)–(d)) and low frequency (5.5 sol: (e)–(h)) states of the winter circulation during the GCM simulations of Collins *et al.* (1996) without a diurnal cycle.

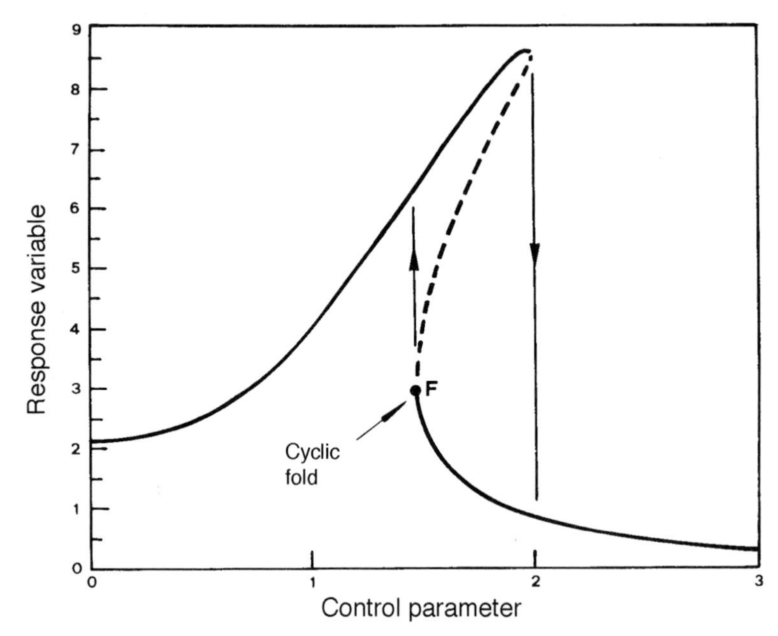

Figure 6.20. Schematic illustration of the phenomenon of hysteresis in an intransitive non-linear system with two possible distinct metastable states, in this case produced by a 'folded resonance'. Note that two possible stable solutions are found in the region $1.5 \leq x \leq 2$, depending upon whether x is decreasing from above or increasing from below.

the context of longer term weather and climate regimes on Earth (e.g., Charney and DeVore, 1979; Plaut and Vautard, 1994).

6.6.3 A laboratory analogue – storms in a teacup?

Some of the clearest examples of hysteresis and intransitivity have been found in various thermal convection experiments in the laboratory. Baroclinic instability has been studied in the laboratory for many years, using a rotating cylindrical tank of fluid in which the outer cylinder is heated and another cylinder, placed at the centre of the first cylinder (see Figure 6.21), is cooled. Heat is transferred from the outer cylinder to the inner by convective motions of the fluid in the annular gap in-between, whose motions can be visualized and measured.

If the whole system is rotated rapidly about the axis of the two cylinders, then rotation affects the dynamics of the convecting fluid in much the same way as in a planetary atmosphere at mid-latitudes, with the warm outer cylinder playing the role of the tropics and the cool inner cylinder that of the polar regions. As a result, the experiment can exhibit a vast range of different kinds of flow regime, some of which may emulate the baroclinically unstable regimes found in the atmospheres

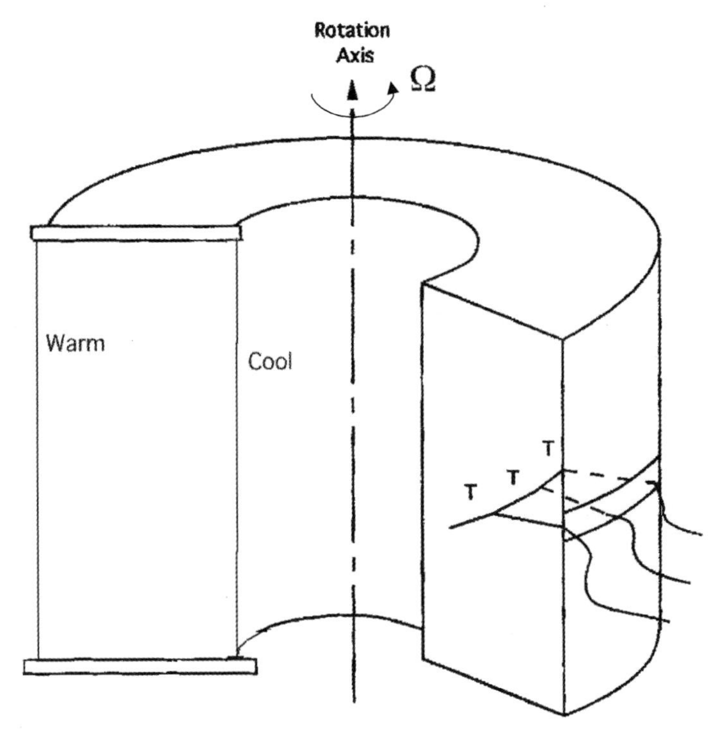

Figure 6.21. Schematic diagram of a differentially heated, rotating annulus experiment, used to study fully-developed baroclinic instability in the laboratory.

of Earth and Mars (see Figure 6.22). In particular, if the experiment is not rotated too rapidly, so that the system is not too strongly unstable, then baroclinic waves are found which equilibrate to regular wave states, which may be steady in amplitude, or with more complicated behaviour (including periodic modulations and chaos).

Depending upon the precise experimental conditions, the dominant wavenumber 'chosen' by the flow can be changed, so that for a fixed temperature contrast between the two cylinders, the faster the system rotates the higher the typical wavenumber (cf. Figure 6.22). For the lower rotation rates, the wave flows are typically regular, 'eastward'-propagating waves of nearly steady amplitude, much as we see in the Martian GCM simulations. Hysteresis is found as the experiment is run while slowly changing the rotation rate Ω. If the experiment is switched on at a particular Ω and allowed to settle, it might develop a regular wavenumber m (say). If Ω is then slowly reduced, keeping everything else the same, then the wavenumber m state will persist only to a critical value Ω_c, beyond which the flow switches to wavenumber $m - 1$. If Ω is then gradually increased, the wavenumber $m - 1$ state is often found to persist well beyond Ω_c, demonstrating a large overlap in parameter space where either m or $m - 1$ can exist as alternative states.

The rotating annulus experiment has been studied for more than 50 years, because of its value as a source of insight into the dynamics of baroclinic instability.

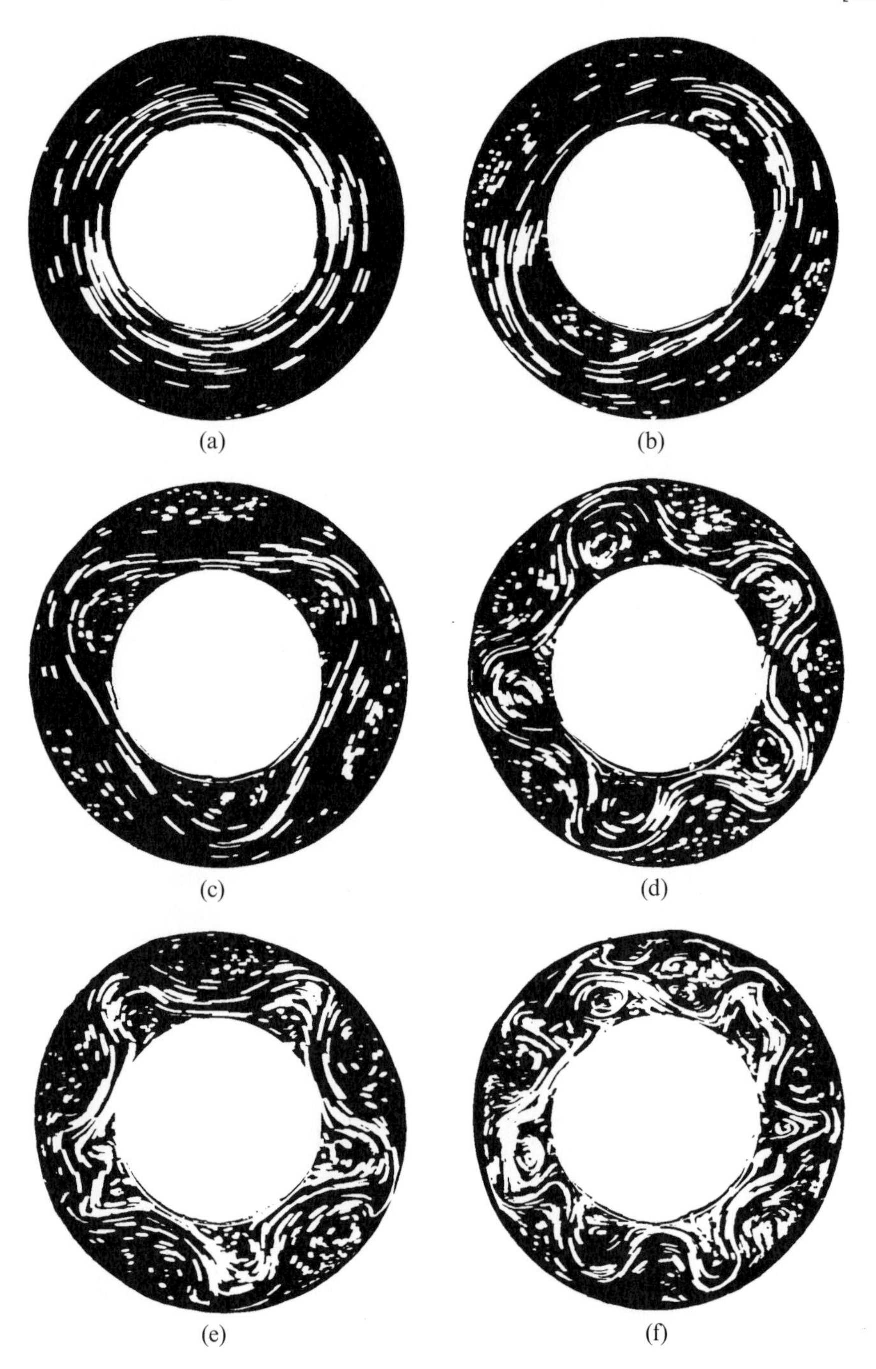

Figure 6.22. Series of streak photographs showing snapshots of typical flows in the baroclinically unstable regimes of the rotating annulus experiment as a function of rotation rate, keeping the temperature contrast fixed.

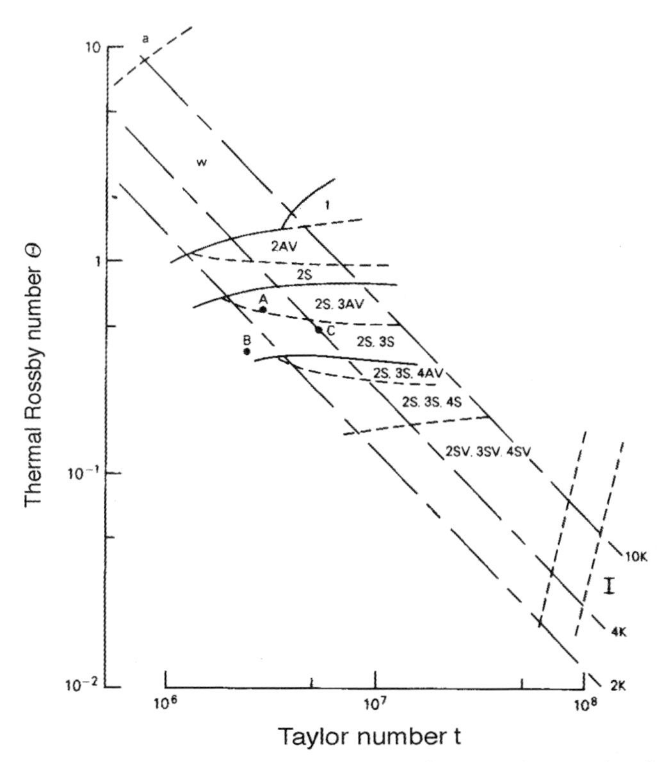

Figure 6.23. Regime diagram from the rotating annulus experiment, showing regions of parameter space where either the flow is baroclinically stable (axisymmetric state with no waves observed), or baroclinic instability is observed with 1, 2, or more possible wavenumber states indicated by regions with lists of numbers denoting the different wavenumber states observed in ensembles of experiments run with different starting conditions.

From Hignett *et al.* (1985).

As a consequence, the detailed behaviour of phenomena (amongst other things) such as hysteresis, intransitivity, and chaos, have been thoroughly documented (e.g., Hide and Mason, 1975; Früh and Read, 1997). The results of such investigations are typically reported in the form of a *regime diagram*, a map showing the different flow regimes as a function of the main control parameters of the system (related in this case to the thermal contrast and rotation rate of the system). A particular example is shown above in Figure 6.23, which illustrates the location of regions where either no waves are observed or regular waves with different wavenumbers are found, with hysteretic transitions in between.

6.6.4 Weather regimes in simplified Mars GCMs

Given the way large-scale numerical models of Mars' atmosphere are formulated (e.g., see Chapter 2), one might have thought it reasonably straightforward to put

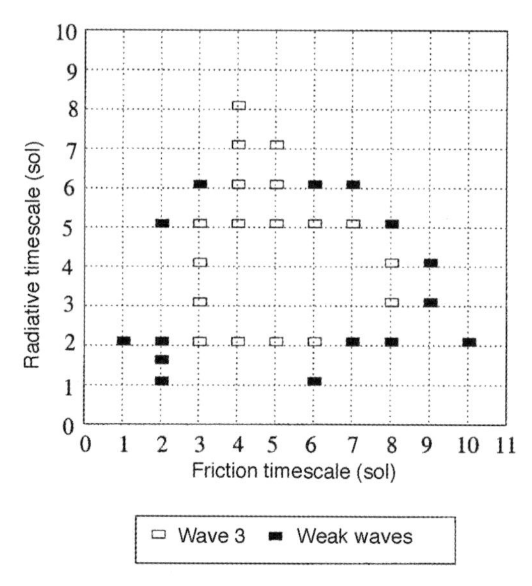

Figure 6.24. Regime diagram from a simple Mars GCM as a function of Rayleigh drag and radiative relaxation timescales τ_d and τ_r.

From Collins and James (1995).

together a regime diagram for Martian meteorology in much the same vein as has been compiled over the years for the rotating annulus experiment. In practice, however, this has never been done, and is unlikely ever to be undertaken, at least for the comprehensive GCMs in current use for accurate simulations of Mars' atmospheric circulation. The reason for this is because, unlike the rotating annulus, Mars GCMs do not have just a few simple external control parameters for the experimenter to change (like the thermal contrast or planetary rotation rate), but have many different parameters available which control the properties of the many different and complex parameterization schemes within the model. As a result, 'thought experiments' using comprehensive GCMs have to be designed very carefully to ensure that parameter changes are made in a controlled way which address the question being investigated.

It is possible, however, to simplify a GCM by using much simpler forms of parameterization for thermal forcing and dissipation (e.g., using linear surface drag and thermal relaxation towards a 'radiative equilibrium' temperature field). Such a model, known as a 'simple GCM' or SGCM, is another useful device in the atmospheric scientist's armoury, which can yield valuable insights. In this case, the model can be formulated to have just a few control parameters, and a regime diagram may be constructed in a similar way to the laboratory experiment. Figure 6.24, for example, shows one such regime diagram, in which a series of simulations were run using a model in which only the surface drag and thermal relaxation timescales could be varied. No diurnal variations were included, and

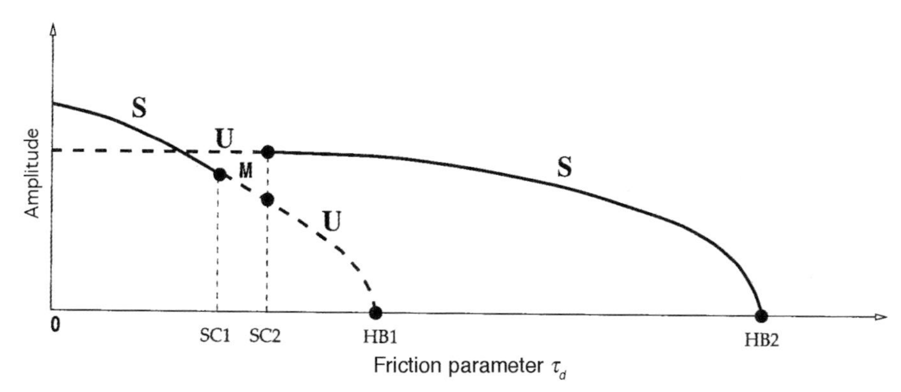

Figure 6.25. Schematic bifurcation diagram of baroclinic instability in a highly simplified model of the atmosphere of Mars, showing the amplitude of the two permitted baroclinic modes as a function of the friction parameter τ_d.
from Whitehouse (1999).

the model was further simplified by eliminating any topographic variations at the surface.

From this diagram, various transitions between different kinds of wave behaviour are seen, including regular baroclinic waves and a weaker chaotic wave state. Although the layout of the regime diagram is somewhat different in detail from Figure 6.23, there are many qualitative similarities, though the investigators found no direct evidence in their experiments of hysteresis or intransitivity. However, subsequent experiments (using the Oxford MGCM) suggest that these phenomena are indeed present in the model.

In an even more simplified model, in which the flow is formulated to filter out almost all components except for the mean zonal flow and two baroclinic wave modes, it is possible to overcome this practical limitation of running the GCM and reduce the problem to one where more sophisticated mathematical tools can be brought to bear, such as those from the theory of non-linear dynamical systems (or 'chaos theory'). For this kind of model, it can be demonstrated (Whitehouse, 1999) that baroclinic instability under Martian conditions can be thought of as representing a supercritical Hopf bifurcation from the zonally symmetric state. With two possible modes, then two possible Hopf bifurcations are found, usually at different values of the control parameter (in this case the surface friction parameter τ_d).

From a detailed analysis of this system (Whitehouse, 1999, see Figure 6.25), the existence of a small region of parameter space where either wavenumber can exist is found, confirming the possibility of hysteretic transitions between the two states. Of course this does not prove that Mars itself will operate in the same kind of regime, since many simplifications and assumptions were made to formulate the model illustrated in Figure 6.25, but does at least suggest the possibility of a clear dynamical similarity with the rotating annulus experiments. Such a similarity was suggested soon after the discovery of the regular character of Mars' baroclinic waves, by Leovy

(1985), for example, though it has only been through recent theoretical work, such as that mentioned briefly above, that the nature of this similarity is becoming clear.

6.6.5 Tidal perturbations and chaos: shaking the table?

Although the baroclinic oscillations found by the VL spacecraft were reasonably regular for intervals of up to 50 sol or more, they were still found to be substantially more erratic than was found in the Mars GCM simulations discussed above in Section 6.6.1. It was found there, that the waves on Mars itself would switch sporadically between roughly periodic states for no apparent reason. What could be causing the system to make these switches?

A clue again comes from further Mars GCM simulations, in which the diurnal cycle was put back into the model and another set of simulations was carried out under otherwise identical conditions over several Mars seasonal cycles. The result of such simulations (Collins *et al.*, 1996) was to restore the model's ability to recapture the same kind of erratic variability in baroclinic wave activity as found by the Viking landers. This version of the model would also exhibit apparent switching between two roughly periodic states, each one of which was dominated by either $m = 1$ or $m = 2$. The net result of restoring the diurnal cycle was to stimulate the system during northern winter to switch erratically between two possible regular wave states. However, because the diurnal tide was always present, the flow could never settle permanently into one state or the other but was continually being perturbed between the two states.

This would suggest that Mars' weather behaves somewhat like a ball bearing rolling around on a horizontal tray with at least two depressions, as shown schematically in Figure 6.26. Left to itself, the bearing would eventually end up in one depression or another, corresponding to either of the two regular wave states. However, if the tray were continually being vibrated (in Mars' case, by the regular thermal tide), then the bearing could not settle permanently in either depression, but would keep on being jogged out of its equilibrium to wander around the tray until it

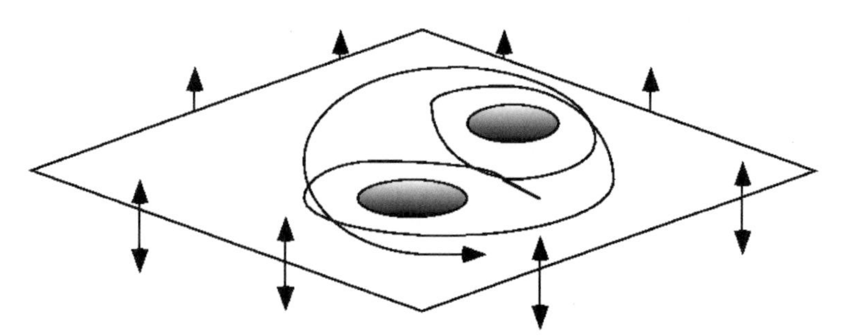

Figure 6.26. Schematic representation of the effect of the thermal tide on baroclinic waves on Mars as a noisy perturbation, jiggling the system between two equilibrium states, represented here as two potential wells.

could fall into one of the two depressions again – only to be disturbed yet again by the vibrating tray.

This behaviour represents a rather special kind of dynamical chaos, caused by the enforced merging of two coexisting periodic attractors through the action of an external perturbation. This phenomenon has been extensively studied in the context of noise-induced chaos found, for example, in laboratory experiments on electronic oscillators using Josephson Junctions in condensed matter physics (e.g., Sommerer *et al.*, 1991; Hamm *et al.*, 1994), and in periodically-forced non-linear oscillators (e.g., Moon, 1979). A somewhat similar mechanism has also been suggested as an origin for chaotic variability in the Earth's climate system, either between different weather regimes (the stimulus coming from baroclinic activity itself – see Plaut and Vautard, 1994) or in relation to the El Niño Southern Oscillation (where the annual cycle of the seasons stimulates irregular switching between El Niño and La Niña states on a timescale of 3–6 yr; Jin *et al.*, 1994). The situation here, however, where the diurnal tide plays such an important role in generating the chaotic variability of baroclinic weather systems, is unique to Mars.

6.7 OVERVIEW

Weather systems on Mars occur mainly during the seasons either side of winter in each hemisphere, and the dominant source of atmospheric variability which is not directly related to the seasonal or diurnal cycles is due to a mixed baroclinic–barotropic instability of the circumpolar winter jetstream. When they occur, these instabilities cause large transports of heat and momentum into the polar regions, and are a major factor affecting Mars' climate system.

By comparing evidence from observations and GCMs, it is seen that weather systems on Mars are only weakly chaotic compared to their terrestrial counterparts, and were it not for the diurnal tide, this chaotic character might disappear altogether. Such a result is quite remarkable, and may have important implications for future weather forecasting operations on Mars. It may also explain in part why data assimilation methods seem to work so well for Mars, even in circumstances (such as during the aerobraking phase of the MGS mission) when the observational information is relatively sparse and has many gaps.

Other sources of variability than baroclinic instability may also be important in some regions of Mars during certain seasons. Various sources of instability in Mars' tropics have been hinted at in GCM simulations, such as inertial instability (e.g., Barnes *et al.*, 1993), though observations are as yet very limited. However, the possibility of these alternative sources of variability should not be forgotten as potential stimuli for affecting the weather at the surface of Mars.

7

Dust storms, devils, and transport

7.1 A DESERT PLANET

From a casual glance at many of the images of Mars obtained at the surface by the Viking Lander (VL) and Mars Pathfinder (MPF) spacecraft (e.g., see Figure 7.1, colour section), one might easily mistake the landscape for that of many desert regions on Earth, especially those in the subtropics at high altitude (such as the Atacama desert in South America) where vegetation is very sparse. The landscape seems fairly flat and strewn with rocks and boulders of varying size, though with some low hills and mountains in the distance. No vegetation is visible. The whole area seems barren and inhospitable, and the soil between the boulders looks dry, loose, and sandy. Indeed the resemblance to terrestrial deserts is so strong that some extreme sceptics have even cast doubt on whether the spacecraft ever left the Earth!

Look more closely, however, and a number of significant differences from the Earth are clear. On closer inspection, it is apparent that there is absolutely no sign of any vegetation, in contrast to even the most barren deserts on Earth. Also, the soil colour is quite distinctive, being a pale reddish sandy colour, which is unusual on Earth (though by no means unknown). Most telling, however, is the sky, which is usually cloudless, with little evidence of the white, fair weather cumulus clouds typical of the Earth's deserts – and not blue, but pink or yellow! This rather startling colour of the sky on Mars led to considerable confusion in the early days of Martian exploration, causing spacecraft engineers to worry whether the calibration and colour balance of their cameras was defective. It is now clear, however, that there was nothing wrong with the cameras, and that the sky on Mars is typically coloured pink, suggesting that Mars' atmosphere has some quite different optical scattering properties from that usually found on the Earth.

On Earth, the cloudless sky is generally a deep blue colour because light from the Sun is strongly scattered by air molecules. Because the air molecules are much smaller in size than the wavelength of light, the scattering efficiency varies strongly with wavelength (as λ^{-4}) and independently of the atmospheric composition (so nitrogen molecules scatter as efficiently as oxygen or CO_2). As any undergraduate physicist will know, this is characteristic of *Rayleigh scattering*. If Mars' atmosphere was as clear as that of the Earth, then we would expect Rayleigh scattering also to occur on Mars, in which case the sky ought to be a dark blue or indigo in colour – deeper in colour than on Earth because Mars' atmosphere is much thinner and more tenuous (much as at high altitude on Earth).

Mars' atmosphere is not clear, however, but virtually always has a thin veil of dust particles suspended in the air. These particles are much larger than air molecules, and are typically larger than the wavelength of visible light (most commonly with a radius of around 1–2 μm). Such particles scatter light very effectively, but with a dependence on wavelength which is quite different to that of Rayleigh scattering. They also absorb some of the sunlight (becoming warmer in the process), and typically scatter very anisotropically with a strong projection of scattered light into the forward direction – much as seen, for example, in shafts of sunlight through a window, rendered visible by dust particles in the air on Earth. Such shafts of light are brightest and most clearly seen if one looks almost into the Sun along the shaft, so that the forward-scattered light can enter the eye.

Although the sites chosen for landing spacecraft on Mars have so far been rather special regions, limited to low-latitude, relatively flat plains in the northern hemisphere, we now know that the conditions encountered by the landers were reasonably typical of the rest of Mars outside the polar regions, in having a relatively loose, sandy soil and with a substantial amount of dust suspended in the atmosphere. This high level of atmospheric dust is an important feature of Mars' atmosphere, indicating a continuous interaction between the atmosphere and surface, and strongly affecting the transfer and absorption of solar and infrared radiation in the atmosphere itself.

The amount of dust in the atmosphere evidently varies significantly with season, though some dust is always present. Figure 7.2 shows the variation of dust optical depth found by the VL 1 spacecraft during an entire Mars year (Martin and Zurek, 1993). The clearest seasons were found to be around northern summer, when optical depths fell to around $\tau \sim 0.3$–0.5. During northern winter, however, peaks were found of up to $\tau \sim 2$–5 or even greater, corresponding to the development of large-scale dust storms which produced significant obscuration over large tracts of the planet. At other times when the atmosphere is clearer, dust hazes are frequently seen, especially towards the limb of the planet (e.g., see Figure 7.3).

It has been known for several centuries now that Mars undergoes periods of widespread obscuration by 'yellow clouds', on scales ranging from small, localized regions to that of the entire planet itself. Most of these obscurations would seem to be due to massive dust storms, in which large amounts of dust are lifted into the atmosphere at times when the surface winds can become strong enough to disturb surface sediments.

Thus, the behaviour of dust and its long-range transport within the atmosphere are major components of the atmospheric environment and climate system on Mars.

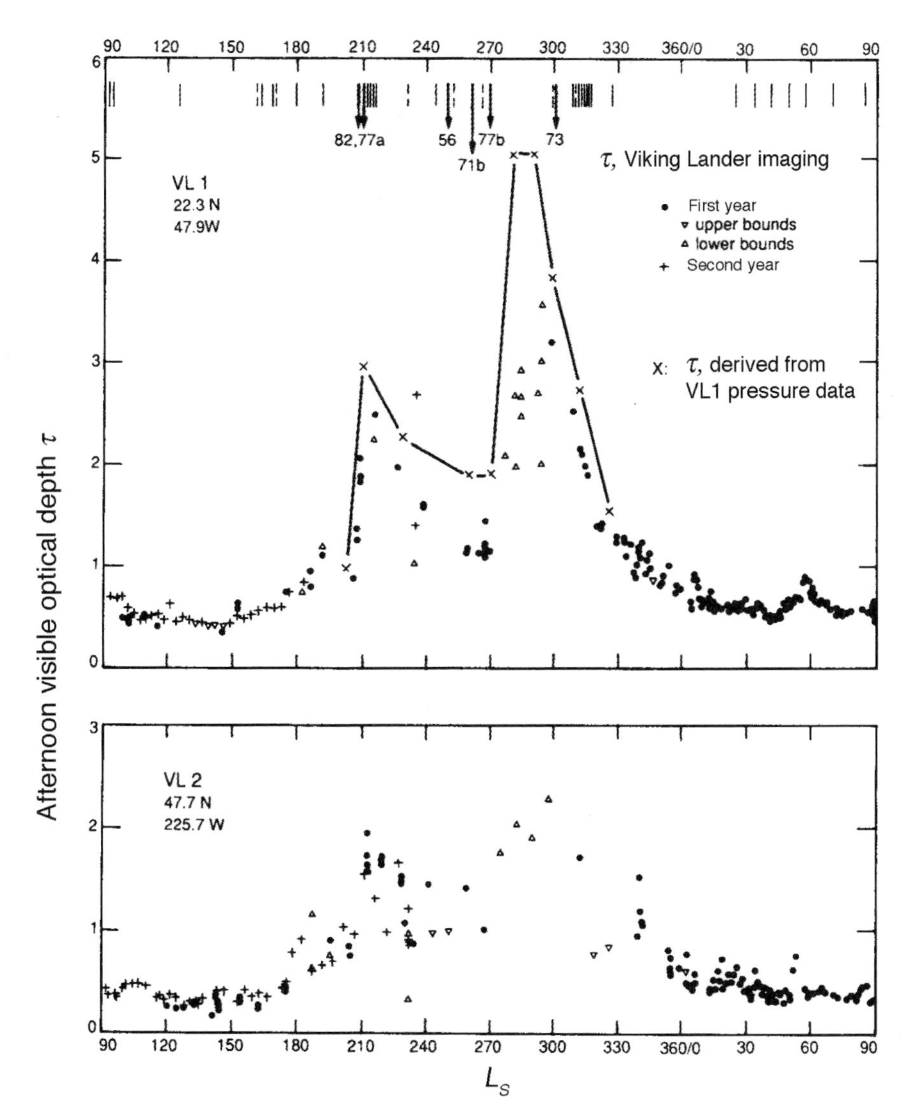

Figure 7.2. Variation in visible optical depth during the Martian year, as found by the Viking Lander spacecraft from measurements of the absorption of solar radiation at the surface of Mars. The upper frame shows variations at the Viking Lander 1 site while the lower shows variations at Viking Lander 2.

Adapted from Martin and Zurek (1993).

In this chapter, therefore, we explore various aspects of dust on Mars, its physical properties, how it finds its way into the atmosphere, atmospheric phenomena which interact strongly with surface sand and dust, and the effects of dust transport on the landscape.

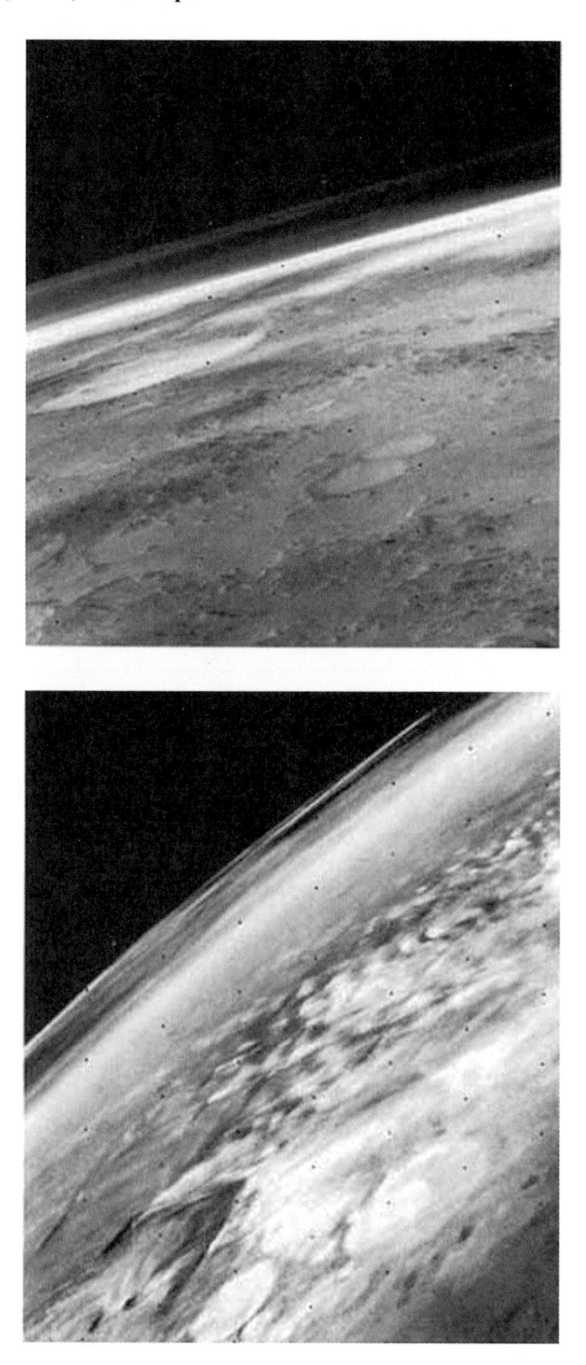

Figure 7.3. Viking Orbiter images of Mars showing the presence of thin layers of atmospheric dust in the atmosphere near the limb of the planet.
From NASA/JPL.

7.2 THE NATURE OF MARS' SAND AND SUSPENDED DUST

7.2.1 Martian soil and sand

The composition of the soil on Mars has not yet been determined with a great deal of confidence, since the amount of *in situ* analysis carried out to date has been quite limited. The VL spacecraft performed some direct chemical analysis of soil samples, but primarily with the objective of seeking evidence for living organisms rather than basic mineralogy. The MPF Lander also analysed the composition of nearby surface rocks and soil using X-ray techniques (Wänke *et al.*, 2001), though results were generally similar to those found at the VL sites. Much of the information on a global scale, however, continues to be derived largely from infrared spectral measurements (e.g., Bandfield *et al.*, 2000).

Based on infrared spectra (Toon *et al.*, 1977; Bandfield *et al.*, 2000), the dominant component of the soil is a form of silicate mineral containing at least 60% SiO_2, equivalent to a mixture of terrestrial basalt and clay. The reddish colour of the surface sand, and most likely the atmospheric dust, is widely believed to reflect a fairly large component of iron oxide (FeO and Fe_2O_3) in proportions up to 18% by mass (Banin, 1992; Wänke *et al.*, 2001) based on analyses of the surface material at the VL and MPF sites. Toon *et al.* (1977) obtained their best fit of the spectra at wavelengths $\lambda \leq 15\,\mu m$ to a terrestrial clay mineral sample, montmorillonite 219b, and this result has dominated subsequent work ever since, although there is no strong mineralogical reason to expect Martian soil actually to consist of montmorillonite (or indeed any other form of clay) itself. The spectral fit is far from perfect, and other features of the spectrum, which have been inferred by extrapolation of the measured properties of montmorillonite, especially in the infrared, are still controversial.

7.2.2 Suspended dust in the atmosphere

The properties of dust suspended in the atmosphere are also highly uncertain in detail, and are again based on quite limited measurements, mainly from remote sensing. The optical scattering properties of solid particles depends strongly on the particle composition, size, and shape. Without direct access to the particles themselves, however, the properties can only be inferred indirectly, typically from various optical measurements of extinction, scattering intensity, and polarization over ranges of wavelengths and/or phase angle (the angle subtended at the particle by the lines connecting it with the Sun and the observer). These measurements may then be compared with the predictions of Mie theory (e.g., Hanel *et al.*, 1992), which enables the integrated optical properties of homogeneously dispersed, uniform particles of idealized shapes (e.g., spheres, disks, or needles) to be computed, or other semiempirical relationships for more general shapes of particle (e.g., Pollack *et al.*, 1979).

The optical scattering properties are generally interpreted in the context of the delta-Eddington two-stream model of radiative transfer (e.g., Joseph *et al.*,

1976). In this model, radiation is assumed to scatter not more than once from any single particle, and the transfer of radiation through a plane-parallel layer of dust-laden air is decomposed into two beams – upward and downward. The net transfer of energy is then governed by three main (Henyey–Greenstein) parameters, the single-scattering albedo ω_0, asymmetry parameter g (representing the tendency for forward-biased scattering), and the extinction efficiency parameter Q_{ext}. These parameters can be linked via Mie theory to the complex refractive index of the particle material, $n_r + in_i$, which may be strongly dependent on wavelength.

In the visible region of the spectrum, n_i for Martian dust is strongly dependent on λ, and is affected substantially by trace constituents in the dust, such as iron-bearing oxides like magnetite. This tends to make the dust particles strongly absorbing in the blue, leading to a reddening of scattered light and helping to account for the pinkish colour of the 'clear' Martian sky (Huck *et al.*, 1977; Maki *et al.*, 1999; Thomas *et al.*, 1999).

As well as affecting the colour of the daytime sky, suspended dust provides a significant source of direct absorption of sunlight distributed within the atmosphere, which can have a strong effect on the distribution of solar heating between the atmosphere and surface. This is clearly illustrated in Figure 7.4, in which we show two calculations, using the delta-Eddington approximation, of the vertical distribution of direct heating due to absorption of sunlight by dust suspended in the Martian atmosphere. The calculations were carried out for local noon at a latitude of 45°, representing typical mid-latitude conditions. The optical parameters of the dust were taken as representative of the average values across the visible spectrum, and the dust mixing ratio was assumed to be uniform in height from the surface up to an altitude of around 20 km, above which it decayed exponentially with height. This distribution is similar to that inferred by Conrath (1975), based on observations of dust from the Mariner 9 mission, but seems to be broadly consistent with recent observations and has been used, for example, in defining the dust distribution for the European Space Agency (ESA) Mars Climate Database (MCD) (Lewis *et al.*, 1999; see Section 2.5.2.3 and Appendix A).

Under relatively 'clear' conditions, the visible optical depth is typically around 0.3 (see Figure 7.2), and this leads to a moderate heating rate of around 20 K sol^{-1} within the atmosphere, whose distribution with height largely reflects the distribution of dust. Thus, there is a more or less uniform heating rate all the way from the surface to around 100 Pa. When the dust layer is more substantial, however, the heating becomes confined to upper levels (Figure 7.4(b)) because the sunlight gets strongly absorbed before it reaches the lower part of the atmosphere. This has the effect of substantially reducing the amount of sunlight reaching the surface, and producing enhanced heating near the top of the dust layer. Such changes to the distribution of heating lead to an enhancement of the static stability in the lower atmosphere, which in turn might be expected to reduce the intensity of vertical transport processes and hence, perhaps, to sow the seeds of the decay of the original dust storm which lifted the extra dust into the atmosphere. This will be discussed in more detail below.

In the infrared, the optical properties of the dust are dominated more strongly

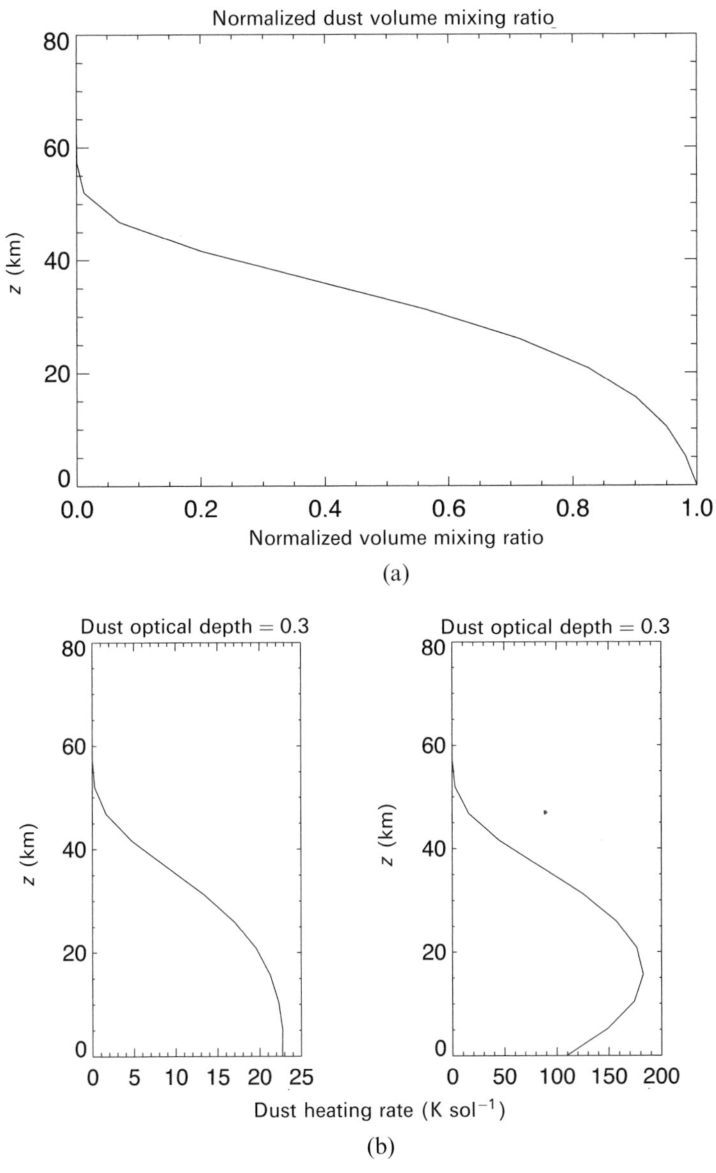

Figure 7.4. Calculations of the heating rate due to absorption of solar visible radiation by a layer of suspended dust in Mars' atmosphere. Conditions correspond to local noon at latitude 45°, values of optical parameters ($\omega_0 = 0.86$, $g = 0.79$, $Q_{ext} = 2.74$), and surface albedo ($= 0.24$) taken to represent mean values across the visible spectrum. (a) illustrates the profile of dust mass mixing ration, showing it to be uniform up to $p \sim 100$ hPa and then decaying rapidly with height, (b) shows two examples of the heating rate with a total optical depth of just 0.3 (left frame), representing typical 'clear' conditions on Mars, while the right frame shows the heating rate for an optical depth of 3, more representative of conditions during a substantial dust storm.

by the silicate content of the dust particles, which leads to a strong absorption band around 9 μm wavelength. Detailed investigations have made use of various measurements of dust hazes and layers in the Martian atmosphere under a variety of conditions, and using measurements both from the surface (using VL and MPF) and from orbit, for example, via the Mariner 9 InfraRed Interferometer Spectrometer (IRIS) (Toon *et al.*, 1977; Pollack *et al.*, 1979, 1995), Viking InfraRed Thermal Mapper (IRTM) (Kahn, 1980), the Phobos mission, and ground-based measurements (Clancy *et al.*, 1995; Tomasko *et al.*, 1999). These demonstrate significant variability in the main parameters derived, probably owing to factors such as geographically variable composition (though the ubiquity of atmospheric dust and efficient transport suggests that the composition of Mars' fine surface soils is probably relatively uniform across the planet), variations in the spectrum of particle sizes, and the possibility of occasional surface coating of dust particles with water ice deposits. An intriguing, and as yet unexplained, result of these investigations, however, is that the size distribution seemed to change little during the development of large-scale dust storms, such as the 1977 storms during the Viking period.

An important overall effect of the dust is to remove radiation from the direct solar beam and lead to an extinction coefficient, which may vary strongly with wavelength. Dust content in the atmosphere may be conveniently measured from orbit using absorption in the 9 μm band. Based on comparisons between VL visible and orbiter infrared measurements (e.g., Clancy *et al.*, 1995; Ockert-Bell *et al.*, 1997), the ratio between mean extinction in the infrared and that in the visible is typically assumed to be around 1:2–1:2.5, though uncertainties are fairly substantial and may vary with location and season. The neglect of multiple scattering is also a major source of uncertainty in understanding the impact of dust on the optical properties of the atmosphere. Computations of radiative transfer including multiple-scattering are difficult and expensive to carry out, and not many cases have been attempted for Mars so far.

The net effect of a layer of dust in the atmosphere is quite complicated, but may be interpreted in certain respects as a 'negative greenhouse effect', in the sense that the presence of dust enhances the opacity of the atmosphere in the visible compared with the infrared. This leads to a net warming of the atmosphere in the vicinity of the dust layer itself and a net cooling of the surface during daylight hours. However, the overall increase in infrared opacity due to the dust layer also affects the rate at which the surface and lower atmosphere can cool during the night. Figure 7.5 shows an example of the predicted effect of a substantial dust layer on the diurnal variation of temperature at the low-latitude landing site of the Beagle 2 lander. The effect of the increased level of dust in the DS5 case in Figure 7.5 shows a strong reduction in the diurnal temperature variation at the surface from around 40 K to less than 10 K. At the same time, the diurnal mean surface temperature is reduced by around 10 K, consistent with the 'negative greenhouse effect' suggested above; day-time temperatures are much cooler, but night-time temperatures are actually slightly warmer than in the clearer case.

In this respect, the layer of dust behaves rather like a layer of stratocumulus

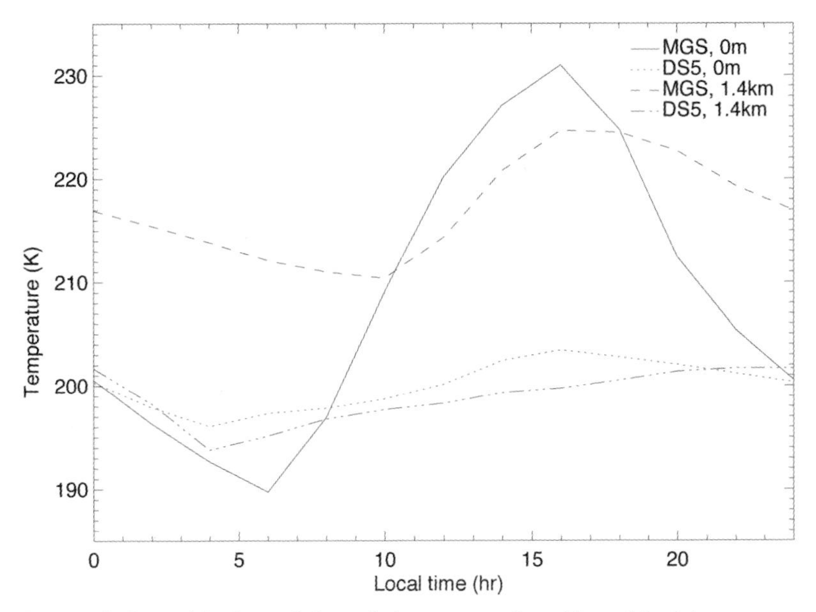

Figure 7.5. Variation with time of day of the near-surface (5-m altitude) temperature and atmospheric temperature around 1.4-km altitude for the cases (a) with low dust optical depth ($\tau \sim 0.3$) and (b) with large optical depth ($\tau \sim 5$) during a major dust storm. Variations were computed in the European Space Agency (ESA) Mars Climate Database (MCD) (Lewis *et al.*, 1999) for the Beagle 2 landing site (longitude 90°W, 10°N) during northern autumn conditions ($L_S \sim 322°$).

clouds on Earth, 'blanketing' the surface and almost eliminating diurnal temperature variations near the ground (cf. Figure 7.6(a) and (b)). Gardeners are familiar with the observation that a cloudy night on Earth in winter may prevent an overnight frost, for example, though cloudy skies will also limit the peak in temperature in early afternoon if the Sun does not break through. The analogy between Martian dust and stratocumulus clouds is not quite complete, however, since terrestrial water clouds are less efficient at absorbing sunlight and causing local heating in the atmosphere near the top of the cloud. Instead, the incoming sunlight is strongly scattered by the highly reflective droplets in the cloud back out to space, and the interaction has rather less effect on the static stability in the atmosphere than is the case for thick dust clouds on Mars.

The resulting radiation fluxes are illustrated in Figure 7.6, in which (a) and (b) compare the upward and downward short wave (visible) and long wave (infrared) fluxes under relatively clear and very dusty conditions. The strong reduction in downward visible radiation at the surface is clearly seen in Figure 7.6(b) compared with (a), yet the infrared fluxes do not change much. This is bound to lead to a net cooling of the surface compared with clearer conditions, as observed in Figure 7.5.

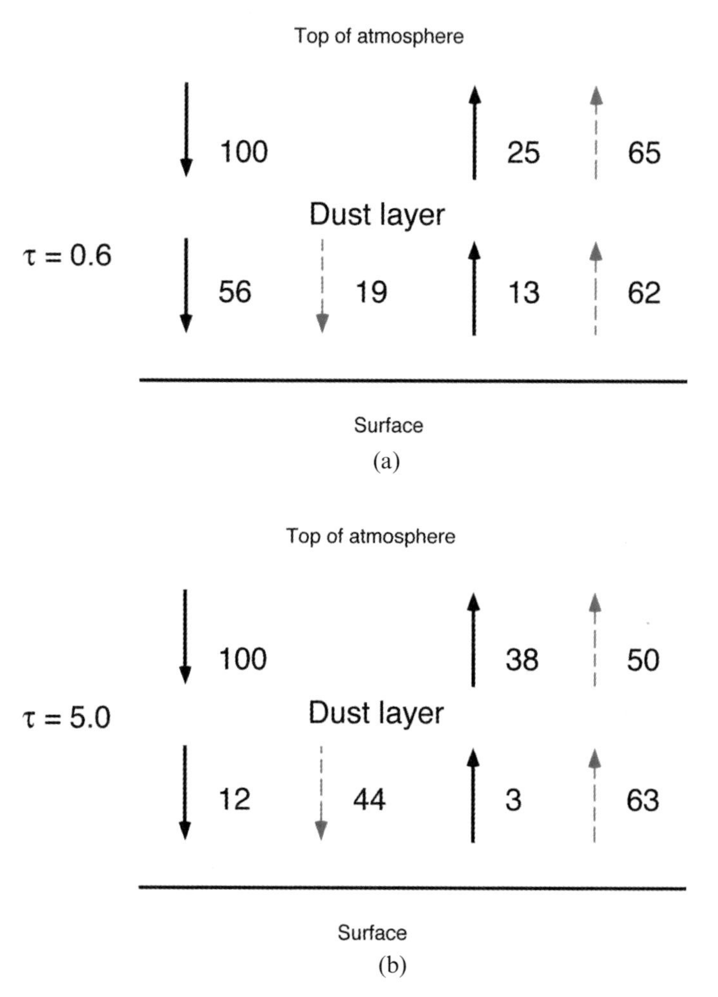

Figure 7.6. Schematic diagram comparing visible and infrared radiation fluxes computed at the top and bottom of the Martian atmosphere under roughly typical mid-latitude conditions (where net fluxes are approximately in balance). Solid arrows represent visible upward and downward fluxes, including scattered fluxes. Dashed/hatched arrows indicate infrared fluxes. Conditions are for a 'typical' atmospheric temperature profile with an assumed surface albedo of 0.2 and zenith angle of 75°, and correspond to (a) dust optical depth $\tau = 0.6$ and (b) $\tau = 5$, the latter representing strong dust storm conditions.

7.3 HOW DOES DUST ENTER AND LEAVE THE ATMOSPHERE?

7.3.1 Near-surface wind stress

Most people probably have experience of being caught in a strong gust of wind on a dry, sandy beach or dusty yard, from which there might seem to be no great difficulty in lifting dust into the air. Strong winds on Earth seem to be able to lift clouds of dust easily into the air around us. However, the lofting of dust into the free atmosphere, beyond the lowest few cm – m of the boundary layer, is far from straightforward. This is because most of the dust particles which are caught in strong near-surface gusts are relatively large in size ($\geq 20\,\mu m$) and, although they may blow around at low levels in local eddies for a while, they quite quickly fall to the ground and don't get very high. On Earth too, surface moisture plays an important role in consolidating soil and causing particles to stick together, preventing erosion by the wind, although this may have been less effective in the distant past, such as during earlier glacial maxima in previous ice ages (Jossaume, 1990; Reader *et al.*, 1999).

Moisture and ice probably play no significant role on Mars in consolidating the surface soil, except perhaps at the polar ice caps, but other factors (such as electrostatic forces, intermolecular forces, and even weak magnetic interactions (Madsen, 1999)) can lead to dust particles in the soil sticking together. These factors generally make it more difficult for the smaller particles ($\sim 1\,\mu m$), which would be sufficiently small to remain in suspension once in the free atmosphere, to be lifted by surface winds than larger ones.

Based on his observations of wind-blown sand and dust in the north African desert during World War II, Robert Bagnold (Bagnold, 1954) suggested that the resistance of surface particles to being lifted by the wind could be represented quantitatively by defining two threshold wind speeds: the fluid threshold and the impact threshold. The fluid threshold is the speed at which wind stress alone enables dust particles to be lifted directly from the surface. The impact threshold takes into account the possibility of *saltation*, in which large particles are temporarily lifted into the air by surface winds, but then quickly fall out by sedimentation. On impact with the surface, however, they may dislodge some smaller particles and lift them into the air. The impact threshold is then the minimum wind stress component of the total surface stress (including particle impacts) which leads to the lifting of small particles in the presence of impacting larger ones. The process is schematically illustrated in Figure 7.7.

The fluid threshold is generally represented by defining a threshold drag velocity, u^t_{drag}, which must be exceeded by the actual drag velocity u_{drag} for lifting to occur. u_{drag} is related to the near-surface wind stress ζ and atmospheric density ρ:

$$u_{\text{drag}} = \sqrt{\frac{\zeta}{\rho}} \tag{7.1}$$

and may be determined from near-surface wind velocities in the surface boundary layer (e.g., Garratt, 1994). Air velocities are generally very close to zero in a very thin layer next to the ground surface, and just above this layer, there is a sublayer within which velocities vary approximately logarithmically with height:

$$u(z) = \frac{u_{\text{drag}}}{k} \ln\left(\frac{z}{z_0}\right) \tag{7.2}$$

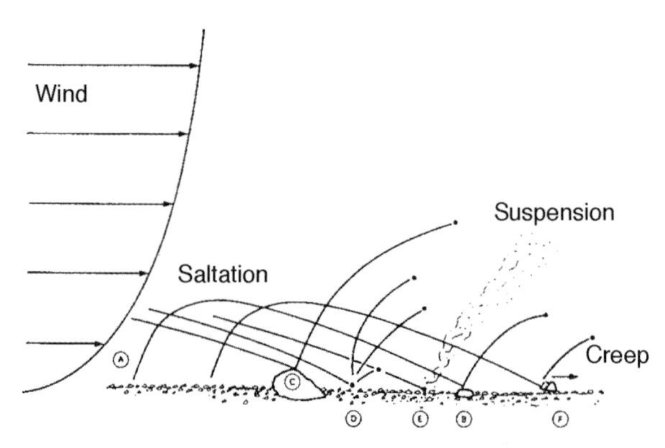

Figure 7.7. Schematic illustration of the process of saltation, in which large particles are lifted directly by surface winds, fall back to the ground and dislodge smaller particles which can remain suspended.

From Greeley and Iversen (1985).

where k is von Kármán's constant ($\simeq 0.4$) and z_0 is the so-called roughness length (probably ~ 1 cm for Mars). Hence:

$$u_{\text{drag}} = \frac{ku(z)}{\ln(z/z_0)} \tag{7.3}$$

The dependence of the threshold velocity u^t_{drag} has been determined semi-empirically from a range of experiments carried out in the laboratory, covering conditions applicable to both the Earth and other planets (including Mars – (Bagnold, 1954; Greeley and Iversen, 1985)), and found to take the form:

$$u^t_{\text{drag}} \simeq A\sqrt{gD_p \frac{\rho_d - \rho}{\rho}} \tag{7.4}$$

where ρ_d is the density of the material comprising the dust particles and D_p is their diameter. This is clearly consistent with the notion that larger, heavier particles are harder to lift than smaller, less dense ones. However, some important details are concealed by the factor A, which in turn depends on factors such as the interparticle cohesion I_p, particle size D_p and the friction Reynolds number at threshold $R_{*t} = u^t_{\text{drag}} D_p/\nu$, where ν is the kinematic viscosity of the air. From numerical solutions of the relations determining A (e.g., see Greeley and Iversen, 1985; Lorenz *et al.*, 1995; Newman *et al.*, 2002a), the final dependence of u^t_{drag} on particle size and interparticle cohesion can be determined, given the appropriate parameters for Mars' atmosphere and surface ($\rho_d = 2.7 \times 10^3$ kg m^{-3}). Some typical results are illustrated in Figure 7.8.

The actual value of I_p applicable to Mars is not known with certainty, but a typical value used for both Earth and Mars (e.g., Greeley and Iversen, 1985) is $I_p \simeq 6 \times 10^{-6}$ N m$^{-1/2}$. For this value, the 'optimum' particle size for direct lifting by the wind (i.e., with the lowest threshold velocity) is around $D_p \sim 90$ μm, requiring a wind around 5-m altitude of around 30–40 m s^{-1}. For much smaller particles,

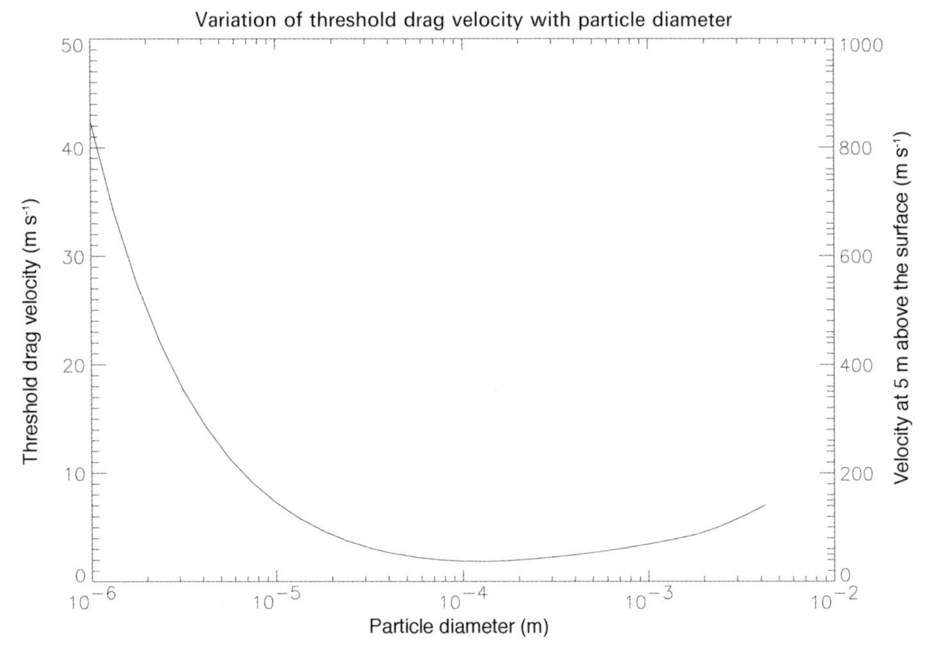

Figure 7.8. Variation of threshold drag velocity (left axis) and corresponding wind magnitudes at 5 m altitude (right axis) with particle diameter, using interparticle cohesion parameter values of $I_p = 1 \times 10^{-6}\,\mathrm{N\,m^{-1/2}}$.

After Newman *et al.* (2002a).

however, closer to the size typically found in suspension in the free atmosphere on Mars ($D_p \sim 1\,\mu\mathrm{m}$), the threshold velocity is found to be extremely high, requiring enormous wind speeds ($>500\,\mathrm{m\,s^{-1}}$) at 5-m altitude which would never occur. Thus, saltation must be crucial to the lifting of very small particles into the air.

Bagnold (1954) and White (1979) found that the upward flux of small particles lifted once the fluid threshold for large particles had been exceeded could be related to the horizontal flux of large (sand) particles, F_H, given by the empirical formula:

$$F_H = \max\left[0, 2.61\frac{\rho}{g}(u_{\mathrm{drag}})^3\left(1 - \frac{u^t_{\mathrm{drag}}}{u_{\mathrm{drag}}}\right)\left(1 + \frac{u^t_{\mathrm{drag}}}{u_{\mathrm{drag}}}\right)^2\right] \tag{7.5}$$

This is now starting to be used by general circulation modal (GCM) modellers to parameterize the lifting of dust in numerical simulations of Mars' atmospheric circulation (e.g., Newman *et al.*, 2002a), in which the small-diameter (dust) upward particle flux is set proportional to F_H. Other effects, such as the gustiness of the near-surface wind, may also be important in determining the mean rate of dust lifting close to the ground. Once lifted into the lowest region of the atmospheric boundary layer, turbulent eddies might then be expected to mix the dust upwards into the free atmosphere.

7.3.2 Convection and 'dust devils'

The other main way that surface dust can be drawn up into the atmosphere is by entrainment into regions of strong vertical motion, which can occur as a result of thermally-driven convection in the atmospheric boundary layer. Convection will occur readily during summer days on Mars, driving a convective layer which can reach altitudes of over 10 km before late afternoon, when the lowering of the sun in the sky reduces the surface heating and causes the convection to collapse in the evening and overnight. Even so, near-surface velocities in such convection are unlikely to exceed a few $\mathrm{m\,s^{-1}}$, so simple convection is likely to suffer from the same obstacles to dust raising as direct wind-stress effects.

In clear, strongly heated conditions, however, the convection is sometimes observed to organize itself into an intense vortical structure, with strong updrafts and relatively low pressure at its centre. Such structures are frequently found in desert regions on Earth, where they are commonly known as 'dust devils'. This is because the low pressure region at the centre of the vortex can help to lift dust particles into the air much more efficiently than simple horizontal wind stress. So dust rapidly gets entrained into the vortical circulation creating a coherent column of dust, which can have a typical lifetime of several minutes or more before the vortex is disrupted. On Earth these 'dust devils' can have sizes from $\sim$1 m in diameter to over 100 m, with heights of up to several hundred metres or more.

The equivalent structures on Mars can apparently reach much larger proportions, commensurate with the deeper convective boundary layer on Mars during summer, with features up to several hundred metres across and 8 km high having been seen in Mars Orbiter Camera (MOC) images from Mars Global Surveyor (MGS).

Figure 7.9 shows an example of a dust devil seen from Mars orbit by the MGS spacecraft, in which the dust column and its shadow, and often a dark trail, can be seen clearly. Such features are evidently common in many places on Mars, at least at certain seasons and times of day, and move steadily across the ground gathering surface dust and lifting it into its vortical column. Dust devils were first identified from Viking Orbiter images (Thomas and Gierasch, 1985), though these features were often seen at the limit of the resolution of the Viking Orbiter camera.

Dust devils were experienced by the VL spacecraft (Ryan and Lucich, 1983), and also seen by the MPF Lander (e.g., Metzger and Carr, 1999), both as tenuous dust columns on the horizon and from their signature in wind and pressure as the feature passed over the spacecraft itself. Figure 7.10 shows the traces in pressure and wind measured by the MPF Lander during the passage of a dust devil, from which some of its characteristics could be inferred (Schofield *et al.*, 1997). This investigation was somewhat limited, however, by the lack of an instrument to measure wind speed accurately, though the rapid change in wind direction as the low-pressure centre of the vortex passed by is clearly seen. The relationship of the variation of p and wind direction above to the passage of the dust devil can be found in Figure 7.11, which schematically illustrates this event relative to the Lander.

The fact that both the Pathfinder and Viking spacecraft emerged unscathed from

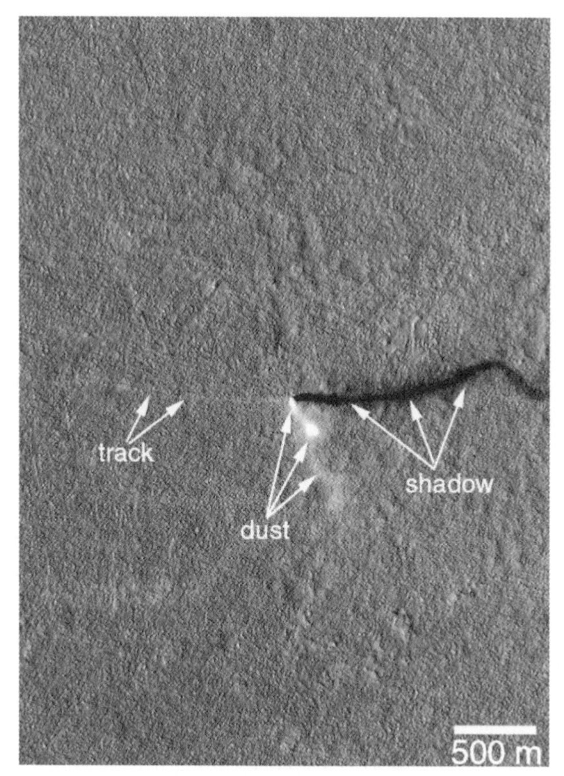

Figure 7.9. Mars Orbiter Camera (MOC) image of a large dust devil on Mars.
From NASA/JPL and Malin Space Science Systems.

the experience at least indicates that the wind cannot have been too strong or destructive, suggesting that such features are (as on Earth) relatively mild phenomena, simply lifting dust and blowing it around. The VL instruments could measure wind speeds during the passage of dust devil vortices, and experienced gusty winds as the vortices went past with wind speeds (at a height of 1.6 m) exceeding 25 m s^{-1} in 7 out of 120 cases. Three vortices were even thought to have produced gusts of 36–44 m s^{-1}. These wind speeds would correspond to a strong gale on Earth, though the much lower density of the Martian atmosphere would mean that the wind stress on obstacles would be much less – roughly equivalent to winds on Earth around 10 times weaker. Even given the low density of the atmosphere at Mars' surface, however, these speeds would almost certainly be enough in themselves to lift dust into the air by the action of direct wind stress and saltation. But the additional effect of the low pressure in the vortex probably means that even much gentler vortices would be likely to lift fine dust particles into the vortex core.

Figure 7.12 shows a collection of streaks on the surface of Mars, almost certainly produced by the passage of a number of dust devils which lift the available dust layer on the ground and scour it clean, revealing darker subsoil underneath. The impact

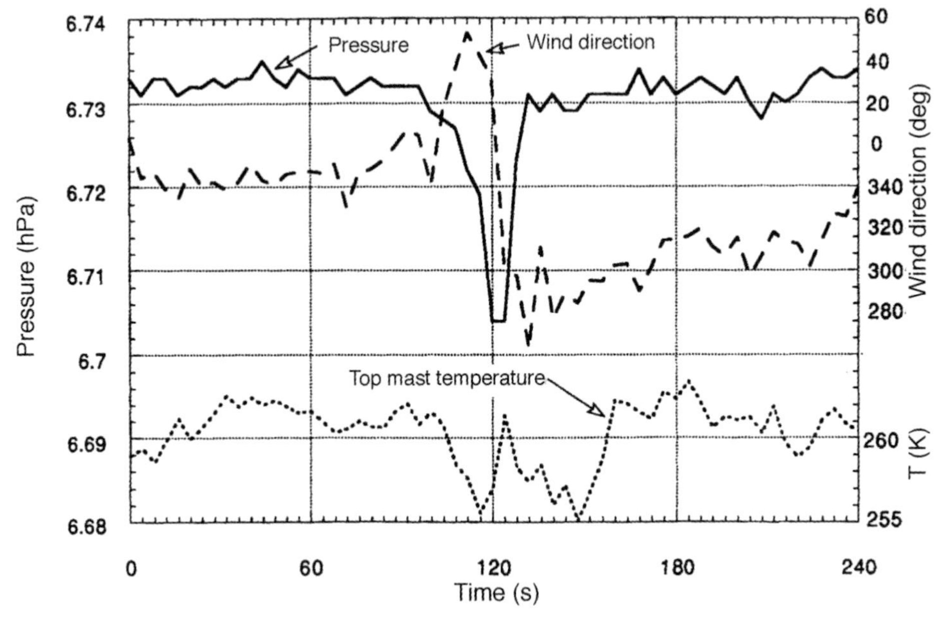

Figure 7.10. Pressure and wind direction variations measured during the passage of a dust devil over the Pathfinder Lander spacecraft

From Schofield *et al.* (1997).

on the landscape of dust devils is probably limited to these temporary perturbations to the dust covering at the surface.

7.3.2.1 Dust lifting by dust devils

The above observations suggest strongly that dust devils may provide an important mechanism for lifting dust into the air, at least in localized regions. Because large dust devils occur relatively rarely, however, they don't lead to very strong obscuration over large areas, but may well make an important contribution to maintaining a continuous component of background dust loading in the atmosphere, even under relatively clear conditions when no dust storms are occurring nearby. The large-scale efficiency of dust devils as a source of dust in the atmosphere has been estimated in several studies, which have made use of a simple theoretical model of a dust devil as a convective heat engine (Rennó *et al.*, 1998, 2000). Such a model seems capable of predicting where dust devils are most likely to occur, and also to give some indication of their likely relative intensity, from which an estimate of the average flux of dust into the atmosphere can be obtained.

The critical parameter for these predictions is the so-called 'dust devil activity' index, Λ, which is defined as the flux of energy available to drive dust devils near

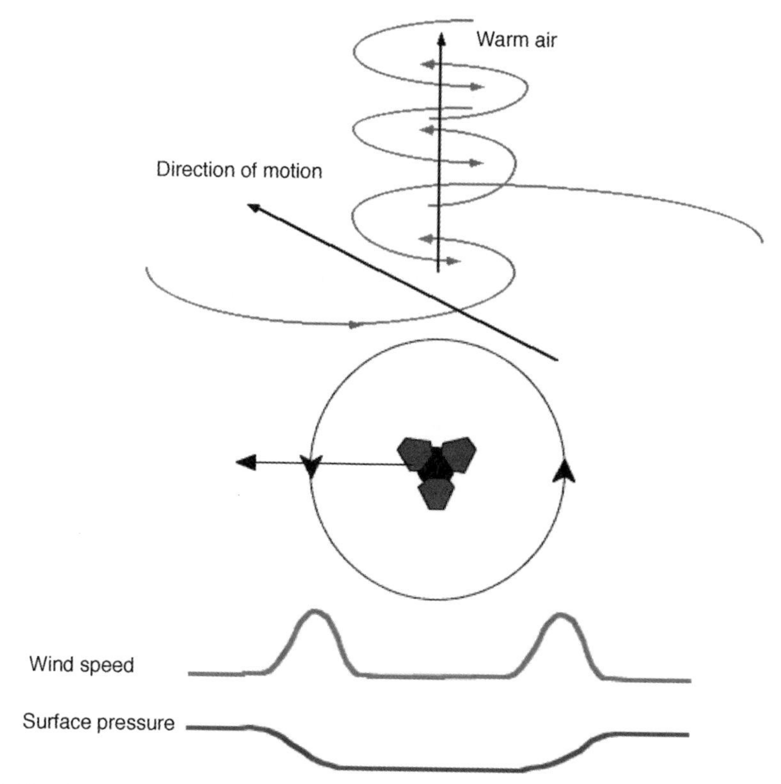

Figure 7.11. Schematic pressure and wind speed and direction variations during the passage of a dust devil over the Pathfinder Lander spacecraft.

the surface. It is roughly related to the basic upward sensible heat flux at the surface, F_s, by:

$$\Lambda \simeq \eta F_s \tag{7.6}$$

where η is the thermodynamic efficiency of the convective dust devil heat engine (the fraction of input heat energy transformed into mechanical work). η generally increases with the depth of the convective boundary layer, while F_s, determining the heat input to the base of the dust devil, increases with the difference between surface and air temperatures at the ground and is also affected by the surface wind stress. Rennó *et al.* (1998) suggested that η may be given approximately by $1 - b$, where:

$$b = \frac{(p_s^{\kappa+1} - p_{\text{top}}^{\kappa+1})}{(p_s - p_{\text{top}})(\kappa + 1)p_s^{\kappa}} \tag{7.7}$$

and where p_s is the surface ambient pressure; p_{top} is the ambient pressure at the top of the convective boundary layer; and $\kappa = R/c_p$.

Figure 7.13 (see colour section) shows the results of computing Λ, averaged over a Martian 'month' (where a 'month' is defined as 1/12 of the Martian year, measured

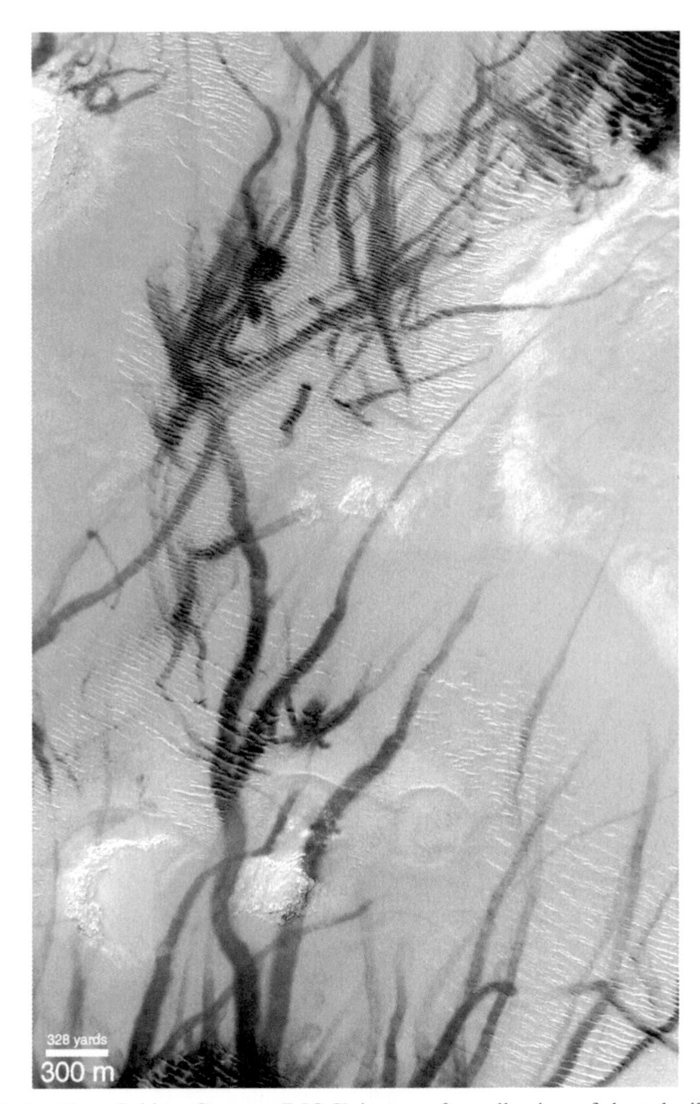

Figure 7.12. Mars Orbiter Camera (MOC) image of a collection of dust devil tracks.
From NASA/Malin Space Science Systems.

in L_S) from the ESA MCD (see Appendix A) by Newman (2001). From these maps, it is clear that the most intense dust devil activity is predicted to occur in fairly low-lying plains during summer, with the largest activity occurring in a band in the southern subtropics. The index is quite inhomogeneous and patchy, however, suggesting that the occurrence of dust devils is likely to be highly intermittent and variable. More detailed calculation of these maps at different local times of day also clearly reflects the tendency of dust devils to reach their peak activity shortly after local noon, when the intensity of surface heating has just passed its daily peak and the thermal contrast between the surface and atmosphere is greatest. Such behaviour is well known for terrestrial dust devils, and is probably consistent with the observations of Martian ones too.

7.3.3 Dust deposition and removal processes

The main mechanism by which dust is likely to be removed from the atmosphere is via gravitational sedimentation. Dust particles in general are denser than the rest of the atmosphere and will therefore tend to fall under gravity. Their terminal fall or 'sedimentation' velocity may be found by equating the buoyancy-adjusted downward force (due to gravity) and the upward drag force (due to viscous friction between the falling particles and the atmosphere, given by Stokes' Law). An additional correction factor is also required in principle in the Martian atmosphere due to its comparatively rarefied nature. This factor is known as the Cunningham 'slip-flow' correction (Rossow 1978), and is required because the typical radius of the dust particles is less than the mean free path between molecules of the atmosphere. Thus, dust particles will tend to 'slip' through the air with reduced viscous resistance. The effective sedimentation velocity, w_{sed}, is then given by:

$$w_{\text{sed}} = \frac{2g(\rho_d - \rho_a)}{9\eta_d}\left(\frac{D_p}{2}\right)^2\left(1 + \frac{8\zeta_f}{3D_p}\right) \tag{7.8}$$

where ρ_d (which is $\gg \rho_a$, the density of the surrounding air) is the density of material making up the particle; g is the gravitational acceleration ($\sim$3.73 m s^{-2} for Mars; see Table 1.1); η_d is the molecular viscosity of the air ($\approx 1 \times 10^{-5}$ kg m^{-1} s^{-1} for Mars), D_p is the particle diameter; and ζ_f is the mean free path ($= 1.6 \times 10^{-5} \times T/p$ metres for Mars).

For typical dust grains of density $\rho_d \sim 2.5 \times 10^3$ kg m^{-3} and diameter $D_p \sim 1$ μm, Eq. (7.8) indicates a sedimentation velocity w_{sed} on the order of 1–2 mm s^{-1}. In the absence of large-scale vertical motion and active dust-lifting, this would suggest that dust clouds extending to altitudes of up to 10 km would settle out by dry deposition on a timescale of around 1–2 months. For sand-sized particles with $D_p \sim 100$ μm, however, sedimentation velocities on the order of 50–80 cm s^{-1} are anticipated, leading to a timescale for suspension in a cloud of altitude 10 km of only a few hours. Such timescales are reasonably consistent with the observation, for example, that local sandstorms on Mars are relatively shortlived, and tend to die out overnight when the wind stress near the surface due to convective sources decays in the late afternoon. Small (1 μm) dust particles, on the other hand, can remain suspended for long periods of time in the atmosphere, leading to persistent veils of dust across large tracts of the atmosphere following major dust storms or other dust-lifting events.

The removal of dust may also be enhanced in the presence of water vapour by dust particles acting as condensation nuclei (particularly in the winter polar regions where large amounts of CO_2 and water vapour condense to the surface). This process is very important in determining the lifetime over which dust remains suspended in the Earth's atmosphere (e.g., see Jossaume, 1990), and may also be a significant factor in the Martian dust cycle. It has even been proposed that dust scavenging by water vapour outside the polar regions may be a vital part of the mechanism to shut down large dust storms on Mars (see Michelangeli *et al.*, 1993, Clancy *et al.*, 1996). Dust scavenging may also occur due to CO_2 ice formation, and even snowfall, in the polar hood clouds which form during polar winter at high latitudes, and attempts have been made to parameterize this process in some MGCMs. But further studies are required to determine the sensitivity of dust transport to these processes, and more observations are needed to identify their importance on Mars.

7.4 DUST STORMS

The presence of suspended dust in the Earth's atmosphere is most clearly seen during the passage of dust storms, which take place from time to time in desert regions. On Earth, such storms are typically associated with gust fronts and windstorms (e.g., see Figure 7.14, colour section) in which winds near the ground are locally enhanced by intense turbulent eddies and organized by larger-scale weather systems.

Dust storms have been seen on Mars almost since the advent of telescopic observations of the planet from ground-based instruments in the 18th century. Transient yellow clouds can sometimes be seen in images of Mars, even from relatively modest telescopes, and occasionally these clouds are observed to extend over large regions of the Martian surface (e.g., see McKim (1999) for a comprehensive review of ground-based observations of Martian dust storms). Nowadays, with the benefit of observations from the Hubble Space Telescope (HST), and from Mars orbit itself, it is clear that dust movement is one of the most active aspects of Mars' meteorology. Dust storms evidently occur over a wide range of scales, with clouds appearing from a few 10 s of km across to the scale of the planet itself. It is now well documented (e.g., Martin and Zurek, 1993; see also Figure 7.21) that major, planet-encircling (or even global) dust storms have been seen to form on Mars roughly every 2–3 Mars years, while smaller, local and regional events seem to occur sporadically every year. Indeed, it is likely that dust storm activity may constitute the main source of intraseasonal and interannual variability in Mars' present climate.

So what is it like to experience a Martian dust storm? The Viking landers recorded measurements for more than a complete Mars year, which included the incidence of two major, planet-encircling dust storms during 1977, as well as smaller, local events. Although such huge dust storms appear dramatic from the vantage point of an orbital platform, at the ground the conditions appear to have been windy but otherwise reasonably benign. VL 1 experienced wind gusts of around $26\,m\,s^{-1}$ at 1.6-m during the arrival of both the 1977 great dust storms (Ryan and Henry, 1979) although VL 2 only recorded wind gusts around $14\,m\,s^{-1}$. A further local dust storm was experienced at VL 1 during the clearing of the 1977b storm, with winds at 1.6-m altitude rising to peak values of 25–30 $m\,s^{-1}$ (with an hourly mean around $15\,m\,s^{-1}$). This local storm seems to have been associated with the passage of a baroclinic wave system. Such wind speeds were probably sufficient to produce local lifting of dust by saltation, but none of these storms seems to have led to major changes to the local landscape or surface in the vicinity of either lander.

7.4.1 Local and regional storms

With the establishment of semipermanent observational platforms in orbit around Mars, starting with MGS, it is at last becoming possible to carry out reasonably comprehensive surveys of dust activity on Mars, both as a function of season and areography (e.g., Cantor *et al.*, 2001). Images from the MOC instrument on MGS can easily resolve dust clouds arising from even quite small events (e.g., see Figure 7.15, colour section), and the regular orbital coverage allows us to see the evolution of these individual events from initiation to final decay, at least in the visible portion of the spectrum.

Figure 7.16 (see colour section) shows the result of one such survey by Cantor *et al.* (2001), in which the locations of nearly 800 local dust storms which occurred during 1999 are plotted with respect to location and season. From this figure, it is clear that the latitude bands close to the polar cap edge in each hemisphere are fertile regions for small dust storms to be initiated. At more equatorward latitudes, however, it would appear that topography plays an important role in focussing dust storm activity into preferred places. Strong activity in 1999 was found in the Arcadia plains northwest of Olympus Mons and northwest of Elysium, and also on the eastward flanks of the Tharsis Plateau and other low-latitude topographic features in the north. In the southern hemisphere, dust storms tended to remain close to the edge of the retreating polar ice cap.

Cantor *et al.* (2001) apparently saw a general southward drift of dust storms as the seasons progressed, but with storms which originated in the north crossing the equator predominantly around longitudes corresponding to the Western Boundary Currents discussed in Chapter 4. This indicates a strong connection between some of the principal meteorological systems discussed in previous chapters and the organization and initiation of dust storm activity on Mars, which we will discuss in more detail below. The low-latitude and subtropical dust storms were also found to be much more common in the southern hemisphere during northern winter, close to Mars' perihelion when diurnal heating in the southern summer is most intense. This largely confirms the observation of increased mean optical depth of dust in the atmosphere around $L_S \sim 250°$ from the VL spacecraft (e.g., see Figure 7.2).

The most common type of storm was the small, local event (see Figure 7.17), with the frequency increasing almost inversely with the size of the storm (measured by the area of the visible dust cloud, see Cantor *et al.* (2001)). The distribution found by Cantor *et al.* (2001) seems remarkably regular, and consistent with a frequency proportional to (storm area)$^{-0.98}$ both globally and in either the northern or southern hemispheres, over a range in area up to around 10^6 km^2.

The duration of dust storms was also found to be highly variable, though with some clear trends. Small, local dust storms tended to be quite short-lived, typically lasting no more than a day or two. This suggests that many of these features may be initiated by local winds which may be intensified by diurnal effects, such as slope

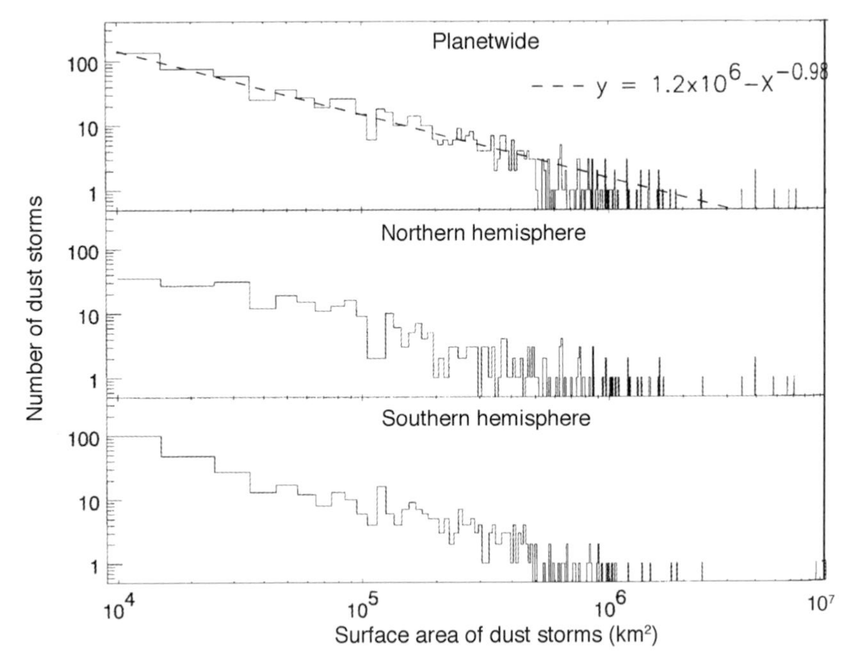

Figure 7.17. Histogram showing the relative frequency of occurrence of dust storms during 1999 as a function of their size.

From Cantor *et al.* (2001).

winds and tides, and focussed by local topography. The majority of small dust storms in the north were found to occur in relatively low-lying regions, where the increased atmospheric density lowered the wind speed threshold for dust lifting. The short duration of the observed small storms also perhaps suggests that these storms are mainly lifting quite large particles into the air, and are therefore akin to terrestrial sandstorms. The large particles settle out relatively quickly without rising high into the air, almost as soon as the enhanced wind activity dies out during the night, so that the return of the diurnally-enhanced winds the following day frequently leads to the initiation of a new storm rather than a regeneration of the one from the previous day.

On a larger scale, however, storms tend to be more long-lived, leading to the possibility of mergers between smaller storms into larger, regional events. From the statistics of dust storm events (e.g., see Figure 7.17), Cantor *et al.* (2001) suggested a size criterion to distinguish between 'local' and 'regional' events corresponding to an area of the visible cloud of around 1.2×10^6 km^2, where the power law form of the histogram starts to break down. In contrast to an earlier suggestion of Martin and Zurek (1993), this criterion seemed to be effective regardless of the shape of the dust cloud. The larger 'regional' storm would then persist significantly longer than smaller events ($\geq$4 sol) and have a somewhat different life cycle.

Dust in a regional storm, if it occurs in an appropriate location, may even lead to the entrainment of dust into the main branch of the Hadley circulation. In this case, enhanced dust levels could be carried over much larger distances and begin to affect the circulation of the whole planet. A clear example of such an effect was seen in 1999 during the early mapping phase of the MGS mission. A regional dust storm was observed to develop from a smaller storm in the Hellas region, producing a larger event in the Noachis Plain to the west. As the regional storm developed, infrared mapping of the thermal structure of the atmosphere showed a pronounced warming in the middle atmosphere around $L_S \sim 240^\circ$ (see Figure 7.18(b), colour section), which affected much of the planet. This happened, even though the enhanced dust loading was only apparent near the location of the storm itself (see Figure 7.18(a)).

Subsequent analyses using data assimilation techniques (Lewis *et al.*, 2003; see also Section 2.6) show the development of the dust cloud over a period of nearly 4 weeks, with the most intense storm period lasting around 10 days (Figure 7.19, colour section), and its influence on the rest of the atmospheric circulation. This clearly demonstrated (Figure 7.20, colour section) that, as dust was entrained into the large-scale meridional circulation, there was a rapid enhancement of the Hadley circulation within one day of the eruption of the dust storm, which spread the effects of the storm planet-wide. The injection of dust into the Hadley circulation led to an almost immediate enhancement of solar heating, due to the low thermal inertia of the Martian atmosphere. As soon as one component of the Hadley circulation (the upward branch in the southern summer hemisphere) was enhanced, the rest had to follow as a consequence of mass conservation, leading to increased upwelling near the latitude of the storm and increased downwelling in the northern hemisphere. The enhanced downwelling led in turn to increased adiabatic compressional heating in the northern hemisphere, and hence the entire atmosphere was seen to warm up, at least at low to middle altitudes.

After its initial development, the storm was then seen to move southwards and dissipate around the retreating South Polar cap. At this point, the enhanced temperatures and meridional circulation were also seen to decay back to 'normal' conditions and the storm ended. Such a life cycle seems to be reasonably typical of a regional event, and similar storms are seen almost every Mars year at some stage. The precise location and timing of these storms is quite variable and unpredictable from one year to another, however, and they are a major source of variability within the Martian climate system.

7.4.2 Planet-encircling storms

Although regional dust storms are sufficiently common for several to occur every Mars year, occasionally two or more regional storms may develop around the same time and amalgamate into even larger events. Such major storms can persist for even longer than the typical regional storm, lifting large amounts of dust into the large-scale atmospheric circulation and spreading dust over large stretches of the planet. Such storms are known as 'planet-encircling' storms, and can spread dust globally over the entire planet in extreme events. Planet-encircling storms do not seem to

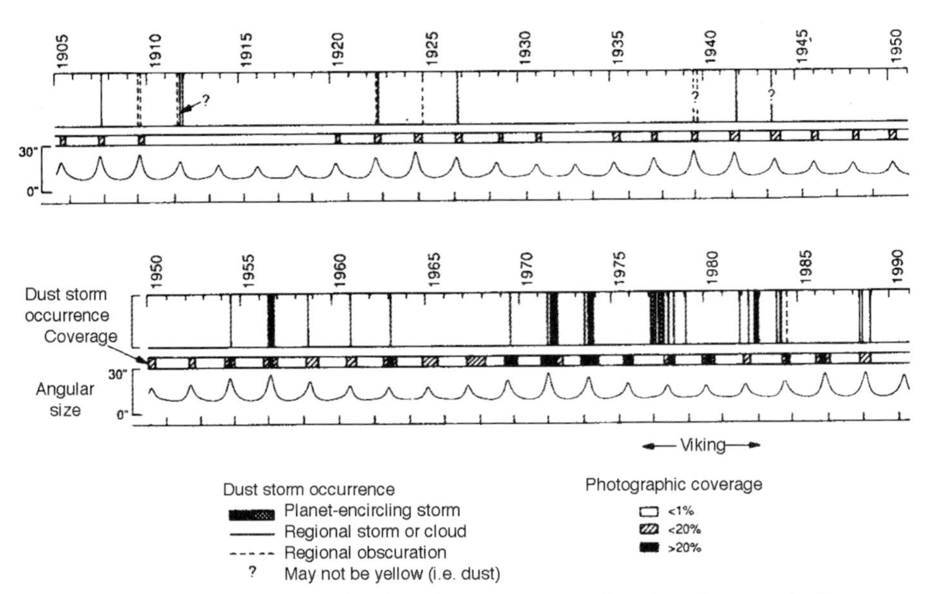

Figure 7.21. Schematic diagram showing the occurrence of major planet-encircling storms during the past 50 years

From an analysis of past ground-based observations by Martin and Zurek (1993).

occur every Mars year but, from ground-based observations over more than 100 years (e.g., Martin and Zurek, 1993; McKim, 1999) seem to happen sporadically roughly every 2–4 Mars years (see Figure 7.21). However, the observational record of this may not be wholly reliable, since observations, especially of smaller, regional events, are only possible around Mars' opposition, when the planet presents a large enough disk to be resolved from Earth.

Such a major dust event occurred around the arrival of the Mariner 9 spacecraft at Mars in 1971. When Mariner 9 completed its orbit insertion manoeuvres and deployed its camera, its first view of Mars was not of craters, mountains, and plains at the surface but the tops of vast dust clouds obscuring almost the entire surface of Mars (see Figure 7.22). Only the peaks of the highest volcanic mountains were visible above the ubiquitous dust obscuration, and this situation persisted for several weeks before the atmosphere finally started to clear sufficiently for surface features to be visible. The dust obscuration also affected infrared observations of the atmosphere and surface, and the IRIS team had to wait until much of the dust had settled out before embarking on its task of mapping the planet.

A similar storm (though of even greater intensity and duration) was observed by MGS in 2001, though in this case the spacecraft was able to observe the initial development of the storm as well as its subsequent evolution and decay (Figure 7.23, see colour section). The 2001 storm apparently grew from a collection of regional

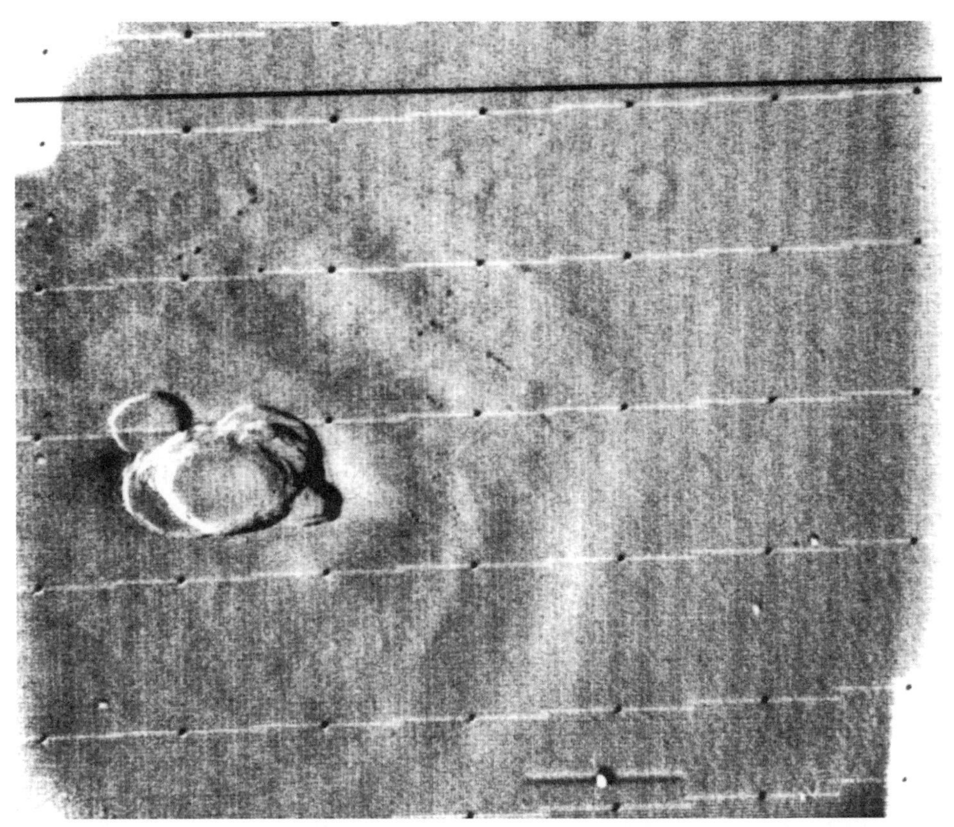

Figure 7.22. Mosaic of images showing the summit of Olympus Mons emerging above the clouds of the great dust storm observed by the Mariner 9 spacecraft in 1971.
From NASA/JPL.

storms in and around the giant Hellas impact basin during June 2001. These storms amalgamated and spread across the southern subtropics of Mars, and began to lift substantial amounts of dust into the main Hadley circulation. This led to a large plume of dust spreading northwest and encircling the planet in the southern hemisphere. By early July, the dust plume had filled the southern skies and extended into the northern hemisphere, where it continued to spread until, by mid-July, the entire planet was covered by a thick veil of yellow-brown dust.

The dramatic change in appearance of Mars during this time was clearly visible from Mars orbit (Figure 7.24, see colour section), and from the HST (Figure 7.25, see colour section), and the change in colour of the entire planet could even be discerned by the naked eye from Earth. This major storm kept a substantial veil of dust across the planet for another 2 months or so before the dust finally started to clear in October/November 2001.

In seasonal terms, the 2001 storm was reasonably typical of large, planet-encircling events. Throughout observational history, storms which grew into

Table 7.1. Large dust storms beginning in 'months' 7–11 ($L_S = 180°–330°$), grouped according to the 'month' in which they were observed to begin.

Mars year	Earth year	Start deg L_S	End deg L_S (month)	Duration in deg L_S	Size	Initial location
						Month 7
12	1977	204	241 (9)	37	Global	Hellas and Argyre
15	1982	208	226 (8)	18	Global?	Unknown
18	1988	189	190 (7)	1	Regional	Hellas
21	1994	201	254? (9)	53?	Global?	Unknown
23	1997	190	201 (7)	11	Regional	Hellespontus and Argyre
25	2001	180?	220? (7)	30?	Global	Hellespontus and Hellas
						Month 8
9	1971	213	220 (8)	7	Regional	Hellespontus
13	1979	212	249 (9)	37	Regional	Unknown
18	1988	214	231 (8)	17	Regional	Hellas
20	1992	230?	? (8)	?	Regional	Unknown
22	1996	230?	? (8)	?	Regional	Unknown
23	1997	224	241 (9)	17	Regional	Noachis
24	1999	224	232 (8)	8	Regional	Chryse
24	1999	228	243 (9)	8	Regional	Amazonis
						Month 9
1	1956	249	304 (11)	55	Global	Hellespontus
9	1971	260	329 (11)	69	Global	Hellespontus
10	1973	244	283 (10)	39	Regional	Unknown
12	1977	268	350 (12)	82	Global	Argyre?
						Month 11
−16	1924	310	350 (12)	40	Global	Unknown
2	1958	309	312 (11)	3	Regional	Syrtis
10	1973	300	336 (12)	36	Global	Solis, Claritas, and Meridiani
18	1988	314	326 (11)	12	Regional	Thaumasia
19	1990	309	312 (11)	3	Regional	Chryse
19	1990	326	333 (12)	7	Regional	Aurorae–Chryse
23	1997	309	319 (11)	10	Regional	North and northwest of Argyre

After a compilation by Newman (2001).

planet-encircling or global events have been found only within the approximate range $190° \leq L_S \leq 310°$, corresponding to the seasons either side of southern hemisphere summer and perihelion. In fact a closer examination even suggests some trends within this period for differences in the character of large dust storms, depending upon the precise season in which they began. Table 7.1 shows a compilation (by Newman (2001), based on the work of McKim (1999), Martin and Zurek (1993), and Clancy *et al.* (2000)) of observations of the largest regional and global dust storms observed during the past 80 years or so, grouped according to the Mars seasonal date in which they were observed to begin. This appears to show a trend in which planet-encircling or global storms have begun within three quite narrow periods: either between $L_S \sim 200$–$208°$ (in late month 7), between $L_S \sim 249$–$268°$ (month 9), or between $L_S \sim 300$–$310°$ (early month 11). These periods also seem to contain the regional storms of longer duration, suggesting a clear tendency for sustained and enhanced storminess. No global storms seem to have occurred between $L_S \sim 208$–$249°$, or during the period $L_S \sim 270$–$300°$ (although major storms developing earlier than these periods would tend to be reaching maturity and decaying during these times of year).

The location of the regions where major storms begin also seems to show some dependence on the time of year. All large storms starting between $L_S \sim 210$–$265°$ or earlier seem to show a preference for originating in the Noachis and Hellespontus regions on the northwest flanks of the Hellas Basin. The 2001 storm appears also to be in this category, having begun around $L_S \sim 180°$ also in the Hellespontus region. The later period storms, in contrast, seem to begin in various regions west of the Argyre Basin, though may also link with storms in the Noachis and Hellespontus regions. In the earlier group of storms, the initial dust cloud was observed to expand first northwest, and then south over Noachis, spreading eastwards and westwards around latitudes ~ 30–$50°$S. Storms from the later period which begin near Argyre, on the other hand, tend to expand eastwards first, and then south, with the main core of the storm in the region 60–240°W until it spreads more widely.

7.5 PROCESSES INFLUENCING DUST STORM INITIATION, EVOLUTION, AND DECAY

From the observationally-based information presented above, we can see that dust storms on Mars originate from a variety of causes and events, though some trends are clear. One important prerequisite is clearly to generate a sufficiently strong disturbance in the air close to the surface to lift small ($\sim$1 μm) dust particles, either directly or (more likely) through saltation or vortical suction (in dust devils).

But simply producing some local lifting of dust into the lowest layers of the atmosphere may not be enough to result in a full-blown dust or sand storm. Additional processes may be needed to organize strong surface winds and consequent dust lifting over a large region necessary to initiate a storm, and also provide some kind of positive feedback to enable the increasingly dusty air in the cloud to strengthen the surface winds and generate renewed and enhanced dust lifting, thereby enabling the storm to spread and sustain itself against losses due to sedimentation and other processes.

7.5.1 Meteorological organization of dust lifting

7.5.1.1 Hadley circulations

The Hadley circulation itself was one candidate (Schneider, 1983), based on a simple model, similar to the Held–Hou model discussed in Chapter 3, with an asymmetric circulation produced by displacing the subsolar latitude away from the equator. Schneider's model concerned itself not so much with the initiation of a major storm, but more with its ability to spread globally. He found that if the strength of the diabatic heating (representing that due to extra dust in the atmosphere) were to become greater in intensity than a critical value, then the dominant Hadley cell in the summer hemisphere could spread across the equator and transport material (including dust) into the winter hemisphere. This would provide some clues as to why some large planet-encircling storms do not eventually become global, though left a number of other issues unaddressed. GCM simulations suggest that the dominant Hadley cell crosses the equator for most of the Martian year in any case, though it does become stronger in the presence of enhanced dust.

7.5.1.2 *Resonant free modes*

An alternative idea concerning the triggering of a major storm, due to Jim Tillman of the University of Washington (Tillman, 1988), invoked a resonant condition which might occur as the thermal structure of the atmosphere changed slowly during the seasonal cycle, such as discussed in Section 5.2. In this situation, a resonant free mode (such as a planetary Kelvin mode) might get strongly excited because its resonant frequency shifts close to either the diurnal or semidiurnal frequencies. This might be expected to lead to the rapid growth in amplitude of the free mode as it comes into resonance, hence amplifying surface winds and triggering large-scale dust lifting.

Evidence in favour of this came from the observations during the 1977 Viking dust storms with frequencies close to (but not necessarily equal to) the diurnal and semidiurnal frequencies, prominent in measurements at the surface by the Viking landers. However, detailed calculations suggested that the modes most likely to come into resonance were short-period Kelvin modes, whose response was likely to be global in nature (with a maximum close to the equator) and not localized as the observations of precursors to the Viking and other major dust storms indicated. Moreover, models suggested (Zurek, 1988) that the resonant effect would be most effective in the cold northern winter hemisphere, contrary to the observed initiation of virtually all major dust storms.

7.5.1.3 *The 'multiple causes' hypothesis*

One of the earliest discussions of this issue (Leovy *et al.*, 1973) proposed that it was only when the superposition of many of the different seasonally-varying components of Mars' atmospheric circulation reinforced each other to produce locally very strong surface winds that a major storm could begin. This might involve adding together the effects of the changing Hadley circulation, atmospheric tides, and local slope winds to achieve sufficiently strong winds to produce wholesale lifting of surface dust. Leovy *et al.* (1973) also emphasized the potential importance of the seasonal maximum in solar heating in the southern hemisphere, close to Mars' perihelion, to account for the seasonal and areographical location of major storm precursors. Interference effects between enhanced tidal modes and the zonally-varying surface topography were expected to lead to strong constructive interference at certain longitudes and destructive interference at intervening longitudes (e.g., where the VL 1 observed a sharp decrease in tidal amplitudes shortly before the onset of the 1977b major storm (Leovy, 1981)). Enhanced winds were expected in the southern subtropics around longitudes of 135° and 320°W, close to the preferred location of major storm precursors.

In practice, it now seems likely that a variant of the original ideas of Leovy *et al.* (1973) may be closest to what really happens on Mars. Various processes may initiate local storms, some of which may be able to merge and grow into larger systems, though elements of all of the above models may also play a role. Additional factors beyond those discussed above almost certainly include the importance of circumpolar baroclinic waves as a major source of local storms at high latitudes, prominent in the MOC images (Cantor *et al.*, 2001), and the equator-crossing western boundary currents in focussing strong cross-equatorial winds into certain longitude bands and coupling the two hemispheres.

7.5.2 Radiative-dynamical feedbacks

The other important ingredient in the development and spread of an initial dust cloud into a major storm is some kind of feedback by which the dynamical behaviour of the atmosphere may be modified by the presence of extra dust in the air. We have already seen that the direct heating of the atmosphere by solar radiation is strongly affected by the amount of dust (e.g., see Figure 7.4), increasing as the optical thickness of the atmosphere increases. The increased heating rate is likely to result in strong increases in the induced vertical velocity (roughly as $w \sim Q_T/N^2$, where Q_T is the solar heating rate). In turn, this would draw heated air upwards and cause cooler air to rush in underneath from elsewhere. Close to the ground, this would lead to stronger surface winds, which in turn could increase the effectiveness of dust lifting, and hence reinforcement of the storm.

7.5.2.1 The 'dusty hurricane'

The potential for this feedback was realized early on, and even led to the suggestion (Gierasch and Goody, 1973; later elaborated, for example, by Houben, 1981) that dust heating could play a role analogous to the release of latent heat in moist convection during the development of tropical storms and hurricanes on Earth. This process, known as Convective Instability of the Second Kind (CISK), occurs when latent heat release reinforces a pre-existing circulation pattern, such as a large-scale vortex, leading to its growth in intensity. Thus, in this concept, a major dust storm would begin with a weak vortex, produced by some other meteorological phenomenon (perhaps initially in geostrophic balance), with light surface winds. As air converges towards the centre of the vortex, it concentrates dust into the vortex core and is heated by the sun. This pulls in air from low levels, which also spirals in and speeds up as it tries to conserve its angular momentum, helping to lift more dust into the circulation. At higher levels, the dust-laden air is envisaged to spiral outwards (in the opposite sense to the circulation at lower levels).

While this model was received with interest at the time as an imaginative solution to the problem of storm generation, it has many shortcomings, not least of which is the expectation (at least for the original Gierasch and Goody (1973) model) that dust storms should develop a prominent and coherent spiral organization. Such an organization in either local, regional, or planet-encircling storms simply is not observed, at least on these large scales. Indeed the only large-scale spiral clouds observed on Mars seem more likely to be due to developing baroclinic waves, visualized either by thin ice clouds or dust entrained by the low-level circulation. The above concept is, however, a reasonable description of the much smaller convective vortices, introduced above as dust devils. In this case, the vortex has to shrink to a comparatively small radius (~10–200 m) before the coherent system can stabilize itself. The application of the concept to larger scale, non-spiral structures (in the same vein as models of CISK produced by moist convection in the Earth's tropics, cf Houben (1981)), however, is still quite controversial.

7.5.2.2 Decay processes

It is important to understand not only how major dust storms may be initiated, but also what processes may lead to their eventual decay. For small dust storms, this is almost certainly a result of the removal of heating by sunlight at the end of the day. Without direct solar heating, the wind very near the surface is likely to drop below the threshold for dust lifting and dust suspended within the storm can begin to settle out overnight.

For larger regional storms, however, they can develop a substantial local circulation that can persist overnight. In these cases, the basic decay of the storm cannot begin until enough dust is lifted into the air to substantially reduce the solar heating close to the ground. In this case, surface winds will reduce even during daylight hours, shutting off the source of new dust lifting, except perhaps near the edges of the dust cloud. It may even be that major regional storms have to grow to sufficient size to modify the large-scale atmospheric circulation before decay can begin.

Local effects undoubtedly also play a role. A spreading storm may rely on dust being lifted vigorously in a gust-front at the edge of the storm. But if local topography breaks up the front, or it runs into the edge of an ice field, then dust lifting may be interrupted and the storm can begin to disperse. Another factor may be the availability of surface dust to be lifted. If a storm runs into a location which is covered relatively sparsely by dust, the rate of dust lifting will reduce and the storm will decay.

Such factors may be hard to quantify in detail, however, and are difficult to study from available observations.

7.5.3 Dust storms in GCMs

7.5.3.1 Radiative-dynamical interactions

In recent years it has become possible to represent the transport of dust tracers in GCMs, and even the parameterization of sedimentation and dust lifting, opening up the possibility of conducting fully non-linear simulations of the dust cycle on regional and global scales. Such simulations make it straightforward, for example, to demonstrate the potentially strong effects of radiatively active dust on the atmospheric circulation and transport. Figure 7.26 (see colour section) shows zonally-averaged latitude–height maps of dust concentration from one example of such a demonstration (adapted from the work of Newman (2001)), in which two simulations were run with dust injected into the lowest levels of the atmosphere and then advected by the winds predicted by the model. In the first case (shown in the left-hand sequence of Figure 7.26) the dust injected is not allowed to affect the radiative transfer in the model ('passive dust'), while in the second case (right-hand sequence of Figure 7.26) it is ('active dust').

From Figure 7.26 (see colour section) it is clear that the plume of dust rises higher and much more rapidly in the 'active dust' case than in the 'passive dust' one. The dust optical depths produced in both cases are reasonably realistic, but in the 'passive dust' case the additional solar heating which would result from the extra dust in the air is not allowed to change the circulation, so the dust tracer continues to be advected in the original circulation pattern, computed assuming a constant, climatological, amount of dust. When the extra heating is allowed to interact with the model dynamics, however, we see an immediate response to the additional feedback, showing the potential for dust injection to affect its environment and transport.

7.5.3.2 Parameterization of dust cycles

In the most recent work (Newman *et al.*, 2002a, b) it has been possible to combine the simulation of dust transport with a range of sources and sinks of atmospheric dust, including loss by sedimentation at realistic rates *and* parameterizations of dust lifting due (a) to direct, near-surface wind stress and saltation, and (b) to the effects of dust devil activity. This enables the GCM to couple a complete simulated dust cycle with the winds predicted by the model, offering the possibility of simulating the onset of model dust storms and forming a 'numerical laboratory' for testing ideas concerning their generation and evolution.

In practice, the lifting of dust into the model is parameterized by setting the flux of dust into the bottom level of the atmosphere proportional in (a) to the horizontal sand flux produced by saltation (via Eq. 7.5 above), and in (b) to the dust devil activity parameter (cf Eq. 7.6), both computed locally at each model gridpoint. The efficiency of each process is controlled by an adjustable parameter in each case, α_S or α_D, whose values have to be set semiempirically. Whilst this may involve some arbitrariness in the calibration of the parameterizations, once a choice is made, the character of the simulated dust cycle, and the nature of the active feedbacks in the system, can be studied in detail.

7.5.3.3 Radiative feedbacks on dust lifting

One of the key results from this type of study has been to show how the radiative-dynamical feedbacks, produced whilst applying either direct wind-stress/saltation lifting or dust devil lifting, affect the efficiency of that lifting process. This was determined by running sets of experiments to assess the sensitivity of dust lifting rates to the respective efficiency parameter, α_S or α_D.

The outcome of this was to show quite clearly that the near-surface stress mechanism exhibited a strongly positive non-linear feedback, in that a small increase in α_S would lead to a much larger increase in the amount of dust lifted. This indicates that when winds reach sufficient speeds to begin lifting large amounts of dust, the radiative-dynamical feedback will enhance the lifting of further dust, thereby helping to sustain and spread the storm in just the way suggested from the more idealized models and discussion above.

For dust devil lifting, however, the opposite result was obtained, in that a negative feedback was found to operate. This means that if dust devils encounter conditions which result from greater amounts of dust being lifted, the efficiency of further lifting is reduced. The reason is straightforward to understand, since the activity index for dust devils depends strongly upon the thermal contrast between the ground and the air near the surface. As more dust is lifted into the atmosphere by dust devils, the air is warmed by the increased absorption of sunlight while the ground tends to be cooled (because it receives less sunlight). Eventually, therefore, the thermal contrast between air and the ground collapses, the activity index reduces sharply, and the dust devils can no longer be sustained. Hence, the parameterized dust lifting is shut off and the whole process decays.

7.5.3.4 Simulation of regional dust storms

The ability of modern GCMs to represent this range of processes in the dust cycle has at last enabled such models to begin to simulate the spontaneous initiation, evolution, and decay of dust storms under Martian conditions. When such a simulation is carried out using a full representation of the seasonal and diurnal cycles, then various dust-lifting events are found to occur spontaneously during the model run, which have shown some remarkable promise in emulating features of observed dust storms on Mars.

Figure 7.27 (see colour section) shows a sequence in time of maps of dust column amounts, during the development and eventual dissipation of a regional storm which passed through the Chryse region in the south of Mars. The seed storm evidently begins as a small circumpolar event, one of many linked to the passage of a baroclinic wave system moving eastwards around the North Polar cap edge. By $t = -0.5$ sol, the small dust cloud becomes distorted in the north–south direction and large-scale winds begin to draw it southwards, into the western boundary current system on the flanks of the Tharsis Plateau around longitude 60°W. At this point, the storm appears to decrease in dustiness until it crosses over into the southern hemisphere. By $t = 2$ sol, however, it begins to grow again as its surface winds are reinforced by those of the boundary current itself, reaching their maximum intensity in the Chryse region around $t = 5$ sol. After this, the cloud spreads out southwards and dissipates around the South Polar cap, the whole process taking around 2 weeks from start to finish.

Such a life cycle is remarkably similar to the Chryse dust storm observed by MGS in 1999, during the science phasing operations (Smith *et al.* 2000; Cantor *et al.*, 2001). The MOC images at this time showed that the storm also began as a small local storm close to the North Polar cap edge before being drawn southwards along a similar track to the simulated storm shown above. The evolution of the dust cloud is clearly seen in the infrared maps from the TES instrument (Smith *et al.*, 2000), shown here in Figure 7.28 (see colour section).

The main development shows a remarkable resemblance to the simulated storm, except towards the end of the sequence when the observations show the development of a second storm which was not present in the simulation in this case.

From a careful study of the model fields during the simulated Chryse storm, and other similar events, it is clear that such storms in the GCM arise primarily through the action of wind-stress lifting and its associated positive feedbacks, rather than through dust devil activity. Although the model system makes many approximations and compromises, it seems likely that similar processes are dominant on Mars itself, suggesting that dust devils do not play a major role in the initiation of major storms. This is broadly consistent with the available observations, in that the majority of dust devils found in the VL observations occurred *after* the main global dust storms had passed overhead and were into their decaying phase. None were found during the development of either the 1977a or 1977b storms, which approached the Lander spacecraft as a 'gust-front' – a burst of gusty winds presaging the advance of the main dust cloud.

7.6 DUST TRANSPORT, CLIMATE, AND INTERANNUAL VARIABILITY

7.6.1 Transport, erosion, and deposition

The ability of low-level winds to lift dust into the atmosphere and blow it about in coherent clouds on local and regional scales leads naturally to the notion that wind action can produce systematic regional and global movements of dust and sediments from place to place. Amongst other things, this can lead to changes in the visible appearance of surface features as they become coated with dust or a layer of dust is removed by the wind.

7.6.1.1 Albedo variations and wind streaks

Ground-based observers, for example, have documented over many years the seasonal and interannual variations in the albedo of various classical Martian features (e.g., Slipher, 1962; McKim 1999). More recent observations from spacecraft have shown that many of these changes can be linked to the transport of bright dust particles by the wind into and out of regions by the action of major dust storms. The Syrtis Major feature is one such classical albedo region whose appearance may change significantly during the Martian year (e.g., see Figure 7.29), especially when a major dust storm occurs. It is usually a dark feature with relatively low albedo (around 0.12). Following a major dust storm,

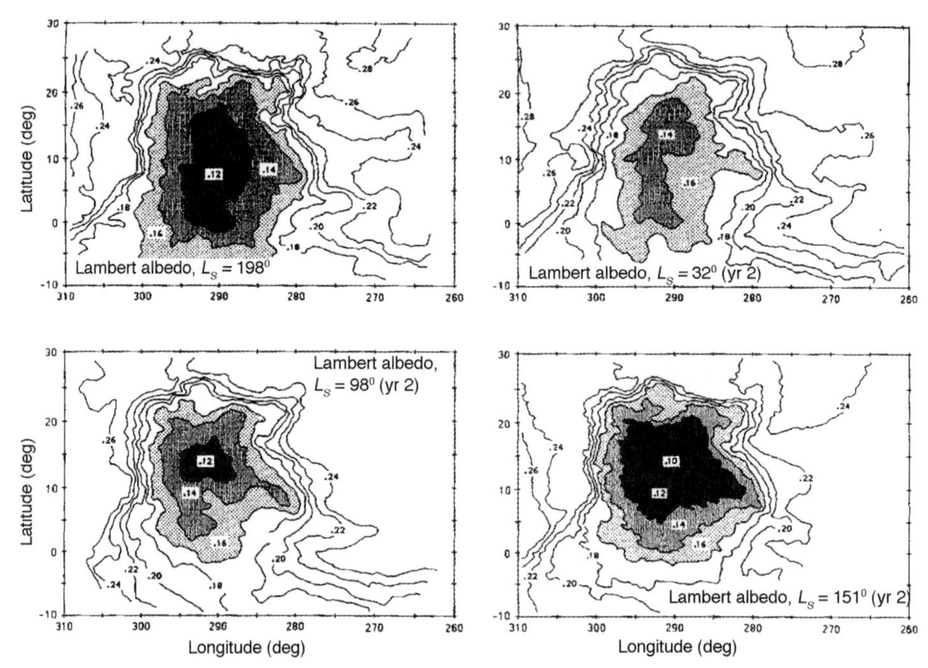

Figure 7.29. Maps of visible albedo in the vicinity of Syrtis Major, showing seasonal variations during the 1977 Viking year.
After Lee 1987.

such as the global storms during the 1977 Viking period, the albedo is found to increase noticeably, as a result of dust deposition from the passage of the storm. During the subsequent months, the albedo then gradually decreases towards its prestorm values, as local winds erode the deposited dust layer.

Another prominent example of seasonally varying albedo features are found in the Solis Planum region, which is most distinct as a low albedo feature during southern spring and summer, and less distinct (with higher albedo and less of a contrast with its surrounding regions) in southern autumn and winter. This again probably reflects the seasonal deposition and erosion of dust layers by the changing pattern of winds.

Some of the most striking examples of variable albedo features affected by wind-blown dust are found in the lee of topographic features such as craters and small, isolated mountain peaks. These can take the form of either bright or dark streaks extending typically a few tens of km from the vicinity of the topographic feature, presumably in the downwind direction (e.g., see Figure 7.30). The dark streaks appear to be regions where winds have stripped away bright dust particles to reveal a darker, coarse-grain underlying surface. In contrast, bright streaks represent regions of deposition of bright dust particles. Such features are typically found to vary in contrast with season, but their orientations often suggest a coherent pattern of erosion by the prevailing wind, and provide strong clues as to the direction

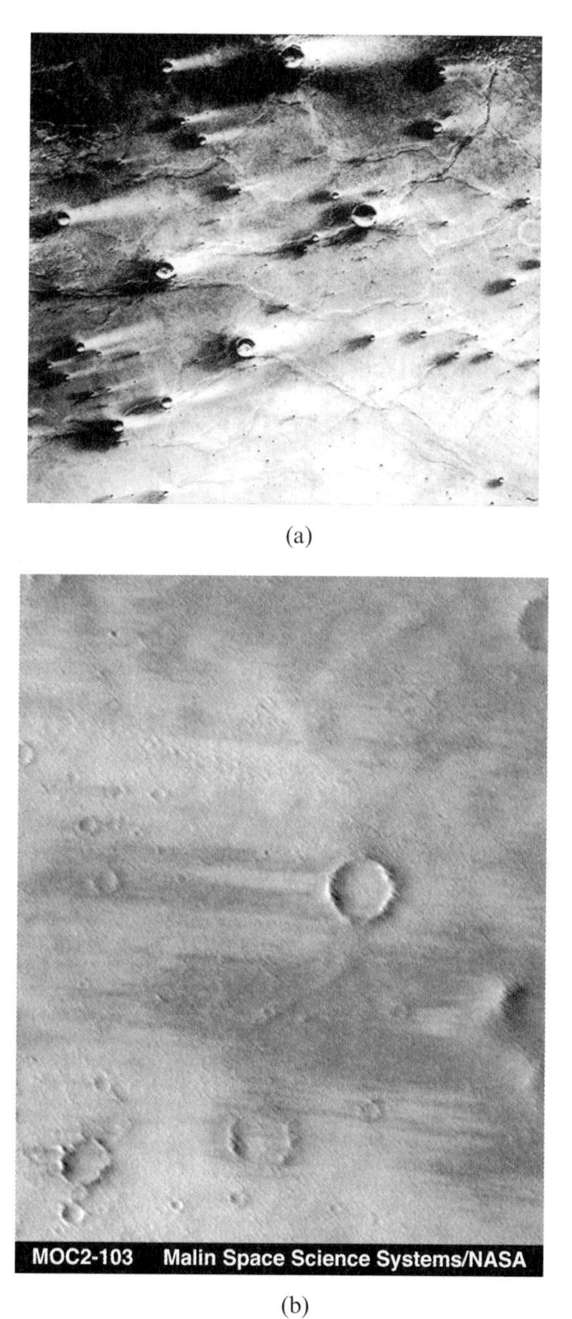

(a)

(b)

Figure 7.30. Two examples of Martian wind streaks, taken (a) from the Viking Orbiter spacecraft and (b) from the Mars Orbiter Camera (MOC) wide-angle camera on Mars Global Surveyor (MGS), showing dark erosional streaks and bright depositional streaks.
From NASA/JPL and Malin Space Science Systems.

and strength of the wind during the season immediately preceding the time of observation.

It is important to appreciate that these changes in surface albedo can be brought about by the movement of very thin layers of dust, perhaps only a few tenths of a μm thick (e.g., see the experimental studies of Wells (1984)). A deposition of only around $10^{-3}\,\mathrm{kg\,m^{-2}}$ of bright dust can change the albedo of a dark underlying area by several tens of per cent. Thus, the typical dust deposition rates prevalent even under fairly clear conditions on Mars could lead to variations in albedo comparable to those observed, even outside the much more intense activity of major dust storms, though dust storms can clearly lead to even stronger albedo variations by depositing layers of dust up to several micrometres thick in some places. In practice, however, typical variations of the classical albedo regions are only of the order of a few tens of per cent, so the amounts of dust must be typically quite small.

7.6.1.2 Sources and sinks of dust

Even though the action of wind can evidently redistribute dust globally, especially during large-scale dust storms, it is also evident that the amount of dust, sand, and other sediments varies considerably from place to place on Mars. The classical low-albedo regions, and in locations where dark dust streaks are found, must presumably have typically quite small amounts of dust covering the surface. In other regions, however, the surface is evidently covered by relatively thick deposits of sediment. This can be seen, for example, from differences in the type of landscape. Large fields of dunes and yardangs, for example, are found in the subpolar plains surrounding both polar caps, and in other areas closer to the equator (e.g., see Greeley *et al.*, 1992). Such regions show clear evidence of the long-term action of wind on the shape of the landscape, much as may be seen in regions of many terrestrial deserts.

The thermal inertia of the surface may be measured and mapped from orbiting spacecraft, from infrared measurements of the thermal response of the surface to the passage of the diurnal wave of solar heating (e.g., Paige *et al.*, 1994; Paige and Keegan, 1994; Mellon *et al.*, 2000). This measurement from MGS TES observations was shown earlier in Figure 4.4 (see colour section), from which it was clear that the thermal inertia varies substantially from place to place on Mars on scales comparable to the major topographic units. As well as being of importance in defining the boundary conditions acting on the atmosphere, such a measurement is also useful as a diagnostic of the underlying surface soil, since thermal inertia is strongly anticorrelated with the typical grain size in the uppermost few cm of the surface. Thus, regions of relatively low thermal inertia are comparatively rich in small dust particles, whereas regions of high thermal inertia are dominated by relatively coarse, sandy, or rocky soil.

7.6.1.3 Longer term transports?

The effectiveness of major dust storms in shifting large quantities of airborne dust long distances across the planet opens up the question of whether such events may be bringing about a systematic transport of material from place to place in Mars' atmosphere. Under present circumstances, where the strongest seasonal heating occurs during southern summer, there is a strong preference for major dust storms to begin in the southern hemisphere. The dust lifted by such initial storms comes from the main source regions in the south (notably near the Hellas Basin near Noachis and Hellespontus, and near the Argyre Basin) and, in planet-encirling storms, is lifted into the solsticial Hadley circulation. At this time of year, the Hadley circulation is predominantly northward at upper levels, so the dust collected in the south is then likely to be carried into the northern hemisphere, where it will be deposited in the vicinity of the descending branch in the northern subtropics.

Thus, if major planet-encircling storms are the main events leading to large transports of dust, the present epoch would seem to favour a systematic transport of dust, generated close to the retreating South Polar cap, from south to north during southern summer. It remains an open question, however, whether there is a corresponding (but less obvious) systematic transport in the opposite direction during the rest of the year? This might entail a steady trickle of dust lifted by dust devils during northern summer (say), which could be carried south by the oppositely flowing Hadley circulation at this time. The problem is that this transport is very difficult to measure directly from existing instruments.

GCM simulations can provide some clues, though this kind of study is very much in its infancy, and results depend greatly on how the various lifting and deposition processes are parameterized. At the present stage of development, it would appear (e.g., Newman, 2001) that a systematic northward transport is probably happening at present, though much more work remains to be done on refining the models before this conclusion can be substantiated.

If such a systematic transport is sustained over interannual timescales, then long-term changes in the surface landscape might be expected to occur, on timescales comparable with the cyclic variations in Mars' orbit and planetary rotation. A possible signature of such long-term variability might be found in the enigmatic 'polar layered terrains', dominating the landscape in the plains immediately surrounding both polar caps on Mars. Ideas concerning the origin of such terrains, and their link with long-term climate cycles on Mars, is taken up in Chapter 9.

7.6.2 Interannual variability of dust storms and climate

The apparent observation that major dust storms occur in some Mars years but not in others, and at certain preferred locations and seasonal dates, poses a substantial challenge to theoreticians and modellers. As the sophistication of atmospheric GCMs for Mars has increased in recent years, it has become possible also to carry out long-term simulations over many Mars years. From the experience of carrying out such simulations under conditions where the seasonal variation of dust opacity in the atmosphere is specified and repeated from year to year (e.g., see Lewis *et al.*, 1999), it is striking how little spontaneous interannual variability is found. Whilst there are always small differences from year to year (e.g., in the wavenumbers and behaviours of baroclinic storms at mid- to high-latitudes during the extended winter seasons in each hemisphere), the same overall weather patterns seem to recur each year with considerable regularity. It is only when the dust loading from year to year is changed that the simulated circulation shows strong changes. This serves to confirm the view, expressed more than 20 years ago (e.g., Leovy, 1985), that large-scale dust storm activity represents the main source of irregular interannual variability in Mars' present climate.

Early attempts to capture this behaviour in models invoked simple non-linear feedback mechanisms forced by an analogue of the annual cycle in highly idealized, *ad hoc* low-order models, much as used elsewhere in meteorology and climatology (e.g., Lorenz, 1963; Vallis, 1986) to demonstrate chaotic behaviour. Ingersoll and Lyons (1993) investigated a couple of such models, and managed to obtain both periodic and chaotic solutions, depending upon the model parameters. This at least lent some credibility to the notion that chaotic intrinsic variability might be a characteristic of the climate system on Mars. However, it left open a number of more detailed questions as to the validity of their particular models.

More recently, Pankine and Ingersoll (2001) have attempted to develop a more 'rational' low-order model of interannual variability due to dust storms, using a highly simplified representation of the atmospheric circulation in the NASA Ames GCM. The model was constructed by deriving a projection of the 2-D meridional transport circulation in the Ames model onto a truncated set of basis functions, and obtaining evolution equations for the amplitude of each basis function. This led to a somewhat more realistic representation of the seasonally-varying circulation, both with and without dust storms. The result, however, was that unless some form of external stochastic forcing was included, the simulated behaviour was always periodic, with every year in the simulation showing the same behaviour. Of course, this may have been partly due to an excessive simplification of the atmospheric circulation as axisymmetric, suggesting perhaps that inclusion of non-axisymmetric processes such as baroclinic waves and regional variations of dust sources might play the role of stochastic forcing in a more realistic model.

With the advent of more complete, fully 3-D GCMs capable of simulating the entire dust cycle (Newman *et al.*, 2002a, b), it is at last becoming possible to study the origins of interannual variability due to the varying occurrence of dust storms. Even so, at the time of writing, a fully realistic multiannual simulation of successive years showing the irregular occurrence of major, planet-encircling storms remains tantalizingly elusive. Under some conditions, the GCM is capable of generating nearly periodic behaviour, with very little interannual variability (e.g., see Figure 7.31(a), colour section), much like the low-order solutions of Pankine and Ingersoll (2001). In other cases, however, there are some intriguing results which suggest the possibility of spontaneous interannual variability (e.g., see Figure 7.31(b)), although in these cases the behaviour of the dust cycle at times outside the main southern hemisphere dust storm season is not particularly realistic, and generally too dusty. Moreover, the results are often highly sensitive to model parameters and resolution.

For the future, it may be that further improvements to the GCM may enable a more realistic representation of dust storm-driven interannual variability will emerge. The effects of 'scavenging' of dust particles by water ice condensation onto dust particles might turn out to be important, especially outside the main dust storm seasons, and lead to more realistic dust loadings during those seasons in the chaotic simulations shown above. It is clear, however, that this aspect of the Martian climate is proving the most challenging to capture in modern climate models.

8

Water, climate, and the Martian environment

8.1 WATER AND MARS

One of the most distinctive features of the Earth as a planet is its abundance of water in liquid form, both covering the surface of the planet (in oceans, rivers, and lakes) and in various guises in the atmosphere (vapour, clouds, and precipitation). This is widely understood to be a critical element in making the Earth, or any other planet, habitable by complex living organisms – at least those relying on the carbon/hydrogen/oxygen/nitrogen-based biochemistry which underlies all life on Earth.

Water not only enables living organisms to exist and operate on the surface of the Earth, but also plays a major role in shaping the landscape. The continuous recycling of water from the oceans, evaporating into the atmosphere, then falling to the surface as precipitation to feed rivers and subsurface aquifers, which in turn return the water to the oceans, leads to the slow but inexorable process of erosion and deposition of rocks and sediments. Over millions of years, this leads to many features of the landscape which make the Earth such a familiar and benign environment, such as rivers, smooth valleys and canyons, lakes, and shorelands. One has only to look at images of alien environments from planetary bodies, such as the Moon, which have always been dry and virtually waterless (except perhaps near its poles), to see what a difference water will make to the surface landforms. The Moon's landscape is evidently dominated by ancient impact features (craters and ejecta blankets) and the results of dry volcanism (smooth mounds, lava tubes, etc.), but with no sign of fluvial erosion (e.g., in the form of small-scale sinuous channels) which may be attributable to water.

The association of water with Mars in the human consciousness goes back far into the history of astronomy. Early observers, such as Huygens, Cassini, and Herschel, had noted and measured the advance and retreat of Mars' polar ice caps in the 17th and 18th centuries, and had even reported transient brightenings on the Martian disk as evidence of dust and water clouds. But in some ways early

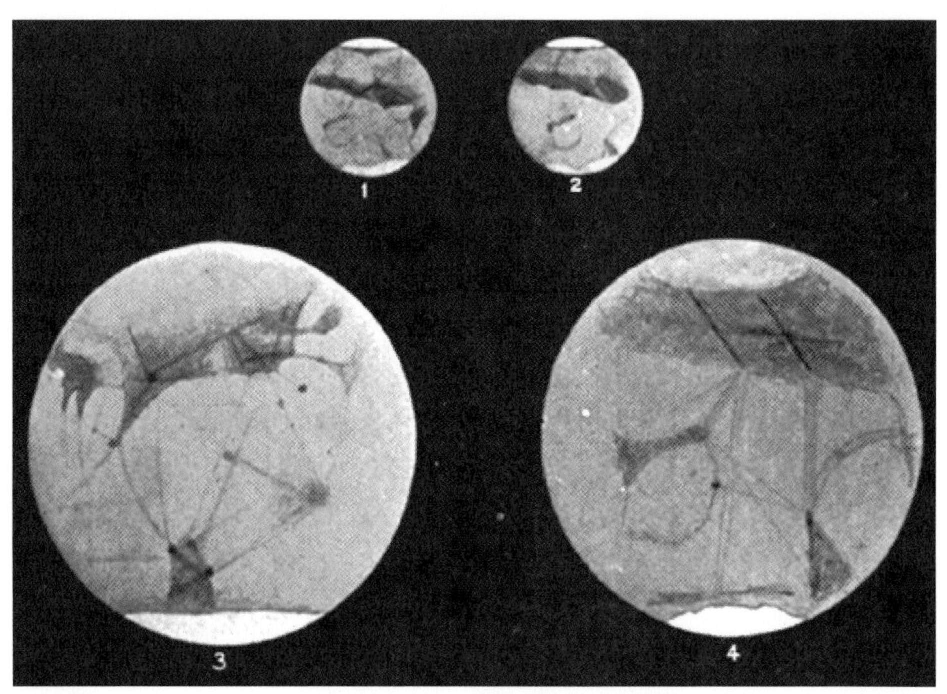

Figure 8.1. Drawings of Mars by the astronomer Percival Lowell, showing examples of the 'canals' which he identified as indicating the presence of a Martian civilization.
Courtesy of Lowell Observatory (Flagstaff, AZ).

observers were often too enthusiastic in drawing analogies between features on Mars and watery counterparts here on Earth. Prominent dark markings were observed to change with the Martian seasonal cycle, and were consequently believed to represent changes in Martian vegetation associated with the movement of masses of water. This culminated in the most famous example of 'mass hallucination' associated with the apparent observations of 'canali' in maps made by Giovanni Schiaparelli (notably during the 1877 opposition of Mars), which led to the widespread belief in intelligent beings on Mars who were somehow managing the sparse water resources of their planet and building irrigation channels (or 'canals') to move water from the polar regions to lower latitudes. The most famous proponent of these ideas was Percival Lowell, the Boston businessman turned astronomer, who founded the Lowell Observatory in Arizona and published widely on the subject around the turn of the 20th century (Lowell, 1895, 1906, 1908; see Figure 8.1). It was not until the efforts of Earl Slipher (Slipher, 1964), working at the Lowell Observatory to obtain detailed photographs of Mars in the 1960s, and finally the advent of the Mariner series of spacecraft missions, that the notion of Mars' canals was eventually put to rest.

But whilst the early Mariner series images clearly disproved the existence of canals, and appeared to show a dry, desolate, lifeless Moon-like landscape, the subsequent missions (most notably Mariner 9 and the later Viking orbiters) showed other features, such as sinuous valley networks, which raised again the possibility of water-related influences on the Martian landscape. Subsequent investigation has led to the prevailing view that something *in liquid form* (which most scientists believe must have been water), or perhaps in the form of mobile glaciers, seems to have played a very substantial role in affecting the landscape of Mars, at least on geological timescales, even though water is very scarce (and mostly in icy form) at present. Indeed this apparent and radical change of climatic conditions, from the time of the formation of Mars' water erosion features to the present time, remains one of the most intriguing mysteries of Martian science. Indeed so remarkable is this change that some scientists have felt moved to question whether water was indeed responsible for the observed features, suggesting that alternatives, such as extensive wind erosion (Leovy, 1999) or liquid CO_2 (Hoffman, 2000) might have been responsible instead.

In this chapter, we will be concerned initially with the role water may have played in shaping Mars' landscape in the past, and will then look at the way water manifests itself in the present Mars environment. We will conclude the chapter by reviewing some recent efforts to model Mars' hydrological cycle and giving an outlook on some of the most important outstanding questions related to water on Mars.

8.2 WATER AND THE FORMATION OF THE ANCIENT MARTIAN LANDSCAPE

Since the first clear images of Mars from the Mariner series spacecraft, a number of surface features have been observed and interpreted as being due to the action of water, either flowing freely in liquid form (as in a river or channel) or in glacial form, or affecting the subsurface properties and resistance to erosion (e.g., leading to mass-wasting). In the following section we will look briefly at some of these types of feature, and the evidence they provide for the role of water in Mars' past climate and environment. It is important to emphasise, however, that we will be viewing this aspect of Mars as atmospheric physicists, rather than as geologists with hard expertise in interpreting geomorphology. As such, the discussion will be fairly qualitative and selective. Those readers wishing for a more in-depth discussion of this aspect of Mars would be well advised to consult a more specialist text, such as those by Baker *et al.* (1992), Squyres *et al.* (1992), Jakosky and Haberle (1992), Carr (1996) and Baker (2001).

One of the major objections to water flowing on the surface of Mars under present conditions is that the ambient surface pressure across most of Mars is below 610 Pa. This value of pressure is significant, since it corresponds approximately to the triple-point pressure of water vapour. Figure 8.2 shows the phase diagrams corresponding to both water and carbon dioxide, covering the conditions appropriate to both Earth and Mars. This shows that if water vapour at a pressure less than 610 Pa or so is cooled below its condensation temperature, it will not condense as liquid but as solid ice. Similarly, if ice on the surface of Mars is

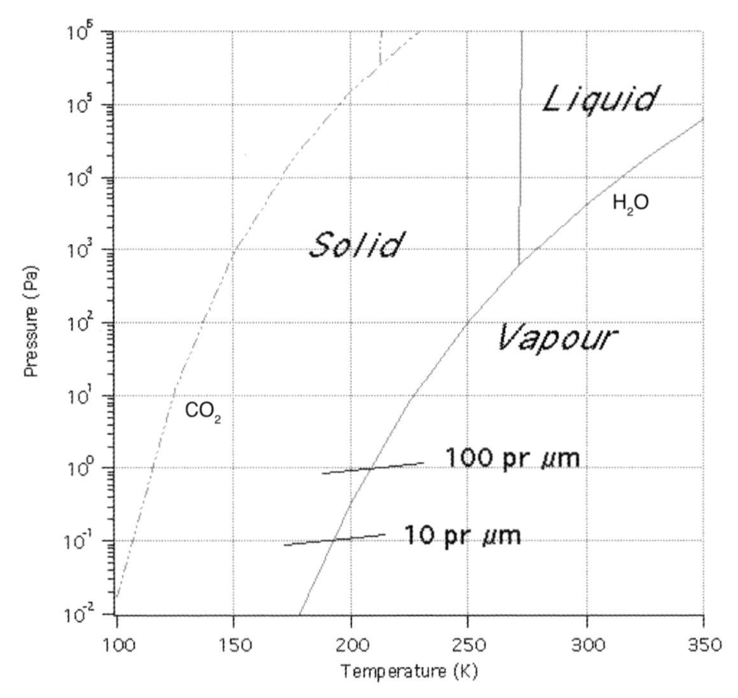

Figure 8.2. Thermodynamic phase diagram for water (solid lines) and CO_2 (dash-dotted lines). The regions indicated 'Vapour', 'Liquid' and 'Solid' refer to the equilibrium states of water, with a phase transition occurring at the solid line when either pressure or temperature is changed. The two lines marked 10 and 100 pr μm correspond to the water vapour partial pressure temperature for a uniformly mixed atmosphere containing the amount of water indicated.

After Carr (1996).

heated, it would not melt into liquid water if the pressure were less than 610 Pa, but would directly sublime as vapour. The conditions suitable for the stable existence of liquid water on present day Mars are therefore quite rare, and are satisfied even just for short periods over less than one third of the planet's surface (Haberle *et al.*, 2001). These conditions may be relaxed somewhat if the water is strongly salty, though the warmer temperatures needed to melt the ice are fairly extreme by present standards. Thus, if water were to have existed in large quantities in the past in liquid form on the surface of Mars, then the ambient conditions must have been both significantly warmer and with a higher ambient atmospheric pressure (>610 Pa). From these kinds of consideration arises the notion that the climate of Mars may have been significantly different in the past from what it is now, though, as we shall see below, the situation is far from being straightforward.

The alternative view, put forward, for example, by Hoffman (2000) as the so-called 'White Mars' hypothesis, would require much higher pressures of several bars for liquid CO_2 to be stable. Such pressures are not inconceivable, especially during early phases in Mars' evolution, or in underground layers which might trap CO_2 ice, though are remote from present conditions at the surface of Mars. Since both the water and liquid CO_2 hypotheses would seem to stretch the imagination somewhat, it is understandable that Leovy (1999) should suggest a radical alternative, invoking long-term wind erosion as the potential cause of flow-like features on Mars. Nonetheless, as we shall see below, many of the features do seem to 'look' very much like features on the Earth which we know to have been shaped by running water. So for the present, we continue our discussion here with the presumption that water was the main source of erosion, but accepting that this interpretation may be subject to change as new observational constraints become known.

8.2.1 Valley networks

Some of the clearest evidence for erosional features produced by flowing water come from the sinuous valley networks on Mars. These occur mainly in the southern highlands, and often take the form of dendritic networks of channels which feed into a single, wide channel. A clear example is shown in Figure 8.3(a), from Viking Orbiter images, showing a region roughly 200 km across. Channels fan out from the wide channel in the lower right of the picture into a broad plateau, giving the impression of a network of valleys draining water towards the southeast, much as one might find on Earth in a mountainous region. On closer inspection, however, it is clear that, although there are many medium-sized channels, there is a shortage of even smaller channels. This would appear to suggest that the channels seen here on Mars may not have been produced in the same way as most river valleys on Earth, by run-off of rainwater. Alternative ways in which liquid water might be stabilized sufficiently to flow along the surface include possibilities either that large amounts of salts were dissolved in the water (changing the physical properties of the water, freezing points, etc.) or that the water layer was covered by a layer of ice. This is the situation prevailing sometimes underneath glaciers on the Earth, for example. As the ice warms close to the surface (perhaps as the result of leakage of geothermal heat from the ground), it can melt into liquid water if the pressure of the overlying ice sheet is sufficiently high. The water could then flow in a 'tube' enclosed by the ground and the ice layer. This is believed to occur on Earth to form the dry valleys in Antarctica, for example, and similar landscapes in the Arctic. The resulting channels then form only on a fairly large scale, but can link up to form networks, much as apparently observed on Mars. An interesting feature of the subglacial channels on Earth, however, is that they do not always flow downhill (unlike any river open to the air), but can have stretches where the water is pushed uphill under pressure for short distances.

With the recent availability of very high-resolution images from the Mars Orbiter Camera (MOC) on Mars Global Surveyor (MGS), however, this picture may not be the whole story. Figure 8.3(b) shows a MOC image of Nanedi Vallis, which illustrates (near the lower right side of the image) a small subsidiary channel in the bottom of the much wider channel, seen in the early low-resolution Viking images. This kind of structure is more typical of river canyons on Earth, such as

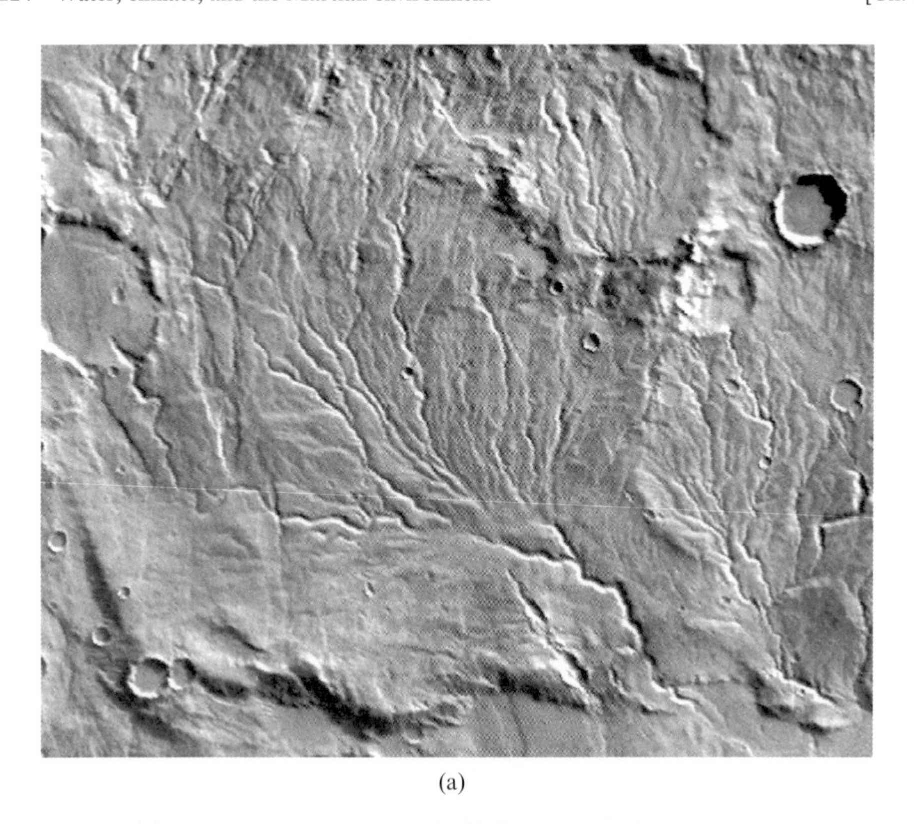

(a)

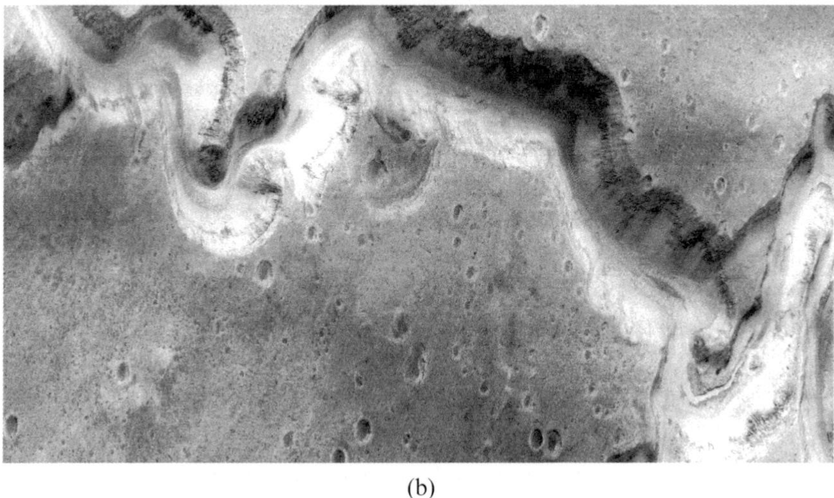

(b)

Figure 8.3. Two examples of sinuous valleys in the southern highlands of Mars: (a) large valley network around 42°S, 92°W from the Viking Mars Digital Map; and (b) a close-up of Nanedi Vallis from the Mars Orbiter Camera (MOC) on Mars Global Surveyor (MGS). (a) Courtesy of Brian Fessler, Lunar and Planetary Institute (LPI), Tucson, AZ. (b) From NASA/JPL and Malin Space Science Systems.

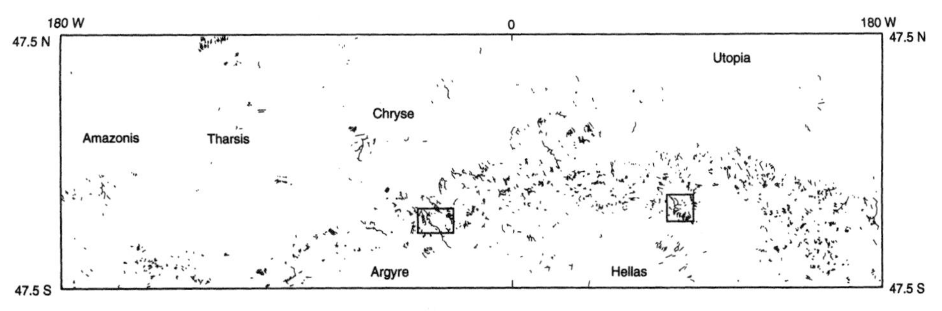

Figure 8.4. Location of valley networks on Mars within the latitude range ±47.5°.
After Carr (1996).

the Grand Canyon itself, in which the Colorado River has cut a narrow channel in the middle of the main Canyon on a much smaller scale, by slow fluvial erosion over long periods of time. The Nanedi Vallis structure is perhaps one of the few pieces of evidence on Mars which might suggest that, at least for a while in Mars' history, open water might have flowed for relatively long periods.

How old are the valley networks? The presence of small craters invading some of the network features, and the fact that they almost all occur in the ancient, densely cratered southern highlands (see Figure 8.4), leads to the conclusion that most are relatively old in geological terms. Estimates for the geological age of the southern highlands range from around 3.5–3.8 Gyr (Tanaka *et al.*, 1992), corresponding to the Noachian era of Martian geological evolution and suggesting that the valley networks primarily formed during the early phases of Mars as a planet. However, it is thought (e.g., Baker, 2001) that up to 25–35% of them could be more recent than this.

8.2.2 Outflow channels and flood plains

Whilst the valley networks give the impression of fairly substantial amounts of water having flowed along the surface of Mars, either as open or closed channels, certain regions of Mars show evidence of at least the transient appearance of prodigious amounts of water flowing from the southern uplands into the northern plains. Such catastrophic flooding has carved wide channels in the ground, inundating the northern plains and sweeping huge amounts of debris some considerable distances. Figure 8.5 shows a close-up of the topography in the vicinity of the Chryse Planitia region, just to the east of the Tharsis Plateau in Mars' northern plains, obtained from the Mars Orbiter Laser Altimeter (MOLA) topographic map. This clearly shows several wide channels along the western and southern margins of Chryse which must have carried huge amounts of water into the low-lying plains. On smaller scales, examples can be seen (Figure 8.6) confirming the impression of channels flooding Chryse, in some cases arising in collapsed aquifers or so-called 'chaotic terrains' where melting water seems to have caused the ground to collapse

Figure 8.5. Topography in the region of the flood plains of Chryse Planitia, showing immense channels caused by catastrophic flooding at some point in Mars' past.

Image adapted from the high resolution Mars Orbiter Laser Altimeter (MOLA) topographic map.

and flow away downstream, perhaps as a result of heat released during volcanic activity on Tharsis, or massive impact events (Segura *et al.*, 2002).

Once the escaping water had emerged onto the northern plains, it evidently continued to flow for some time, causing rapid erosion of the soil around some craters and leading to features such as streamlined 'teardrop-shaped' islands. Figure 8.7, for example, shows some examples of two streamlined islands around two craters, approximately 8–10 km in diameter, near the mouth of the Ares Vallis in Chryse Planitia. The large crater to the right of the rightmost island seems relatively uneroded, suggesting that it may have formed *after* the flooding event itself.

It is important to appreciate the scale of the flooding events which must have led to these topographic features. Some of the outflow channels are immense in size, up

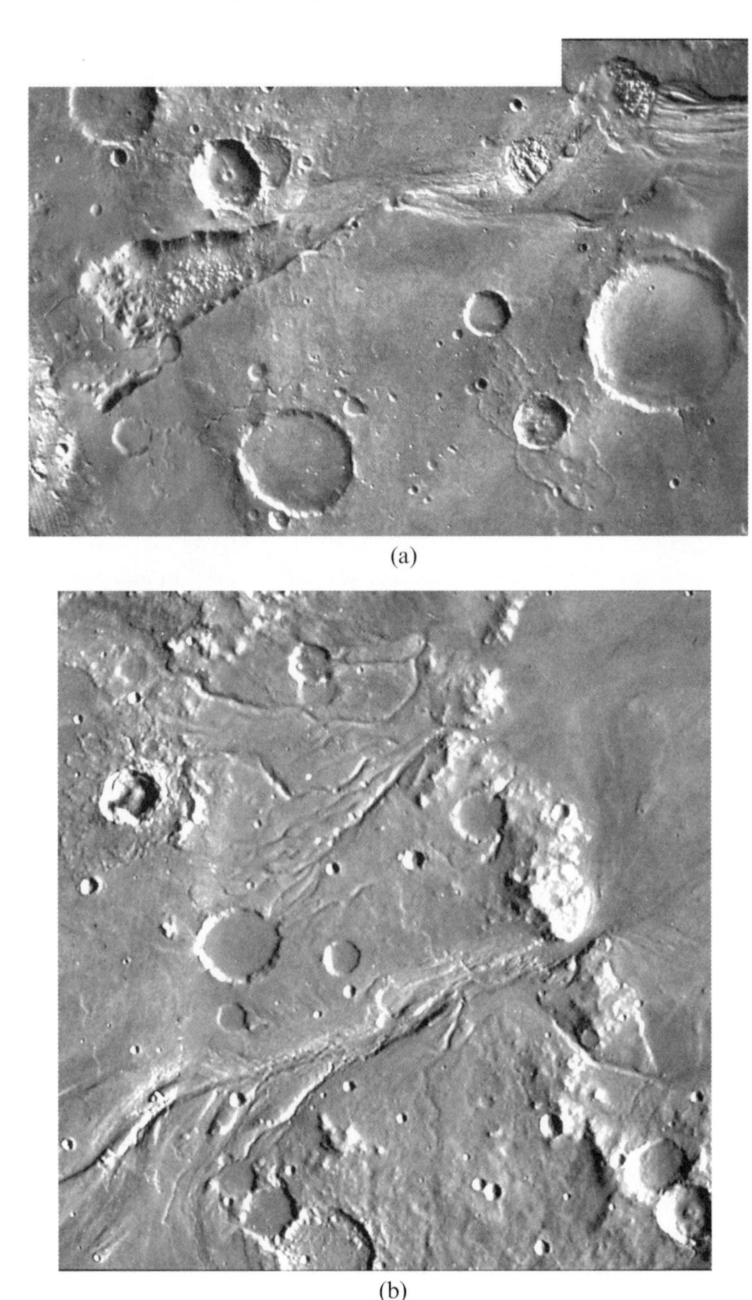

(a)

(b)

Figure 8.6. Two examples of outflow channels which must have formed during major flooding events into the northern plains: (a) a collapsing source region or 'chaotic terrain' associated with the Ravi Vallis outflow channel around 1°N, 42°W, and (b) close-up of some outflow channels into Chryse Planitia, both obtained from the Viking Mars Digital Map.

Courtesy of Brian Fessler and Walter Kieffer, LPI, Tucson, AZ.

Figure 8.7. An example of the formation of streamlined islands during catastrophic floods which invaded Chryse Planitia, obtained from the Viking Mars Digital Map.
Courtesy of Brian Fessler, LPI, Tucson, AZ.

to 25 km across and several hundred metres deep. The flow rate that must have been accommodated by these channels far exceeds any flooding event within human experience on Earth, with flow rates comparable to those in some of the major ocean currents on Earth such as the Gulf Stream in the Atlantic Ocean or the Kuroshio Current in the Pacific Ocean.

The ages of these flooding events is estimated to vary from the early Noachian period through to the late Hesperian and early Amazonian period (perhaps 1.5 Gyr ago). There is even some evidence that flooding events (albeit on a much smaller scale) may have occurred much more recently, perhaps within the past 10 Myr (Mouginis-Mark, 1990).

8.2.3 Oceanus Borealis?

Given the prodigious output of water from these early catastrophic flooding events, it is not surprising that these observations led sometime ago to the suggestion that the northern hemisphere might once have been covered by a shallow, semipermanent ocean, perhaps a few hundred metres in depth (Baker *et al.*, 1991; Parker *et al.*, 1993; Head *et al.*, 1999). The volume of water contained in such an ocean (somewhat fancifully named 'Oceanus Borealis' or the northern ocean by Baker *et al.* (1991), see Figure 8.8, colour section) is by no means out of proportion to the volatile inventory over Mars' history. But its persistence would clearly imply climatic conditions which must have been considerably different from those prevailing today, with much higher prevailing atmosphere pressures and temperatures.

Not surprisingly, this concept of a persistent, semipermanent ocean is still regarded as controversial amongst Mars scientists, though many accept that smaller bodies of open water may well have persisted for some time at various stages in Mars' history, perhaps as large lakes. Questionable evidence for Oceanus Borealis includes some so-called fossilized coastlines, in the form of shallow, scarp-like linear features identified in the northern plains, some of which appear to follow areopotential surfaces (Head *et al.*, 1999) as might be expected for a 'sea level'. This interpretation has been questioned, however, on closer examination of the high-resolution measurements from the MOLA instrument (Malin and Edgett, 2000; Withers and Neumann, 2001), and it may be that these scarp-like structures have a more straightforward (dry) tectonic explanation, rather than as coastline features. Even so, the relative absence of ancient craters in the northern plains, and other evidence of erosion, continues to excite speculation concerning ancient oceans on Mars. Evidence for smaller lakes seems on rather firmer ground, however, with a number of documented cases, for example, of craters which may have been flooded for some considerable time, such as Gusev Crater, see Figure 8.9, which is

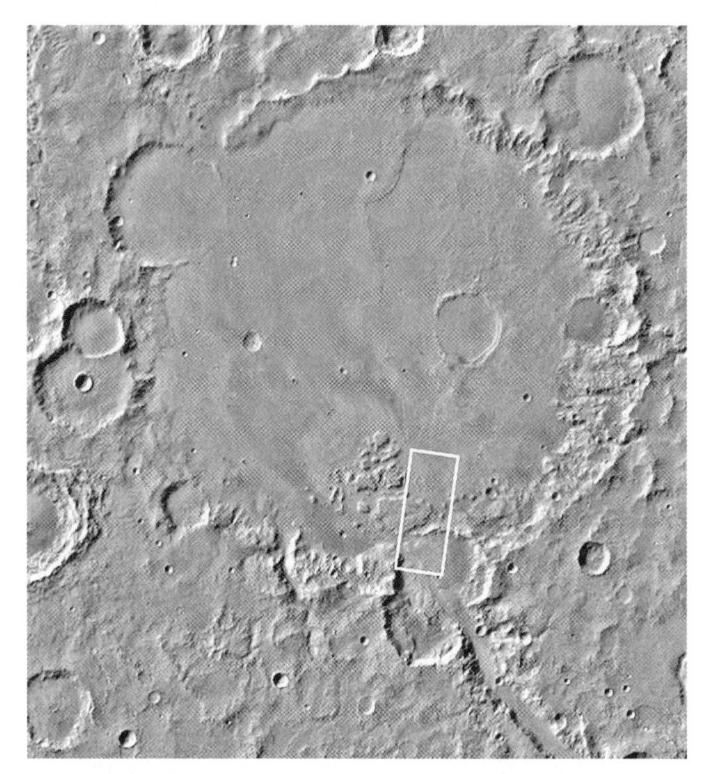

Figure 8.9. Wide-angle Mars Orbiter Laser Altimeter (MOLA) image of Gusev Crater on Mars, which may have been the location of a long-lived ancient lake. Note the inundated and eroded smaller craters within the large structure (which is approximately 60 km across), and the outflow channel toward the southeast.

From NASA/JPL and Malin Space Science Systems.

the planned destination for one of the NASA Mars Exploration Rovers launched in 2003. In these cases, the close-up images and other information seem consistent with the presence of sediments and other features, suggesting persistence of water in such craters for periods up to 10^3–10^4 yr (Baker, 2001). In such circumstances, however, the body of water may have been overlain by a sheet of ice for much of its existence, much (again) as seen in some relatively dry regions in the polar environments on Earth, but kept liquid near the bottom by a combination of areothermal heat and hydrostatic pressure from the overlying ice.

8.2.4 Glacial activity and permafrost

Given the past presence of large volumes of water on the surface of Mars and the relatively cold prevailing temperatures at present, it is reasonable to consider the possible role of ice itself in shaping the Martian landscape, at least in the times following the major floods mentioned above. Direct evidence of glacial activity seems fairly sparse on Mars and remains controversial. There are some features, however, particularly in the southern uplands close to the Hellas and Argyre impact basins where the landscape does show some indications of glacial ice movement in the past which have shaped features such as V-shaped valley profiles, cirques, and troughs (Baker, 2001). In parts of Argyre itself, there are also features which might represent glacial moraines and other structures similar to those found on Earth near the ends of active glaciers. The kinds of glaciers represented may, however, be more representative of debris-covered glaciers or rock glaciers, in which ice plays more of a lubricating and cementing role in amongst rocky fragments, rather than as a river of more or less pure ice.

In the cold conditions typical of Mars at present, the sublimation rate of ice deposits is very small – perhaps as low as 10^{-5} cm yr^{-1} (e.g., Baker, 2001). It would not take a great deal of accumulation, therefore, to maintain a very slowly flowing glacier over very long periods of time, such as observed in rock glaciers on Earth. The annual mean temperature at Vostock on the East Antarctic ice sheet, for example, is around −57°C, and snow accumulates at only around 2 cm yr^{-1}. Such an accumulation rate might have been even smaller during the Last Glacial Maximum, when mean temperatures at these latitudes were some 20° colder. In the dry valleys in Antarctica, the landscape is largely ice-free now, but there is old ice left over from a past glacial era buried some 50 cm below the surface which is subliming into the atmosphere at a rate of around 10^{-5} cm yr^{-1}, much as on Mars. These extreme polar environments on Earth may thus be seen as quite close analogues to present conditions on Mars, apart from the much higher atmospheric pressure at the surface.

So if the water from the catastrophic floods did not form large bodies of frozen water at the surface, what can have happened to it? The most widely accepted explanation is that it most probably would have been absorbed into the surface and then frozen inside the soil to form widespread deposits of permafrost. Evidence for such a permafrost across the surface of Mars can be found from a variety of surface formations which parallel similar features found in permafrost regions on Earth. One of the characteristic formations found in permafrost regions are networks of small-scale polygonal cracks, where the ice in the soil has frozen and produced stresses in the subsoil. The most convincing examples of these polygonal structures are found in high resolution MOC images of the northern plains (see Figure 8.10(a)), where polygons on a scale of around 100 m are visible which appear to be representative of permafrost deposits. A number of these regions are also almost devoid of craters, suggesting a relatively recent formation in geological terms.

Some of the most striking evidence for large deposits of subsurface ice comes from the morphology of the ejecta blankets surrounding certain classes of impact craters. These can be seen (e.g., see Figure 8.10(b)) to take a form quite unlike those on the Moon, for example, but appear to have formed lobate structures resembling a solidified mudslide. This would seem to have been produced by the effect of liquid water, released at the time of the impact itself when the heat of impact would have produced local melting of subsurface ice. Such crater morphology is reasonably common across Mars at low and mid-latitudes.

Most recently, evidence for subsurface ice deposits has begun to emerge from the measurements of the flux of epithermal neutrons from Mars using the Gamma Ray Spectrometer on board the Mars Odyssey spacecraft. This instrument is capable of detecting large concentrations of hydrogen in the near-surface layers of the Martian soil, which is most likely to be due to the presence of water ice in the topmost 1–2 m of soil. The maps obtained so far (the mission is ongoing at the time of writing) do indeed indicate large amounts of subsurface water occurring over large areas of the planet, especially at high latitudes (see Figure 8.11, colour section; Boynton *et al.*, 2002). The maps clearly show substantial concentrations of subsurface water close to the North Pole in northern summer (Figure 8.11(a)), and near the South Pole in southern summer (Figure 8.11(b)) as the CO_2 ice cover retreats, with some weaker concentrations in the more low-lying areas of the equatorial region. The apparent absence of water ice under the surface of the winter pole in each case is almost certainly because its signature is masked when the surface is covered with the seasonal deposits of CO_2 ice; the water ice is almost certainly a permanent feature under both poles. In the Tharsis Plateau and other upland areas, however, concentrations are relatively low. This would appear to confirm the view that water accumulated in the subsoil of Mars in the low-lying and polar regions of the planet before becoming frozen as semi-permanent ice reservoirs.

8.2.5 Warm and wet or cold and dry?

Following the suggestions of Baker (2001) and work cited therein, the evidence for water playing an active role in shaping Mars' landscape in the past would seem to indicate that there was indeed a period when open areas of liquid water may have existed on Mars. But this most likely occurred for any length of time only during Mars' early history, perhaps during the late Noachian era or shortly thereafter. Since then, there may have been brief interludes (perhaps lasting between 10 and 1,000 yr, and occurring at intervals of several hundred Myr or even longer) when violent

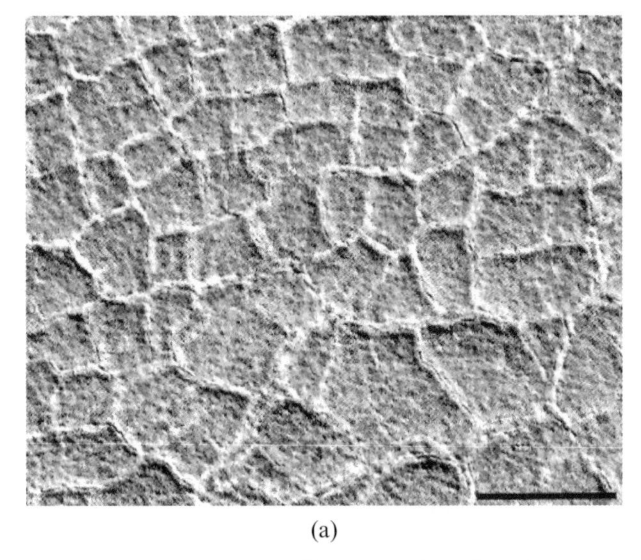

(a)

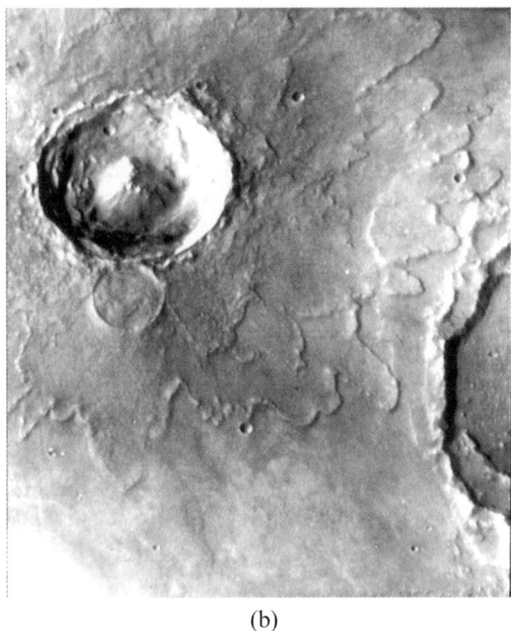

(b)

Figure 8.10. Images indicating the presence of permafrost in the Martian subsoil: (a) small-scale polygonal structures in the floor of a crater in the northern plains of Mars around 67.5°N, 312.5°W (Mars Orbiter Camera (MOC) image M01-00294 – the scale bar indicates 200 m), and (b) a Viking Orbiter image (3A07) of Yuty Crater, an 18-km-diameter representative of the so-called rampart craters on Mars. The lobate ejecta seem to indicate fluidized movement of material, somewhat like a mudslide, which may have been lubricated by the release of water from frozen ice in the soil at the time of the impact.

From NASA/JPL and Malin Space Science Systems.

volcanic outbreaks may have occured, leading to major, but shortlived, flooding events taking place (each perhaps lasting just a few days). But for much of the time during the past 3 Gyr or so the climate may have been much as we observe now, apart from some shorter term, and more modest, climate variability due to cyclic variations in Mars' orbit and obliquity (which will be discussed in more detail in the next chapter). During this more recent time, liquid water may have been only marginally active very gradually under layers of ice, much as found in some semiarid polar regions on Earth, with evidence of frozen lakes (e.g., in the floors of some craters). The majority of Mars' water may now only exist in frozen form underground, though the possibility that some of this water might be in liquid form deep below the surface remains an intriguing, though presently unconfirmed, possibility. Such a possibility has been made all the more intriguing with the discovery of the mysterious gully features observed recently in MOC images (Malin and Edgett, 2000) and discussed in more detail below and in the next chapter.

8.3 EVIDENCE FOR WATER IN MARS' PRESENT CLIMATE

It is clear from the above that water, in either liquid or solid form, may well have played a major role in shaping the environment of Mars, although much of this action seems to have taken place during a much earlier epoch. The present landscape thus represents a relatively old landform compared with the Earth and is currently subject to only very weak erosion. So does water manifest itself in any clear way in the present climate of Mars (at least above ground)?

8.3.1 Residual ice caps

The most obvious appearance of water in any substantial quantity on Mars now seems to be manifest in the very large deposits of water ice comprising the residual polar cap near the North Pole. In contrast to the South Pole, where there appears to be only a very small residual ice cap in summer after the CO_2 seasonal ice cap has sublimed away, the North Polar region has a large body of ice, up to 3–4 km thick and with a total volume around 1.2–1.7 $\times 10^6$ km^3 (Zuber *et al.*, 1998), making it around half the size of the Greenland ice cap on Earth, and which remains intact throughout the northern summer season. Although direct *in situ* measurements of the composition of this body of ice are not yet available, remote-sensing measurement of the ice temperature (e.g., from the Viking Infrared Thermal Mapper (IRTM) instrument), indicate that it cannot be made of CO_2 ice (the surface is too warm in summer, cf Figure 8.2), but is most likely a residual cap of water ice which is covered each winter by a more extensive (though much thinner) veneer of CO_2 ice (up to 1–2 m thick) when the ground temperature becomes sufficiently cold. In the case of the South Polar cap, however, the seasonal CO_2 frost never completely sublimes away, so any residual water ice may only become detectable at the fringes of the seasonal cap during southern summer. Indeed, observations (Titus *et al.*, 2002) from the MGS Thermal Emission Spectrometer (TES) instrument and the THermal EMission Imaging System (THEMIS) instrument on the Mars Odyssey spacecraft have recently revealed evidence of water ice at the fringes of the South Polar cap, probably indicating that the south also has a significant residual cap of water ice which remains relatively isolated from the atmosphere at the present time. The origin of this water ice is not known, but it is presumably an old residue of a much larger reservoir of water from an earlier epoch. Indeed, the long-term stability of the large northern residual ice cap (and the much smaller one in the south) is one of the major outstanding questions in this area of Mars science, with some evidence from MOLA observations (Zuber *et al.*, 1998) that it may have been larger in the past.

8.3.2 Atmospheric water vapour

The direct detection of water vapour in Mars' atmosphere is a comparatively recent achievement, relying on detailed and sensitive spectroscopic measurements (Spinrad *et al.*, 1963). The amount of water vapour in the atmosphere turns out to be extremely small – such that if all the water in a column of Martian air, extending from the surface into space, were to be condensed into a pool of liquid water, it would produce a layer only around 10 μm thick. For comparison, typical total column amounts of water in the Earth's atmosphere are measured in cm – so Mars' atmosphere would seem to be drier than the Earth's by a factor of at least 1000! Note also that this is a much smaller volume of water (by a factor $\sim 10^5$ or more) than apparently present in the north residual polar cap. This small atmospheric concentration of water vapour is largely a reflection of the much cooler temperatures and lower surface pressures on Mars in the lower atmosphere and near the surface, for which the partial pressure of water is correspondingly much smaller than in the warm and humid troposphere of Earth.

Despite such small amounts of atmospheric water vapour, techniques are available to map atmospheric water from space. For many years, the main source of information on variations in atmospheric water vapour during the seasonal cycle on Mars was from the Mars Atmospheric Water Detector (MAWD) on the Viking Orbiter spacecraft (Jakosky and Farmer, 1982). This operated essentially by measuring the absorption of infrared radiation emitted from the warm ground during the day by atmospheric water vapour in characteristic absorption lines around 1.4 μm, and collected data for just over a single Mars year during the main Viking orbital mission period in the late 1970s. Because it relied on measuring absorption of radiation from a warm background, MAWD could only obtain reliable measurements during the daytime. This meant some significant gaps in the data during winter at high latitudes in the polar night. Moreover, the signal-to-noise ratio of individual measurements was fairly low, so the best data set which has been analysed in some detail is of the seasonal variation of the zonally averaged column-integrated water amount. This is shown in Figure 8.12(b) (see colour section) as a function of solar longitude L_S.

Also shown for comparison in Figure 8.12(a) (see colour section) is the equivalent latitude–time map of atmospheric water vapour measured by the TES instrument on MGS (Smith, 2002) during the first mapping year. The TES instrument also measures water vapour by determining the absorption of infrared radiation emitted from the warm ground during the day by atmospheric water vapour, but in characteristic absorption lines around 25–50 μm wavelength (200–400 cm^{-1} in spectroscopic wavenumber units) instead of at 1.4 μm. But since both instruments need warm ground underneath the spacecraft to make the measurement, the TES maps have the same pattern of gaps in the data as for MAWD during polar night in each hemisphere.

Both maps clearly show the mean concentration of water vapour throughout the year at most latitudes to be around 10 pr μm, but with some large seasonal variations, especially in the north. In particular, the vapour maps both show a strong release of water vapour at high northern latitudes during the onset of northern summer ($L_S \sim 30°–150°$), and a local maximum of vapour concentration which moves southwards towards the equator during $L_S \sim 130°–210°$ in northern autumn. The TES data show a somewhat weaker increase in water vapour column amounts during southern summer ($L_S \sim 240°–330°$) close to the South Polar cap, though this does not seem to appear in the MAWD map, which shows only a very weak change around $L_S \sim 270°$ around 60°S from 10 to 15 pr μm. This could represent real interannual variability during the interval from the late 1970s to 1999–2000, but it is now believed that the MAWD observations systematically underestimated vapour amounts during southern summer because of the onset of major planet-encircling dust storms. This produced a marked increase in the infrared opacity of the atmosphere, so that the MAWD measurements could not determine the column absorption due to water vapour all the way to the Martian surface. The TES measurements at this time of year, on the other hand, were under relatively clear conditions with no major storms.

Some similar tendencies have been observed in subsequent observations from the TES instrument (see Figure 8.13, colour section), though during the second mapping year of MGS (when the major planet-encircling dust storm 2001a took place) water vapour amounts were somewhat smaller than during the first mapping year, even when dust levels (see Figure 8.13(a)) were relatively small (e.g., around $L_S \sim 300°$ during the second Mars year). This could well mean that water vapour amounts close to the South Pole in southern summer were actually less in the dust storm year than in the previous year, and not just apparently reduced because of higher atmospheric opacity. Such a tendency might suggest the possibility of scavenging effects associated with the deposition of ice onto suspended dust particles, which would help to remove both dust and water ice from the atmosphere more quickly (due to more rapid sedimentation of larger particles) than due to either dust or cloud ice alone (Michelangeli *et al.*, 1993; Clancy *et al.*, 1996). As discussed briefly in the previous chapter, this is an aspect of the interaction of the two cycles (dust and water) which may well turn out to be of much greater importance than has so far been realized. But future progress is likely to depend on improved knowledge of the microphysical processes responsible for the formation and accretion of ice particles in Martian water clouds.

From the general shape of the latitude–time maps of water vapour from both MAWD and TES, we can already begin to see some indications of exchanges between various reservoirs of water in the Martian climate system. The striking increase in atmospheric water vapour in northern early summer seems to coincide with the time when the seasonal veneer of CO_2 ice evaporates to uncover the residual water ice cap. At this point in the seasonal cycle, the surface temperature of the residual cap can then begin to rise above the CO_2 frost point (around 150 K, cf Figure 8.2) towards the sublimation temperature of water ice (around 185 K for an initial atmospheric column abundance of around 10 pr μm of water vapour). At this point, the northern residual cap can begin to evaporate by direct sublimation, releasing substantial amounts of water into the atmosphere. The subsequent migration southwards in latitude of the maximum of water vapour probably represents actual atmospheric transport of the newly-released water towards the equator.

The corresponding increase in water vapour near the South Pole during southern summer, however, most likely does not primarily reflect the release of large amounts of water from the south residual cap because, as mentioned above, this cap is almost never exposed through the perennial layer of CO_2 ice at those latitudes. In contrast to the north, because the surface temperature of the South Polar cap remains permanently close to the CO_2 frost point, it almost certainly acts as a 'cold trap' for water, immediately freezing it onto the surface of the ice cap from which it cannot later escape. This asymmetry in the action of the two polar caps on atmospheric water seems mainly due to a combination of the high altitude of the South Polar regions and the present eccentricity and timing of the perihelion of Mars' orbit, such that southern summer is relatively short-lived compared with northern summer. Unlike the equivalent in the north, the seasonal increase in water vapour in the south seems to be preceded in both the MAWD and TES measurements by a poleward-migrating maximum in water vapour which then merges with the South Polar maximum itself. This would seem to indicate, perhaps, that the southern peak may reflect a seasonal increase in the transport of water vapour *into* the South Polar region, much of which presumably becomes trapped onto the residual ice cap.

It would appear from these observations, therefore, that the northern residual water ice cap is the dominant source of water vapour in Mars' seasonal hydrological cycle. Moreover, the atmospheric water cycle seems to be strongly asymmetric between the northern and southern hemispheres, with the southern residual cap perhaps acting more as a net sink than a source at the present time. However, a more complete assessment of how Mars' hydrological cycle works requires us to consider other possible reservoirs for water in the Martian environment.

8.3.3 Water ice clouds

When the atmospheric temperature falls sufficiently far to reach the water frost point (cf Figure 8.2), there is the possibility that suspended clouds of condensed ice crystals will form which can be observed as thin, bluish hazes across the planet. Figure 8.14 (see colour section), for example, shows some clear examples of such clouds in

Hubble Space Telescope (HST) images, especially prominent close to the limb of the planet, and close to some of the major topographic features such as Olympus Mons. The association of clouds with topographic features has already been discussed in Chapter 4 in connection with circulations due to forced waves, but tenuous water ice clouds are a common occurrence within the Mars climate system. Figure 8.15 (see colour section) shows a very clear example of clouds associated with the major peaks on the Tharsis Plateau in wide-angle MOC images. Each of the main volcanic peaks seems to have a banner cloud attached to it, although some other, more diffuse, clouds are also present further to the east. Such more general cloud cover seems to be relatively common during northern hemisphere summer, around the time of aphelion when the strength of solar insolation is comparatively weak and the atmosphere is relatively cool at high altitudes. During this time, the TES measurements of water vapour and temperature (Smith, 2002) suggest that the atmosphere reaches the frost point for water ice formation at altitudes of only 10–20 km. This is broadly consistent with other measurements, which also suggest that the most prominent clouds occur around the 10–20-km altitude above the surface.

The observed clouds can take many different forms, not just the simple, uniform veils apparent in the figures shown above. We have already seen in Chapters 4 and 6 some illustrations of clouds associated with topographic lee waves and baroclinic cyclones. But many of the other cloud forms familiar to us can also be seen in some form on Mars. Figure 8.16, for example, shows some streak-like ice clouds which resemble the high-altitude cirrus clouds seen on a fair day on Earth. Water ice clouds have also been observed from the surface of Mars, by the Mars Pathfinder (MPF) Lander, most commonly around sunrise and sunset and in the form of thin veils of cirrus or cirrostratus. Figure 8.17 (see colour section) shows two examples of different cloud formations observed by the Pathfinder camera, illustrating both layered clouds and a field of cirrostratus. Other types of cloud include convective features (equivalent to terrestrial cumulus clouds, e.g., see Figure 8.18), though these are comparatively rare.

While visual imaging can provide a great deal of local detail relating to individual clouds and regions of cloudiness, other types of measurement (especially from orbit) may be better able to quantify the systematic occurrence of cloud within the Martian climate system. The TES instrument on MGS, for example, has obtained maps of water ice aerosol in the Martian atmosphere using similar techniques to those by which the atmospheric water vapour and dust opacity was obtained above (cf Figures 8.13, 8.12, and 7.18 (all in colour section)). Again, because this technique relies on measuring absorption of radiation from a warm planetary surface (Smith, 2002), determination of cloud opacities can only be obtained during the day and the outside regions of polar night.

Thus, Figure 8.13(b) (see colour section) shows a map of zonally averaged atmospheric ice opacity in the 12-μm wavelength band in the infrared which is dominated by water ice absorption. This shows some clear patterns which seem to be roughly reproduced in both the Mars years observed so far by TES. Notable features are the strong increases of cloudiness surrounding the winter polar caps, associated with water condensate clouds on the fringes of the polar hood. The extent

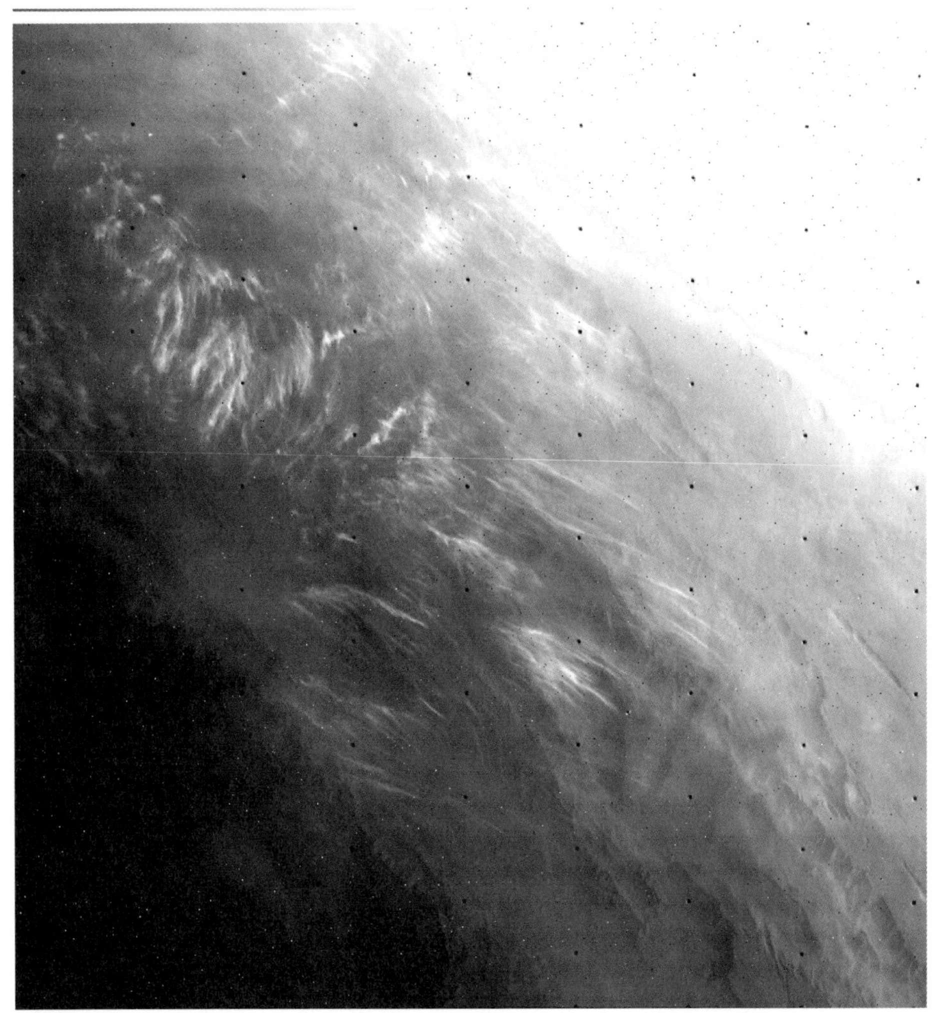

Figure 8.16. Viking Orbiter image showing a field of water ice clouds in the form of cirrus streaks.

From NASA/JPL.

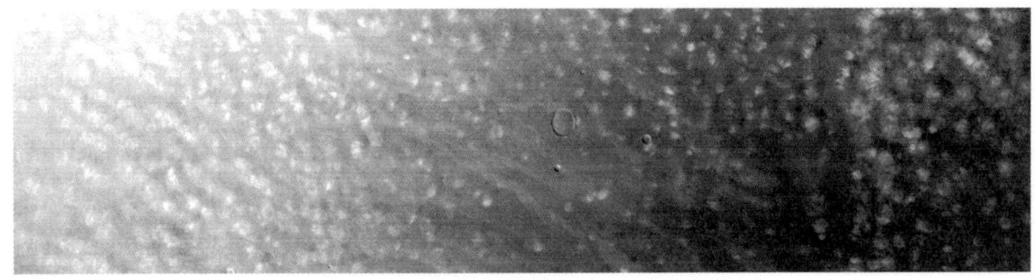

Figure 8.18. MOC image of convective clouds.

From NASA/JPL and Malin Space Science Systems.

to which these features extend into the polar night region is unclear from this kind of map, however. The other striking feature is the prominent maximum of zonal mean cloud opacity close to the equator around $L_S \sim 20^\circ$–160°. This corresponds to the aphelion cloud belt mentioned above, which tends to dominate the appearance of the planet during northern summer. In the first mapping year, the aphelion cloud belt also seemed to persist well into northern autumn/southern spring (until around $L_S \sim 220^\circ$). This turns out to represent persistent topographic clouds in the vicinity of the Tharsis Plateau, rather than more general cloudiness, as in the aphelion period itself. During the second MGS mapping year, however, the onset of the planet-encircling dust storm resulted in a rapid cut-off of low-latitude clouds. The flooding of the atmosphere with dust resulted in a strong warming of much of the global atmosphere at this time, which essentially seems to have evaporated most of the water clouds during this period.

The typical density of these water ice clouds is generally very small, with amounts of water tied up as suspended ice aerosol being equivalent to a few pr μm. This places these clouds in a similar category to the polar stratospheric clouds found on Earth, at altitudes of around 20–45 km. Such tenuous clouds are most unlikely, therefore, to result in precipitation, since this would require the formation of particle sizes of around 100 μm or greater. Such sizes are unlikely to develop because of the low densities and coagulation rates from smaller particles, so that any particle reaching a size sufficient to begin to fall by sedimentation is much more likely to sublime away before it reaches the ground (Jakosky, 1985).

8.3.4 Fog and frost

Another manifestation of water-related weather on Earth is the formation of early morning mists and fog, and the deposition of dew and frost onto the ground. This generally takes place when the surface itself cools to the condensation or frost point temperature or below, causing vapour in the atmosphere either to condense directly onto the surface (forming dew or frost) or onto dust particles forming a cloud in contact with the ground. This commonly happens just before dawn, when the surface is at its coolest during the day.

Such processes can evidently occur on Mars also, at least in the form of ice fogs and frosts. Figure 8.19 (see colour section) shows a striking example of the formation of surface fog shortly after dawn in the Noctis Labyrinthus region of Mars, which consists of a network of interlinked canyons. The fog is almost certainly a cloud of water ice, formed as the surface cooled overnight to the water condensation temperature. Somewhat surprisingly, the formation of fogs on Mars has not been studied extensively so far, either by observation or in models. While all the lander spacecraft so far have been equipped with cameras, none were able to obtain images of the surrounding landscape during the night. However, both the Viking and Pathfinder landers were able to measure increases in the local opacity of the atmosphere during the early hours of the morning (between 02:00 and local dawn) by observing the moon Phobos as it passed overhead (Pollack *et al.*, 1977; Thomas *et al.*, 1999). Colburn *et al.* (1989) made an assessment of the potential for condensation of water at the Viking Lander (VL) 1 site, based on an assumption of uniformly mixed water vapour in the atmosphere with a total water column of 12 pr μm, which suggested that the atmosphere would be saturated above 25-km altitude much of the time, and in the near-surface layer (below 400 m) during wintertime. This kind of behaviour was also found in the boundary layer model of Savijärvi (1991), which indicated the possibility of shallow fog formation overnight provided the water concentration was high enough (around 16 pr μm).

The formation of surface frost has also been observed at the VL 1 site at around 48°N. Figure 8.20 (see colour section) shows an image from VL 1 in which much of the surface close to the lander is covered in a thin, white, granular deposit. Surface temperatures indicate that this could not have been direct deposition of pure CO_2 ice, and is thought to be primarily water ice frost. The close proximity of this latitude to the polar ice cap edge at this time of year, however, suggested the possibility that the deposit could actually be a mixture of water and CO_2 ice in the form of a clathrate when it first formed (Jones *et al.*, 1979), which might account for its quite thick and white initial appearance. Subsequently, as the temperature rose the CO_2 probably sublimed away, leaving a pure water ice frost behind. The frost was observed to persist for a few months during the winter season, persisting particularly in the shadows of boulders, much as winter frosts on Earth.

8.3.5 The regolith

The other main location where water can reside is in the ground itself, as mentioned above. On short timescales, water vapour can diffuse a short distance into the soil through pores and cracks, and become adsorbed onto soil particles. In practice over seasonal timescales, this probably affects a layer a few tens of cm deep close to the surface (see Chapter 9), which is commonly referred to as 'the regolith'. On longer (interannual–geological) timescales water could penetrate much deeper (perhaps hundreds of m or even km into the ground). This region, which has been processed and broken over geological time by volcanic action and impacts, is sometimes known as the 'megaregolith', and may be the ultimate reservoir for the bulk of the water which could have been present during the early warm, wet period of Mars' history.

The take-up and release of water in the Martian soil has not so far been measured directly, though some of the processes have been studied in both terrestrial laboratory experiments and in models (e.g., see Jakosky and Haberle (1992) and references therein). The latter enable the potential of different Martian soil types to adsorb water to be estimated, with equilibrium values ranging from 10–100 g of H_2O or more per kg of soil (depending on the soil type). Models based on these measurements, and representing 1-D diffusion into the soil and adsorption, have then been developed to estimate the daily and longer-term take-up and release of water into the ground under Martian conditions. The most detailed 1-D models have been developed by Zent *et al.* (1993), which seek to model the exchanges between the near-surface atmospheric boundary layer and the uppermost few metres or so of soil (though diurnal exchanges probably only affect the top 10–20 cm or so). Such models take into account the atmospheric conditions (temperature, humidity, etc.) and latitude of the location being computed, together with properties of the soil itself (thermal inertia, porosity, and composition), and can provide estimates of how rapidly, and how much, water is taken up or released by the surface layers, either on diurnal or seasonal timescales.

The diurnal exchanges of water are probably relatively small, with amounts around 1 pr μm or less being cycled in most places (corresponding to around 10% of the typical atmospheric column of water vapour). Although the amounts are small, they can have a significant effect locally on the relative humidity, and hence contribute to the formation of fogs and frost at the surface. On seasonal timescales, however, the amount taken up or released by the regolith might be significantly larger – perhaps as much as 10–20 pr μm between solstice and equinox (Zent *et al.*, 1993). Given sufficient time for the atmosphere to come into equilibrium, the regolith is evidently able to adsorb large amounts of water (consistent with the adsorption densities mentioned above), suggesting that this could be a major reservoir within the water cycle, capable of buffering large variations in atmospheric vapour concentration and transport. Indeed, recalling the need to account for potentially vast amounts of water which might have filled Oceanus Borealis in Mars' geological past, if we extrapolate adsorption densities of several tens of grams of water per kg of soil to the suggested depth of the megaregolith (2–3 km), there would seem to be no fundamental difficulty with the megaregolith accommodating enough water to cover the northern hemisphere to a depth of at least 500 m.

The amount of adsorbed water which may be stably locked up in the soil, however, evidently depends on a number of factors, most notably the soil type and porosity, the surface temperature (and hence the vapour pressure) relative to the frost point temperature of water at the surface, and the temperature at deeper levels. Various models of the regolith-surface interactions have been used to estimate the limits under which a body of ice underneath the surface may be stable to evaporative loss during the Martian year. An example of such an estimate is illustrated schematically in Figure 8.21, which shows those latitudes and soil depths where water ice is either uniformly stable (dark grey regions, close to the poles), unstable (light grey regions), or partly stable (stable during only part of the year, usually around winter). This would seem to indicate that water ice is unstable to at least a few metres depth within 40° of the equator, and so will have evaporated by now to leave dry soil to these depths. Ground ice is evidently stable within 10° of the poles right to the surface (corresponding to the residual water ice caps), and also between about 45–80° latitude at depths of around 1 m or slightly less. In between (i.e., between latitudes 45° and 80° at depths less than 20 cm), water ice may be stable at certain seasons, and it is this region which might be expected to participate as a buffer reservoir in the seasonal exchange of water during the year. This distribution would seem to compare reasonably well with the measurements of ground ice from Mars Odyssey (cf Figure 8.11, colour section), for which epithermal neutrons sampled hydrogen presumed to be associated with subsurface water at depths up to a couple of metres, except perhaps in some of the low-lying regions at equatorial latitudes which also appear to have some subsurface water present.

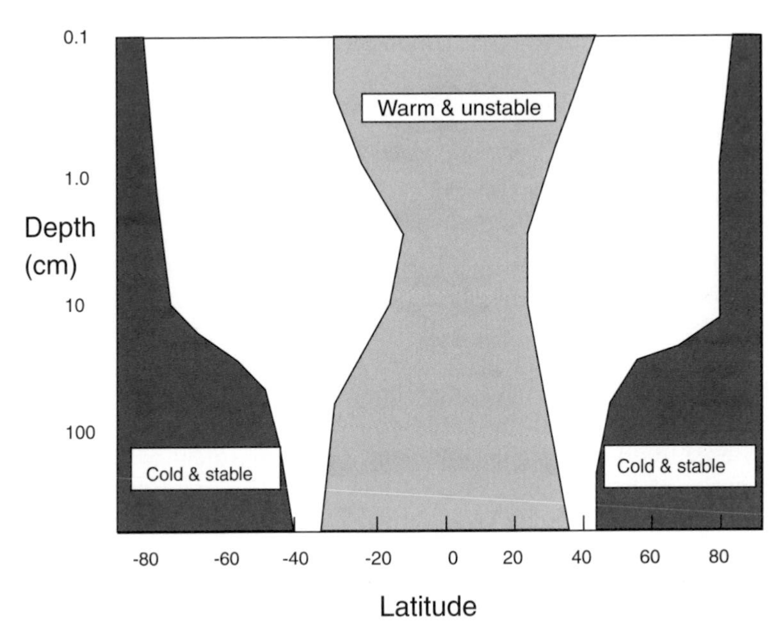

Figure 8.21. Schematic boundaries of subsurface stability of water within near-surface materials on Mars as a function of depth and latitude for present conditions. The dark grey regions indicate permanently stable water ice, while the light grey region indicates where water ice is not stable and will tend to evaporate away on long (interannual) timescales. The white regions in between indicate soil where water ice is unstable for only part of the year.
Adapted from Farmer and Doms (1979)

Such models, however, do not really account for the occurrence of the small-scale gully features discovered recently in MOC images of the Martian surface (Malin and Edgett, 2000) in a number of locations poleward of 30° in both hemispheres. Figure 8.22 (see colour section) shows an example of two groups of these features in the south (poleward)-facing side of a crater in the southern area of Noachis Planitia. The sinuous gullies emerge from small areas of subsidence, from which it appears that a stream of water (or a similar liquid) may have eroded the channel leading from the apron of the subsided region. These features are relatively rare, and appear to be comparatively young geologically (i.e., less than 10 Myr old). If they are produced by the transient release of subsurface water, however, then this raises a number of questions, such as: (a) what causes the water to be released close to the surface on a poleward-facing slope (though by no means all gullies occur on slopes which are poleward-facing) and (b) how can the water emerge and remain in liquid form, given that under present atmospheric conditions the liquid state is inaccessible (cf Figure 8.2)? A possible explanation for (b) could be that the water has a high concentration of dissolved salts. Such a brine solution can have very different properties to pure water, and remain liquid to significantly lower temperatures and pressures. This is by no means the only possibility, however, and alternative possibilities (e.g., involving long-term cyclic climate change) have been suggested (Mellon and Phillips, 2001), and will be discussed in more detail in the next chapter.

8.4 MODELLING THE ATMOSPHERIC WATER CYCLE

From the discussion of the various ways in which water may be stored in the Martian environment, either above the ground in the atmosphere or underground, we can envisage a conceptual picture of the hydrological cycle under the conditions now prevailing on Mars. Figure 8.23 shows a schematic representation of the various storage 'reservoirs' and the exchanges envisaged between them, showing the basic cycle from evaporation of water vapour into the atmosphere to the means by which the vapour may be transported across the planet by the various components of atmospheric circulation (eddies, zonal mean flows, etc.). Some of the vapour may be transformed into cloud ice and also transported in this form, or sedimented back to the surface at a location remote from the poles. Accumulation of ice at the surface leads to seasonal deposits, and vapour may also find its way back to the subsurface by diffusion and adsorption into the regolith.

8.4.1 Diffusive models

In order to quantify these exchanges, scientists have generally tried to construct a range of mathematical models, which seek to relate the transport of water in its various forms to certain properties of the atmospheric flow and local conditions at the surface. Early models, for example, by Davies (1981), Jakosky (1983), and James (1985) tried to quantify the atmospheric transport of water vapour as a diffusion (so

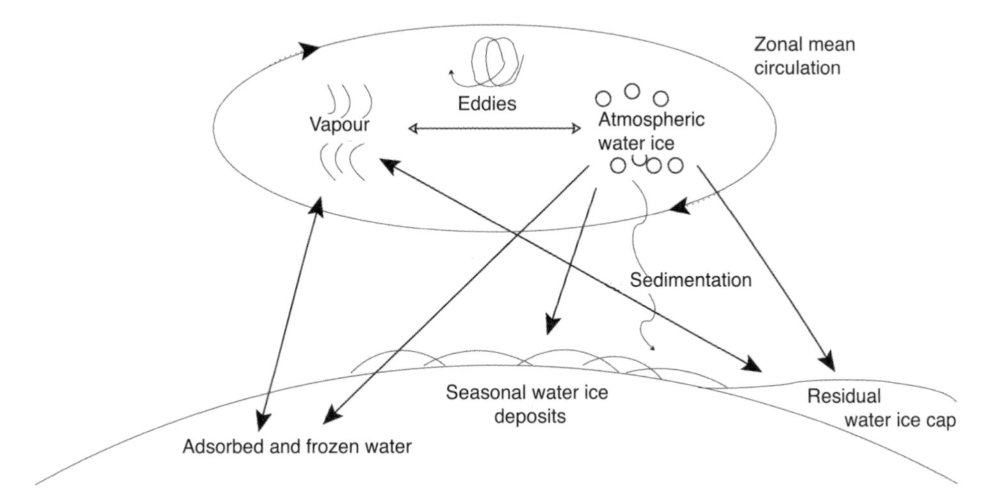

Figure 8.23. Schematic diagram of the interaction of the main water reservoirs on Mars. Adapted from Richardson (1999) and Böttger (2003).

the poleward flux of water vapour per unit area $\sim K_d \nabla q_v$, where q_v is the water vapour mixing ratio and K_d is a diffusion coefficient) and to consider whether the observed atmospheric water vapour (measured by MAWD) could originate solely from the then known residual ice cap in the north. Such a model might be expected to yield atmospheric transports in the appropriate direction, according to season, since the observations indicate that sustained latitudinal gradients of water vapour occur during the observed water cycle. All three studies obtained vapour transports which suggested that most of the observed vapour could indeed be derived by seasonal releases from the northern residual ice cap. Davies (1981) also concluded that the atmospheric water cycle could be closed and stable (i.e., with no net transport of water from the northern hemisphere to the south) provided the transport efficiency (represented by the diffusion coefficient) was substantially different in the two hemispheres (by a factor $\sim$50). This large asymmetry was justified by the assertion that the increased incidence of dust storms in the south during southern summer would radically increase the effective diffusion. Later studies were less convinced by this assertion and its implied rapid mixing in the southern summer (which was not reflected in any other variable), and tended to conclude that, although the seasonal variation of atmospheric water vapour could be accounted for by simple transports from a northern polar source, the cycle may not be closed but could well represent a net transport from the North Polar cap to the south when integrated over the year, of around $1\text{–}4 \times 10^{11}\ \mathrm{kg\,yr^{-1}}$, though this could be alleviated when a parameterization of the CO_2 condensation flow was also taken into account (James, 1985).

8.4.2 Water cycles in simple GCMs

Diffusive models are a rather crude way of representing atmospheric transport which, as we have seen, takes a variety of complex forms. It is clearly desirable, therefore, for a more realistic model to be used to simulate and quantify atmospheric transport, much as described for the lifting, transport and deposition of dust (see Chapter 7). In fact the application of comprehensive GCMs to model the water cycle on Mars has only been achieved relatively recently (Richardson and Wilson, 2002a; Richardson *et al.*, 2002; Böttger *et al.*, 2003; Montmessin and Forget, 2003), though these studies did build upon earlier studies using simplified 2-D and 3-D circulation models (Haberle and Jakosky, 1990; Houben *et al.*, 1997). The 2-D (axisymmetric) model study of Haberle and Jakosky (1990) actually found it quite hard to transport water from one hemisphere to the other, illustrating an important limitation (among several) of axisymmetric models. This is because the tendency of fluid elements to conserve their potential vorticity makes it almost impossible for them to cross the equator (so the ambient planetary vorticity changes sign) without the help of some other process (usually parameterizing a non-axisymmetric effect). The absence of non-axisymmetric eddies also led to a rather sluggish transport close to the polar regions themselves, where baroclinic eddies are typically active and important for large-scale transport of heat and other tracers (see Chapter 6).

The 3-D model of Houben *et al.* (1997) solved for a fully 3-D circulation, forced by some simple drag and relaxation parameterizations but without including detailed representations of topography or the diurnal cycle. It was able to show some significantly more vigorous transports than the earlier 2-D study, but still suffered from some drawbacks and uncertainties which compromised its ability to predict the details of the water cycle.

8.4.3 Comprehensive GCMs

Current versions of Mars GCMs seek to represent the main reservoirs summarized above and shown schematically in Figure 8.23. Direct surface exchanges due to sublimation and condensation of water vapour onto surface ice are represented in much the same way as for CO_2 ice condensation as described in Chapters 2 and 3. The released vapour in the atmosphere is then transported as a material tracer, much as dust (as discussed in Chapter 7), by the fully 3-D time-dependent field of winds predicted by the rest of the model. Current models also take account of the possibility of ice cloud formation within the atmosphere. Cloud ice can then be separately transported as another material tracer, including the possibility of vertical sedimentation at a suitably prescribed rate, though the details of how the clouds are formed by microphysical processes in the atmosphere is the subject of ongoing research at the present time (e.g., see Montmessin *et al.*, 2002). Finally, a representation of the seasonal and diurnal exchange of vapour with the thin regolith may also be included.

The results indicate that such models can produce seasonal distributions of atmospheric water vapour which look remarkably realistic when compared with the MAWD and TES observations discussed above. Figure 8.24 (see colour section) shows results from such a simulation using the Oxford Mars GCM, in which a relatively sophisticated regolith scheme has been implemented (based on that of Zent *et al.*, 1993), but cloud ice is produced without complex representations of microphysics (Böttger *et al.*, 2003). Experiments were typically initialized by assuming a dry atmosphere and a large body of water present in the North Polar residual cap, and no significant southern residual cap. After a few years of simulation, the distribution of atmospheric water settles down to a reasonably stable and repeatable annual cycle. The water vapour map (Figure 8.24(a), see colour section) was obtained after a few years of simulation, and shows a typical cycle in which vapour is released in substantial quantities when the north residual water ice cap is exposed in northern spring. The released vapour is then transported systematically southwards, and eventually some vapour is transported across the equator into the south (around $L_S \sim 200^\circ$). In northern autumn, however, there is some evidence of northward transport of vapour until the cycle begins to repeat in the following year. In the south, there is a small maximum in vapour around $L_S \sim 275^\circ$, much as observed in the TES observations (Figures 8.12 and 8.13, colour section).

In Figure 8.24(b) (see colour section) the formation of cloud ice shows a number of interesting features, several of which resemble observed features in spacecraft images of the planet. These include the formation of a ring of ice clouds surrounding the northern winter polar CO_2 ice cap, the development of a belt of low-latitude clouds around aphelion ($L_S \sim 90^\circ \pm 50^\circ$) and a thin region of ice clouds surrounding the South Polar cap during southern winter. This level of realism seems remarkable, given the simplicity of the schemes used to calculate the formation of water ice clouds (Richardson and Wilson, 2002a; Richardson *et al.*, 2002; Böttger *et al.*, 2003).

Since the water cycle as represented in a GCM explicitly computes the large-scale, fully 3-D and time-dependent transports of water and ice, and the exchanges between the atmospheric and other known reservoirs of water in the Martian environment, it is of considerable interest to use such models to assess in some detail (a) to what dynamic equilibrium state (which may be repeated from year to year) the simulated water cycle eventually settles down, (b) whether the observed water cycle is reproduced accurately, and if so, (c) what such an equilibrated water cycle might imply about the long-term stability of the large surface and subsurface reservoirs of water on Mars. At the time of writing, these issues remain somewhat controversial. Richardson and Wilson (2002a) established that their simulated water cycle (even without an active regolith) would eventually equilibrate to a roughly constant atmospheric state, but which appeared to retain around 2–3 times as much vapour on average in the atmosphere than suggested from observations. The inclusion of a more sophisticated representation of the regolith led Böttger (2003) to suggest that exchanges with the regolith during the annual cycle would exercise some buffering of changes in water vapour in the atmosphere, and lead to a somewhat lower average burden of water vapour during the year – arguably closer to that observed by MAWD and TES. The transport of cloud ice also seems to play a minor though not insignificant role during the annual cycle (cf. James, 1990).

The extent to which the seasonal exchanges constitute a closed cycle, however, is perhaps the most interesting issue which such models may be able to address. The early diffusive models needed significant 'tuning' to produce a closed cycle which did not result in an integrated transport of water from the North Polar cap (as source) to the south residual cap (as a cold-trap sink). The simple GCM study of Houben *et al.* (1997) found a substantial north–south transport of water vapour between hemispheres in their model, though this model left out a number of potentially important processes (surface topography, CO_2 condensation, etc.) affecting both the transport and the interaction of water with the surface. Even so, the more comprehensive models developed recently also seem mostly to indicate that some residual transport from north to south is inevitable, though differ in their assessments of how large the residual transport may be. Certainly including the regolith reservoir and exchanges does tend to bring the cycle more closely into balance, since vapour which is advected southwards from the North Pole in northern spring can then be temporarily absorbed by the regolith, only to be rereleased in northern autumn to be retransported back onto the North Polar cap during winter. This allows the vapour cycle to remain highly asymmetric between the hemispheres without very large net transports between the polar ice reservoirs. In some ways, it would be a rather remarkable coincidence if there were to be precisely *zero* net transport of water from one hemisphere to the other at the present time, given the asymmetries between the northern and southern hemispheres, the ellipticity of the orbit, the permanent presence of a cold-trap associated with the residual southern CO_2 ice cap, and other factors. Such a condition would only have to be the case if the present configuration of insolation were unchanging over very long periods of time. In practice, however, as we shall see in the following chapter, the present conditions for solar heating are not permanent and unchanging, but are simply part of an ongoing long-term cyclic variation in the Martian climate. It may yet turn out that a net transport of water between hemispheres may be a necessary feature of long-term climate variability on Mars in order to explain other features of the Martian environment.

8.5 OUTSTANDING QUESTIONS

At the present time, therefore, the study of the hydrological cycle on Mars has reached an intriguing stage in which there is a substantial observational base, at least for the atmospheric components of water storage, but increasingly for subsurface reservoirs too. Models seem to be capable of simulating quite plausible cycles with a good deal of local detail which may be amenable to quantitative comparison with observations – either now or in the foreseeable future. But a number of critical questions remain to be resolved:

- Given the evidence in Mars' landscape of large-scale features requiring much warmer and wetter conditions on Mars in the distant past, even including the possibility of relatively long-lived lakes and oceans, where has the bulk of this water gone?
- If large quantities of water are stored within the Martian subsurface, is it likely that this water can be released again in the future – either naturally, or by deliberate human action?
- Is the current annual cycle of water transport the result of a closed cycle, or is the hydrological cycle on Mars in a transitional period leading to the systematic transport of water from north to south?
- What are the respective roles of the 'permanent' northern and southern residual ice caps in the present water cycle?
- Can we quantify the relative roles of all the main water sources and sinks on Mars, including the possibility of localized water reservoirs?
- How do the dust and water cycles interact with each other (and is there any link with the polar layered terrains, discussed in the next chapter)?
- What causes the local gullies on Mars, and what is the origin of the water released?
- Did the presence of water on Mars in its early history ever lead to the development of living organisms?

These questions (and doubtless others which the reader may be able to formulate him/herself) are among the most compelling concerning Mars (or indeed any other planet in the Solar System), and continue to exert a strong influence on the priorities for the current and future phases of Mars exploration.

9

Cyclic climate change

Throughout much of this book so far we have been concerned with understanding the circulation and climate of Mars' atmosphere under the conditions we observe at the present time. But in examining the processes which formed and modified the Martian landscape, we have already seen evidence that Mars enjoyed a rather different, and perhaps more hospitable, climate in its distant past, the investigation of which forms a major focus of the current exploration of the planet. The changes to the Martian climate must have been pretty dramatic to transform the water-rich environment which formed the outflow channels and valley networks discussed in the previous chapter. But even in Mars' present state, there are some clear signs that its climate is not static, but undergoes cyclic variations which change the conditions at the surface of the planet quite substantially from those at present.

In this chapter, therefore, we draw together a number of threads which have linked much of our investigation so far of Mars' climate, and consider the potential of Mars to exhibit climate states which may be quite far removed from Mars' present state. We begin by looking at the evidence for natural cyclic variability in Mars' climate, which will turn out to reflect quite strongly the well known astronomically-driven ('Milankovitch') cycles of climate change on Earth, which lead to the cycles of glacial and interglacial periods on timescales of 10^4–10^5 yr or so.

9.1 CYCLIC CLIMATE CHANGE AND THE POLAR LAYERED TERRAINS (PLTs)

When the Mariner 9 spacecraft went into orbit around Mars in 1971, its camera detected some strange linear features at the fringes of the North Polar ice cap which seemed to take the form of parallel stripes, apparently following contours around the edge of the ice itself. The higher-resolution images of the Viking Orbiter camera (e.g., see Figure 9.1(a), colour section) showed these features more clearly, and indicated that the stripes were apparently marking the edges of exposed layers in the soil and ice, some of which were as thin as a few tens of metres. Most recently, the Mars Orbiter Camera (MOC) on the Mars Global Surveyor (MGS) spacecraft has enabled the acquisition of images with a resolution of just a few metres, which show (e.g., see Figures 9.1(b) (colour section) and 9.2) that this layered structure is apparently visible down to a thickness of perhaps 10 m or less in places. Yet such layers appear to be coherent horizontally across hundreds of km of the polar landscape on Mars, apart from some occasional breaks or jumps (known as 'unconformities').

This kind of terrain evidently occurs widely around both the northern and southern residual ice caps, and is clearly visible during the respective summer seasons in each hemisphere. Figure 9.3 shows schematic maps of the northern and southern polar regions which indicate the extent and location of the PLTs, as surveyed during the Viking Orbiter mission. This clearly shows that the layered structures occupy much of the land surrounding both polar ice caps, extending to around 78°N in the north and as far equatorwards as 70°S in the south, though the distribution in the south is more asymmetric.

The precise properties of the icy soil which lead to the contrast in albedo and texture of the layers are not well understood at the present time, and their investiga-

Figure 9.2. A narrow-angle image from the Mars Orbiter Camera (MOC) on Mars Global Surveyor (MGS) of a small region of the southern layered terrains centred at 73°S, 224.5°W. The image shows a region 1.5 km × 4.6 km long, at a resolution of 1.2 × 2.2 m per pixel.
NASA/JPL and Malin Space Science Systems.

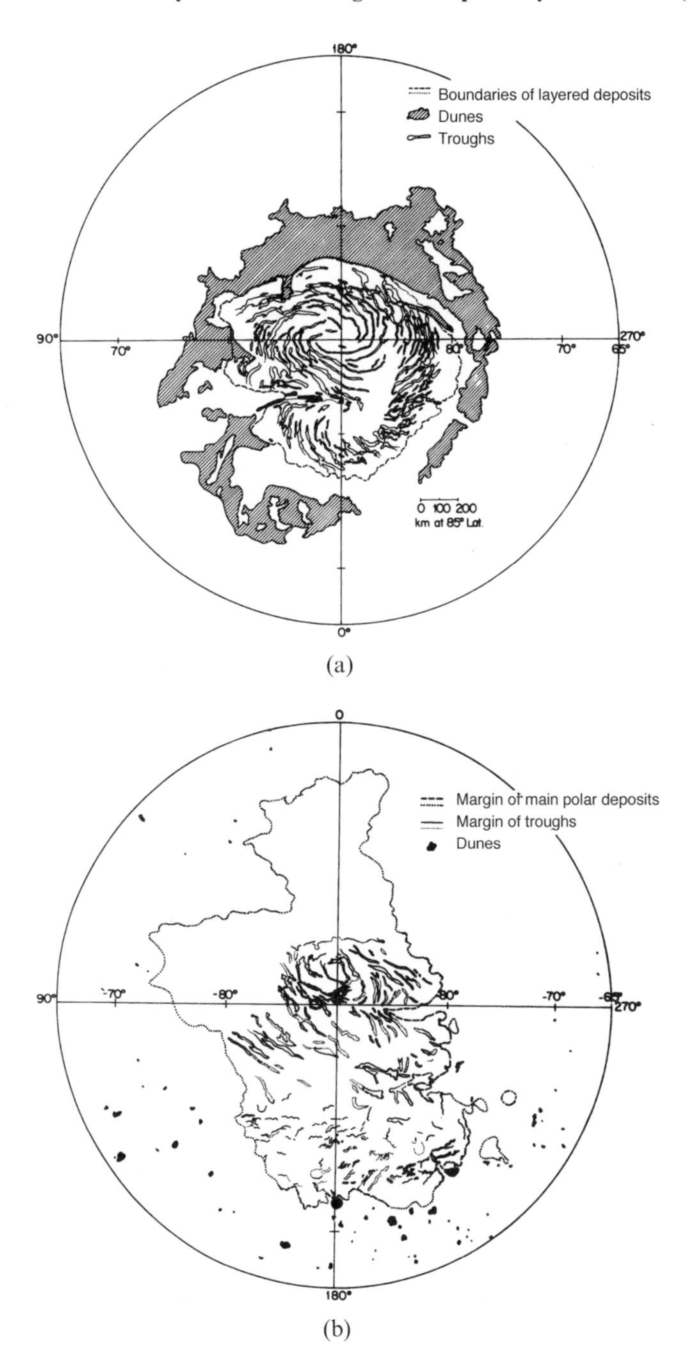

Figure 9.3. Schematic maps showing the extent of the polar layered terrains (PLTs) around (a) the North Polar and (b) the South Polar caps on Mars.

Thomas *et al.* (1992).

tion was a major objective of the ill-fated Mars Polar Lander (MPL) spacecraft. This was due to reach Mars in 1998 but was lost during the final stages of the landing itself. A definitive resolution of this question will have to await a return mission to these high-latitude regions, but for the present it is believed that the layers reflect a varying ratio of dust to ice during the deposition of water ice from the atmosphere. The measured albedo of the layers seems to be approximately similar to that of bright, wind-blown dust, though perhaps with an additional darker component mixed in (e.g., Thomas *et al.*, 1992). From laboratory studies, it would seem that the albedo of dust–ice mixtures is likely to be approximately the same as for the dust itself, even for very small fractions of dust (10^{-2}–10^{-3} by mass), provided the grain size of the ice crystals is significantly larger than that of the dust particles (Kieffer 1990). So variations in albedo within the layers could either reflect different ice–dust ratios or varying ice–grain sizes, or some undetermined mixture of the two.

The current concensus concerning these features is that the layering reflects a cyclic variation in the relative deposition of a mixture of dust and ice over what must be long periods of time. Given present conditions, the column densities of atmospheric dust and water vapour amount to a few tens of microns of material which are likely to be deposited each year. Allowing for the possibility of some concentration of dust and ice into the polar regions (e.g., Pollack *et al.*, 1979), this would seem to suggest that the formation of a layer a few metres thick would take a few hundred thousand years. This would seem to date the layered terrain structures as comparatively recent in geological terms, which also seems consistent with a relative absence of impact craters invading the layers (at least in the north – see below). The contrasting nature of the layers would therefore suggest variations in the ice–dust composition of annual polar deposition on those kinds of timescales. Thus, we are driven to consider the possibility that Mars' climate itself (or at least those aspects which control the dustiness of its atmosphere) may be changing cyclically on a timescale of around 10^5–10^6 yr.

A significant uncertainty concerning the possibility outlined above is whether this cyclic change leading to variable deposition at the poles is ongoing, or was a phase which Mars went through in the more distant past. In this respect, there seems to be conflicting evidence between the North and South Polar regions. From a detailed study of Viking Orbiter images of the southern layered terrains, Plaut *et al.* (1988) found evidence of some impact cratering which had invaded and modified the layered regions, on scales of around 300 m or so. This suggested that much of the southern polar layered terrains are relatively old in geological terms; perhaps up to 200 Myr or so, rather than being recent and fresh. In contrast, in the north there seem to be relatively few craters modifying the layered terrains, suggesting that the surface there is more recent, perhaps even with layers being actively laid down up to the present time. This possibility has received some further support from a recent analysis by Laskar *et al.* (2002), which is discussed in more detail below. The presence of the unconformities in the layers around both poles also suggests the possibility that deposition rates may have varied in the past, to the extent that at some stage erosion of the layers may have taken place in some regions (e.g., Carr 1996).

9.2 QUASIPERIODIC AND CHAOTIC ASTRONOMICAL CYCLES

9.2.1 Spin–orbit cycles on Mars

The timescale of around 10^5 yr is an interesting one in the context of the celestial mechanics of Mars, since it has been known for more than 30 years that the orbital and rotation parameters of Mars (its orbital eccentricity, L_S of perihelion and obliquity) undergo cyclic oscillations in response to various dynamical influences (e.g., see Ward 1992). These can be calculated in detail from the known perturbations to Mars' motion from the other planets of the Solar System (e.g., Ward, 1974, 1979; Touma and Wisdom, 1993). Simpler theories isolate the dominant periodicities (e.g., showing that the eccentricity e varies over the range 0–0.12 as the superposition of two oscillations, one of amplitude 0.05 and period 96 kyr and the other of amplitude 0.1 with period $\sim$2 Myr). The L_S of perihelion follows the precessional period of Mars' spin axis with a period of around 175 kyr, while the normal to the orbital plane fluctuates with an amplitude of around $\pm 3^\circ$ about its mean direction with a period of 70 kyr. Mars' obliquity θ_o is found to vary over the range $24.4^\circ \pm 13.6^\circ$, according to Ward (1979), with periods of 125 kyr and 1.3 Myr. In fact the latter turns out to be somewhat controversial, since more recent calculations (Laskar and Robutel, 1993; Touma and Wisdom, 1993) using full numerical solutions of the governing equations indicate that Mars' obliquity cycle is in fact chaotic. This is mainly due to some previously ignored spin–orbit resonances, in which the period of one of the oscillations affecting the spin axis of the planet comes into a commensurate relationship (i.e., with periods in a ratio close to that of simple integers) with one or more of the orbital oscillation periods. This may have caused the obliquity to vary between $0^\circ \leq \theta_o \leq 50^\circ$ or even greater during the past 10 Myr or more. Figure 9.4 shows a time series from the computations of Laskar *et al.* (2002), which illustrates the large and erratic variations in θ_o and other parameters which may have taken place during the past 10 Myr or so. However, the details of the reconstructed oscillations depend sensitively on the conditions assumed for the present time, and so become less and less certain the further into the past the time series is extrapolated.

9.2.2 Spin–orbit cycles and the Earth's climate

Note that, while these variations have their counterparts in corresponding oscillations of the Earth's orbit and rotation, they are generally much more extreme on Mars. For the Earth, for example, e varies between 0.01 and 0.05 with periods of 100 kyr and 400 kyr, while the precession of the equinoxes occurs with a period of 21 kyr and obliquity θ_o oscillates between 22° and 24.5° with a period of 41 kyr. Even so, it is believed (Imbrie, 1980, 1982) that these cycles exert a strong influence on cyclic variations in the Earth's climate on the 10^4–10^5 yr timescale, associated with corresponding modulations in the annual distribution of solar radiation. The relatively small amplitude of obliquity variations on the Earth is now known to be due to the stabilizing influence of the Moon, which exerts a significant gravitational

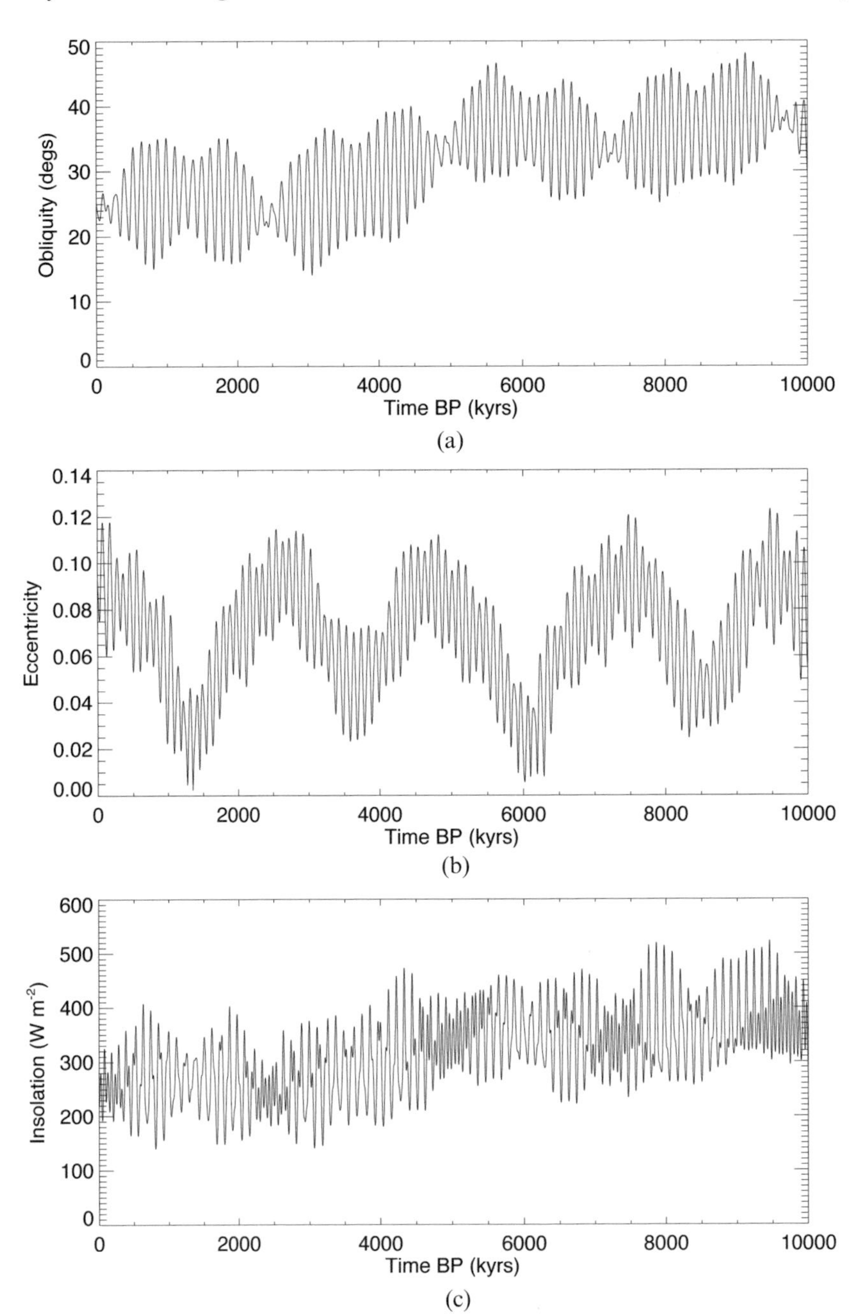

Figure 9.4. Time series of the obliquity (a) and eccentricity (b) of Mars' orbit during the past 10 Myr, as computed by Laskar *et al.* (2002), showing the wild and chaotic oscillations of the planet's obliquity during this period. The corresponding insolation at the summer pole at summer solstice is shown in (c).

From data courtesy of Dr Jacques Laskar.

perturbation of its own on the motion of the Earth. In the absence of any spin–orbit resonance with the Earth, the Moon's influence is to stabilize the Earth's obliquity oscillations, and actually to prevent more substantial climate fluctuations than could be the case without its influence. This is a happy coincidence for human life on Earth, though cannot be expected to last indefinitely. Because of tidal dissipative effects, the radius of the Moon's orbit is gradually increasing, and will continue to do so until it reaches a spin–orbit resonance condition at a radius of ~66.5 Earth radii (Ward, 1982). At this point, the Moon will release its stabilizing grip on the Earth, allowing much more violent oscillations in obliquity (possibly up to 60°). But until this happens (in at least 1–2 Gyr time!), we can continue to enjoy the stability our Moon confers upon us.

9.2.3 Obliquity cycles and insolation

Although variations in orbital ellipticity and spin-axis precession cause some variations in the distribution of insolation during the year, it is the planetary obliquity which exerts the strongest influence on the annually averaged distribution of solar heating. Taking into account both the obliquity θ_o and eccentricity of the orbit, it can be shown (Ward, 1974) that the annually averaged insolation at latitude λ is given by:

$$I_a = \frac{S}{2\pi^2(1-e^2)^{1/2}} \int_0^{2\pi} [1 - (\sin\lambda\cos\theta_o - \cos\lambda\sin\theta_o\sin\xi)^2]^{1/2} d\xi \qquad (9.1)$$

where S is the solar constant at the mean orbital radius of Mars (590 W m^{-2}) and ξ is the solar hour angle measured from midnight (over which an average is taken). The resulting variation of insolation with latitude is illustrated in Figure 9.5, in

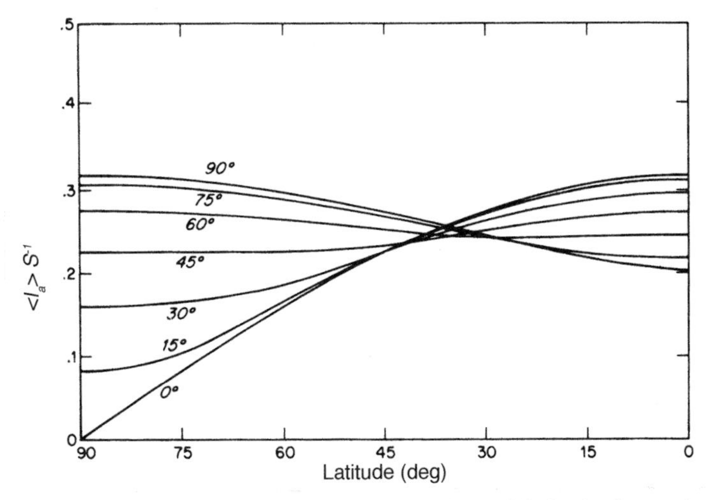

Figure 9.5. Averaged annual insolation I_a as a function of latitude for various values of obliquity. I_a is normalized to the average solar flux $S/(1-e^2)^{1/2}$ taking into account the eccentricity e.

which Eq. (9.1) is integrated numerically for a set of different obliquities spanning the likely range of interest. Major features to notice in this figure are (a) how much more the insolation at high latitudes depends on obliquity compared with low latitudes, with a minimum of influence around the subtropics, and (b) for θ_o greater than about 54° the poles receive more insolation averaged over the year than the equator. This is particularly important for planets such as Uranus (with obliquity 97°), for which solar heating would indeed peak at the poles with a minimum of solar heating on the equator itself. For Mars, however, the planet may approach (but is unlikely to exceed) the 'break-even' obliquity at times in its oscillatory cycle, which must have substantial implications for its atmospheric circulation and thermal state, especially in the polar regions.

At the very least, therefore, obliquity oscillations will lead to a major modulation of solar heating, especially in the polar regions (cf. Figure 9.4(c)), on a timescale of around 30–60 kyr or longer. Such modulations will be compounded by the effects of eccentricity oscillations and spin-axis precession, though the latter are likely to be significantly weaker overall than the obliquity variations, because the amplitude of the latter is so large.

9.2.4 Correlating PLTs with the obliquity cycle

The 'acid test' of the possibility that variations in the pattern of annual insolation, controlled by the obliquity cycle and other orbital variations, govern the deposition of dust and ice in the PLTs would be to determine the extent to which the properties of the layers themselves are correlated with the pattern of insolation variations. This has not been feasible until quite recently because the level of detail available on the observed structure of the individual layers has not been sufficient to correlate individual layers with particular orbital cycles in the past. In principle, however, it ought to be possible to treat the polar layers as a form of 'climate proxy', much as has been done very effectively for the Earth, using measurements of layer thicknesses and isotope variations in polar ice cores, tree rings, and lake sediments as proxy records of climate variability for correlating with the Milankovitch orbital and rotational variations.

Given the availability now of very high-resolution images from the Mars Orbiter Camera (MOC) instrument on Mars Global Surveyor (MGS), with resolutions of a few metres or less, it is finally becoming possible to carry out such studies. Laskar *et al.* (2002) used one such MOC image, together with information on the topography of the region from Mars Orbiter Laser Altimeter (MOLA) measurements, to obtain a profile of surface albedo with depth along a section perpendicular to a region of PLTs close to the Northern Polar cap. Figure 9.6(a) shows the region analysed by Laskar *et al.* (2002), in which they took a MOC narrow-angle image aligned perpendicular to the layers, and averaged the image brightness along the layers to maximize the signal-to-noise ratio. The result is shown in Figure 9.6(b) as a profile of image brightness with horizontal position across the layers.

The brightness profile was then transformed to represent the variation of brightness with depth, covering a range of around 350 m using MOLA measurements of

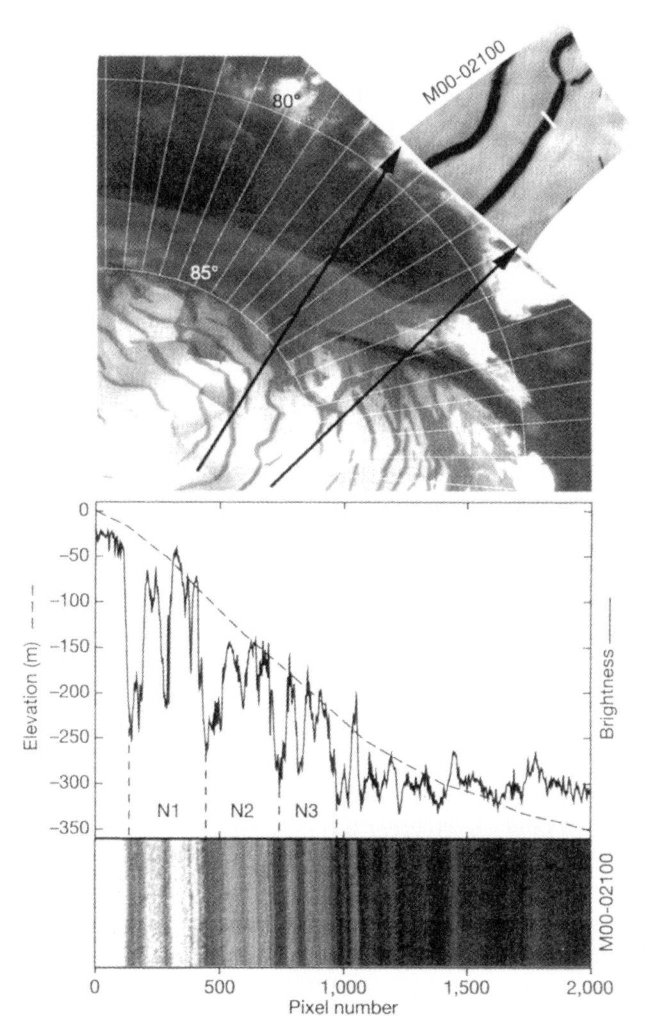

Figure 9.6. (a) Mars Orbiter Camera (MOC) context and narrow-angle images analysed by Laskar *et al.* (2002) to obtain a profile of brightness (shown in (b)) across a region of the northern polar layered terrains (PLTs) on Mars.

From *Nature*, with permission.

the local slope of the topography perpendicular to the layers. Laskar *et al.* (2002) then looked for correlations between their brightness versus depth profile and the variations in annual mean polar insolation during the first 1 Myr or so of Figure 9.4(c). The result is shown in Figure 9.7, in which the brightness profile is shown as a continuous line and the insolation variation is shown as a dashed line (the horizontal coordinate has been scaled in both cases to represent time before present).

Though far from perfect, the correlation is remarkably good over much of the 800 kyr shown, and is at least as clear a correlation as is typically found for most

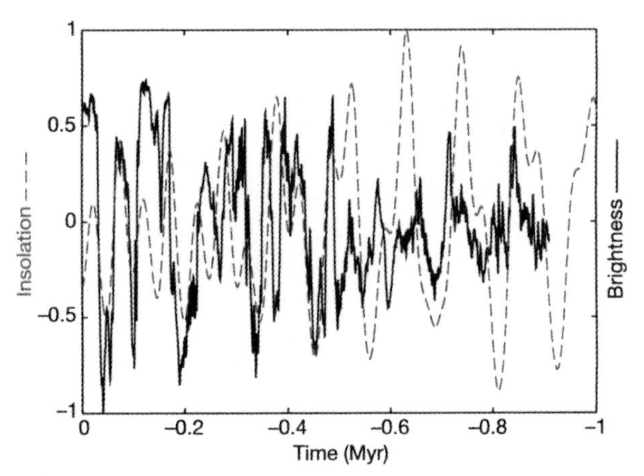

Figure 9.7. Profiles of image brightness across a region of the northern polar layered terrains (PLTs) on Mars (solid line) shown correlated with the variations of polar insolation during the past 1 Myr, as derived by Laskar *et al.* (2002).
From *Nature*, with permission.

climate proxies investigated in terrestrial palaeoclimate studies (e.g., Hays *et al.*, 1976). The scaling of the depth variation with time necessary to obtain this level of correlation suggested that the rate of deposition may have changed during this time, from around 0.5 mm yr^{-1} for the topmost 250 m of depth (representing around 500 kyr) to around 0.25 mm yr^{-1} for the remaining 100 m (or 400 kyr). This represents a relatively rapid rate of deposition, and if sustained continuously would lay down the entire 2.5 km of northern polar layered deposits in around 5 Myr. This is a remarkably rapid process, suggesting that the northern PLTs may be astonishingly recent compared with the age of the planet (and presumably the PLTs in the south). The reason why the northern layers are apparently so recent, having presumably been laid down following an earlier period when the polar cap must have been essentially eroded away completely, remains an intriguing mystery. For the future, however, this kind of 'climate proxy' study needs to be extended much more widely to other regions of Mars, to establish whether the chronology suggested from the study of Laskar *et al.* (2002) is consistent with a global deposition variation following a similar pattern.

9.3 MODELLING THE RESPONSE TO THE OBLIQUITY CYCLE

9.3.1 Qualitative overview

The previous sections have established that there is evidence from the PLTs surrounding both of Mars' polar caps that the climate state may have varied cyclically in the past on timescales of around 10^4–10^6 yr. This would seem to be consistent with the timescales expected from the computed cyclic variations of Mars' orbit and planetary rotation parameters, for which the obliquity is perhaps the most significant. The fact that the obliquity cycle primarily controls the annual mean insolation close to the poles may be especially significant. We have seen from earlier chapters that the CO_2 partial pressure in the atmosphere is in approximate equilibrium with the CO_2 ice caps, at least when averaged over the Martian year. This is consistent with the seasonal variations of CO_2 ice coverage at both poles being determined by a local energy balance between the rate of radiative heating or cooling of the polar regions and the latent heat exchanges associated with the condensation or sublimation of CO_2 from/into the atmosphere. When averaged over the year, therefore, a reduction in mean solar heating, for example, would be expected to lead to increased net cooling and hence a displacement of the atmosphere/surface equilibrium towards lower temperatures and atmospheric partial pressure (and a corresponding increase in the ice mass). Thus, as Mars changes towards a lower obliquity state than the present one, we would expect the poles to experience net cooling, and the mean atmospheric CO_2 content to decrease in favour of larger residual CO_2 ice caps, at least at the colder pole (depending upon the relative phases of the precession and eccentricity cycles).

On moving towards larger obliquities than at present, however, the polar regions would experience a net increase in solar heating, which would tend to evaporate reserves of CO_2 ice presently locked up at or below the surface close to either pole. Thus, in this case we might expect the mean atmospheric pressure on Mars to increase, perhaps substantially if there are large reserves of polar CO_2 ice which can be released on a timescale of 10^4 yr or so.

Even without detailed calculations, therefore, it is fairly straightforward to deduce that the obliquity cycle might lead to substantial modulations of the mean surface pressure and the extent of the CO_2 polar ice cover. This might also feed back onto the mean dustiness of the atmosphere (as would seem to be necessary to account for the formation of the PLTs) since, at significantly lower pressures, the atmospheric bulk stress at the ground produced by winds would be substantially reduced, making it much harder for organized wind systems to lift large amounts of dust into the atmosphere (cf. Chapter 7). Thus, at low obliquity we might expect large-scale dust storms to cease, leading to relatively cold, clear conditions and relatively little aeolian transport of dust. At high obliquity, however, the atmospheric stress would be substantially enhanced from present conditions, perhaps leading to *more* widespread dust storms than today and much greater aeolian transport of dust to the poles. This would presumably be reflected in the deposition of sediments in the polar regions, not only in dust but also in the net transport of water ice and vapour between the poles and lower latitudes, and from one pole to the other. Thus, we have the embryo of a potential explanation for the formation of PLTs, though this has to be made more quantitative if it is to become a credible theory.

9.3.2 Quantitative models: energy balance models

Until comparatively recently, the main way to quantify Mars' response to changes in radiative heating and cooling has been to develop more elaborate versions of the simple 1-D model we used to illustrate the seasonal condensation of CO_2 at the two polar caps in Chapter 3. The overall approach typically used follows the early work by Ward (1974) (see also the review by Kieffer and Zent, 1992). These studies typically make the following assumptions in setting up their model. The total atmospheric CO_2, which determines the annual mean surface pressure, is assumed to be in equilibrium with the colder of the two polar CO_2 ice caps. The radiation budget is calculated for an atmospheric column at various latitudes (though primarily at or close to the poles), for which atmospheric opacity is generally neglected to simplify the problem. Solar insolation is computed in the annual mean, as outlined above (see Eq. 9.1), taking into account both obliquity and orbital eccentricity. Atmospheric advective heat transport is either neglected altogether, or crudely parameterized by a simple constant term V. This is broadly consistent with recent GCM computations (Pollack *et al.*, 1990), which suggest that atmospheric heat transport is relatively small compared with direct radiative processes at the present time (at the level of a few per cent at most).

The surface and subsurface CO_2 reservoirs are also assumed to be essentially unlimited. This is an idea which goes back to the suggestion of Leighton and Murray (1966), though is now regarded as somewhat controversial. Indeed some recent evidence from sequences of high-resolution images of CO_2 ice features at the South Pole (Byrne and Ingersoll, 2003) may even suggest the opposite (i.e., that there might be rather little additional CO_2 ice available at the poles to be released), though the presence of more substantial reserves underneath the surface cannot be ruled out. In any event, if the reserves of CO_2 ice are not unlimited, then some of the conclusions from models of the type outlined by Ward (1974) may need to be modified, especially under conditions of high obliquity.

Thus, taking the current obliquity for Mars ($\theta_o = 25.2^\circ$), and assuming the polar cap albedo $A_c = 0.65$, advective heating $V = 0$, the polar frost emissivity in the infrared $\epsilon = 1.0$, and $e = 0$ leads to a predicted annual mean atmospheric pressure at the surface of $p_{s0} = 7.1$ hPa. This is perhaps a little on the high side for present conditions, though within the bounds of plausibility. By changing each of these parameters within their respective uncertainties, we can gain some insight into the relative sensitivity of the climate state to each parameter in turn. Thus, if we change ϵ to 0.95, to represent the additional 'greenhouse' opacity due to the atmospheric 15-μm absorption band, atmospheric cooling is reduced and p_{s0} increases by around 30%. Perhaps more remarkable is that, by changing the ice albedo A_c to 0.7 from 0.65 produces substantial cooling and reduces p_{s0} by a factor of nearly 2. Despite the relatively small contribution to heat fluxes by atmospheric advective transport, setting V to a plausible value of $\sim 3\,\mathrm{W\,m^{-2}}$ increases p_{s0} by around 70%, while setting e to 0.1 from 0 leads to an increase in p_{s0} of just 3%.

The extreme sensitivity of p_{s0} to A_c is a particular concern, since the factors determining this on Mars itself are not well understood. Measurements indicate that its present value lies somewhere in-between that expected for pure ice and bright dust particles (Thomas *et al.*, 1992), though some evidence suggests that it may vary with time of year and with other factors. Indeed Paige (1985) proposed from an analysis of Viking Orbiter observations that A_c might vary systematically with seasonal insolation itself, roughly as:

$$A_c \simeq 0.52 + 9.2 \times 10^{-4} I \tag{9.2}$$

where I is the local insolation in $W m^{-2}$. If we include this formulation for A_c, and set $e = 0.093$, $V = 3 W m^{-2}$, $\epsilon = 0.95$, and $L_S = 251°$ at perihelion, the new computed equilibrium leads (Kieffer and Zent, 1992) to a new predicted surface pressure $p_{s0} = 5.0$ hPa, which is still within the bounds of plausibility for present conditions. However, the formulation for A_c in Eq. (9.2) cannot span the entire range of θ_o encountered on Mars, because it will eventually lead to $A_c > 1$ at very high obliquities, which is unphysical. This is hardly surprising, since this formula was derived by a heuristic and empirical approach, and is therefore unlikely to represent the true behaviour of the system outside the range of conditions under which it was measured.

The energy balance approach to modelling Mars' reponse to obliquity oscillations and other variable parameters external to Mars thus reveals a range of sensitivities and dependencies on poorly-constrained variables which leave an uncomfortably wide margin for error in settling some of the major issues. It is clearly desirable from all sorts of perspectives to reduce these uncertainties, both by refining measurements in the polar regions of Mars and increasing the sophistication and physical consistency of the predictive models.

9.3.3 Quantitative models: GCMs

As in other areas of Mars' climate research, the ultimate tool in modelling the inner workings of the climate system is the full numerical circulation model or GCM. At the present time, however, the application of this kind of tool to the study of climate variations substantially removed from the present conditions on Mars is still in its infancy. The energy balance models, for example, indicate an important role for a variety of complex feedbacks (e.g., between ice albedo, polar energy balance, the deep regolith, and atmospheric mass and composition). The details of these processes are highly uncertain at present, and it is still premature to contemplate including them in more detailed (and much more expensive and complex) 3-D models such as GCMs. Even so, several studies have been undertaken so far by some of the main GCM modelling groups, focussing on more limited problems related to variations in Mars' obliquity.

Fenton and Richardson (2001), for example, used the Princeton Geophysical Fluid Dynamics Laboratory (GFDL) Mars GCM to investigate the effect of varying obliquity on the prevailing wind direction at the MPF and other lander sites. This study was aimed at evaluating the possibility that changes in the mean circulation due to obliquity variations might account for systematic changes in wind direction at the MPF site, as inferred from ventifacts and wind streaks in that region. The orbital parameters in the model were altered so as to compute the daily and seasonal insolation corresponding to conditions applicable at other epochs when the obliquity and other parameters differed from those of the present day. Simulations for a few Mars years were carried out, though the experiments were not entirely realistic to these earlier epochs. This is because, in order to constrain the number of parameters to be varied in a controlled way, it was necessary to keep the CO_2 mass in the atmosphere and seasonal caps the same as at present. In addition, the other surface parameters (topography, thermal inertia, etc.) were not changed (in the absence of evidence that they should be different from today) and a geographical pattern of dust injection from the surface was prescribed which was not allowed to vary with obliquity. Given these limitations, however, the simulations showed that the intensity of the seasonal atmospheric circulation is likely to increase with increasing obliquity, but that orbital changes do not seem to lead to the changes in the prevailing wind direction needed to explain the various indicators at the MPF site.

Both the NASA Ames and LMD/Oxford Mars GCMs have also been used in a similar mode recently, to simulate various aspects of the atmospheric circulation at other obliquities (Haberle *et al.*, 2003; Newman *et al.*, 2003), including the extent to which patterns of dust lifting and aeolian erosion might vary. These experiments have so far mostly kept the total CO_2 inventory in the atmosphere and polar caps and other surface properties unchanged from the present, and simply varied the orbital parameters, though some recent Oxford MGCM runs have allowed p_s to vary in response to reducing θ_o. Despite these limitations, some fairly consistent and plausible trends in the structure of the circulation with obliquity are clear.

The extent of the changes varies considerably with season, with the equinox circulation being relatively insensitive to varying obliquity. At the solstices, however, the changes in the circulation are considerably more profound. Figure 9.8 (see colour section) shows a series of zonal mean sections of zonal wind and temperature at northern winter solstice for a range of obliquities from 15° to 45°. When the obliquity is significantly lower than at present, even the solstice temperature structure still looks reasonably symmetrical about the equator, apart from a shift in the warmest surface temperature to the subsolar latitude. The corresponding zonal wind remains westerly in both the winter and summer hemispheres (somewhat like the Earth's troposphere). At higher obliquities than at present, however, the summer pole becomes the warmest region at this season, and the polar temperatures over the summer pole get steadily warmer as obliquity increases. This increases overall the thermal contrast with latitude, leading to more intense zonal winds in the middle atmosphere.

These trends in the seasonal variations of surface temperature are even more apparent in Figure 9.9 (see colour section), which shows maps of mean surface temperature as a function of latitude and L_S for several different obliquities. From this figure the shift in both the location and intensity of the maximum surface temperature towards the summer pole, and a tendency for higher values at high obliquity are clearly seen. Note also that for obliquities only slightly greater than at present, the seasonal CO_2 ice caps disappear during summer in both hemispheres but extend a long way equatorward during winter. For an obliquity of 45°, the CO_2 ice virtually reaches the equator from the winter pole. For low obliquities, however, the CO_2 ice caps over either pole become stable, and hardly vary at all with season.

In apparent contrast to some of the early energy balance model studies, the global and annual mean surface temperature and pressure do not vary greatly with obliquity. This may be in part due to the design of the model, in which the boundary conditions were artificially fixed at their present values, and there was an inability of the model regolith to adsorb or release stored CO_2. But it may also reflect some subtle 3-D effects in the structure of the model response to changing obliquity which effectively compensate for some trends found in simple, globally-averaged energy-balance models. Figure 9.10 (see colour section) shows the variation in these parameters from simulations using the Oxford Mars GCM, which show remarkably little change with obliquity in the seasonal variation of surface temperature, except perhaps around northern autumn ($L_S \sim 180°–270°$). The seasonal maximum in mean surface pressure (Figure 9.10(b)) also shows little change with obliquity, though the minimum changes substantially. In particular, the minimum in pressure around northern summer (and to a lesser extent in northern winter) shows a strong sensitivity to obliquity, associated with the formation of increasingly large CO_2 ice deposits during the winter in each hemisphere. Over longer periods of time, however, the mean surface pressure is found to equilibrate to significantly smaller values as θ_o is reduced, much as found in the 1-D models.

The 3-D response of the circulation to varying obliquity is particularly striking in the variation of surface winds and corresponding ablation of surface dust. As the obliquity increases, the strength of the cross-equatorial Hadley circulation increases dramatically at the solstices in the model simulations. This leads to a significant intensification of the western boundary current (WBC) flows crossing the equator with increasing obliquity. Insofar as dust lifting and ablation from the surface are directly related to surface stress, this would indicate these WBC regions as likely sites for strong erosion, while net deposition might be expected at longitudes outside the WBC. From these considerations, Haberle *et al.* (2003) defined a statistic known as 'deflation potential', which measures the rate of erosion and mass ablation due to dust lifting, based on similar near surface windstress lifting parameterizations to those discussed in Chapter 7.

Haberle *et al.* (2003) and Newman *et al.* (2003) noted that this pattern of net 'deflation' (see Figure 9.11, colour section) correlates strongly with the observed pattern of variation of thermal inertia at the present time (see Figure 4.4, colour section). Given that we may interpret thermal inertia (as in Chapter 4) as measuring the extent to which the surface soil is highly porous (and dust-rich) or compacted (and dust-poor), this would seem to suggest that the observed distribution of thermal inertia directly reflects the topographically steered pattern of near-surface winds, averaged over the range of obliquity variations exhibited by Mars. At low obliquity, Haberle *et al.* (2003) and Newman *et al.* (2003) also note that dust-lifting potential is confined largely to the stormy baroclinic regions close to the polar cap edge, while at high obliquity, regions of intense dust lifting shift systematically to the WBCs at low latitudes. Whether this would lead to systematically increased atmospheric dustiness at high obliquity, as suggested from the energy balance studies, remains to be confirmed in more realistic experiments (though preliminary work suggests that it does). For these purposes, systematic changes in mean surface pressure associated with slow exchanges with the regolith would need to be taken into account more accurately, as would changes in the role of the hydrological cycle on Mars, since increased deposition of water ice at the surface, and ice cloud scavenging of dust particles, could well have important roles to play in relation to the net dust level maintained in the atmosphere.

In the context of the water cycle, Richardson and Wilson (2002a) and Mischna *et al.* (2003) have extended the earlier study of Fenton and Richardson (2001) to consider the effects of varying obliquity on the hydrological cycle included in the GFDL model. Again not all of the significant parameters could be varied realistically for past epochs on Mars, and again the surface pressure and/or CO_2 mass were kept at present values. Increasing the obliquity in the simulations up to 45° was found to lead to sustained evaporation of the residual water ice cap in the north and a consequent large increase in the amount of water resident in the atmosphere, with column amounts rising to as much as 1,000 pr μm or more over the residual caps in spring. This suggests that water frosts on the ground could become stable over wide areas on Mars at times of high obliquity, which might then be capable of bonding the surface sediments and preventing dust lifting. Moreover, a large increase in atmospheric humidity might also be expected to lead to increased cloudiness of the atmosphere. This could in turn also lead to an increased tendency for scavenging of suspended dust in the atmosphere by ice accretion and sedimentation (Michelangeli *et al.*, 1993; Clancy *et al.*, 1996), thus leading to a less dusty atmosphere (but with more water clouds?). GCMs are not yet in a position to confirm these speculations, but this is an interesting area for future investigation.

9.4 SCENARIOS FOR MARS' RESPONSE TO OBLIQUITY VARIATIONS

In the previous sections, we have seen something of the qualitative range of effects which obliquity variations might impose on Mars, although it also clear that many uncertainties remain at the present time concerning the quantitative impact these effects may have on the Martian climate system. Even so, these kinds of study have led to at least a broad concensus on at least some of the principal effects of varying the obliquity from its present value. These are illustrated schematically in Figure 9.12 and discussed below.

9.4.1 Low obliquity conditions

The response to reduced obliquity from the present conditions is perhaps more reliably constrained by the existing range of models and observations than for increased obliquity states. Starting, for example, from the present obliquity and decreasing θ_o, insolation at the polar caps will decrease significantly, allowing both poles to cool. This allows the seasonal CO_2 frosts to accumulate year by year, steadily depleting the atmosphere of mass and reducing the mean surface pressure. At values of θ_o somewhat less than the present conditions, permanent CO_2 ice caps

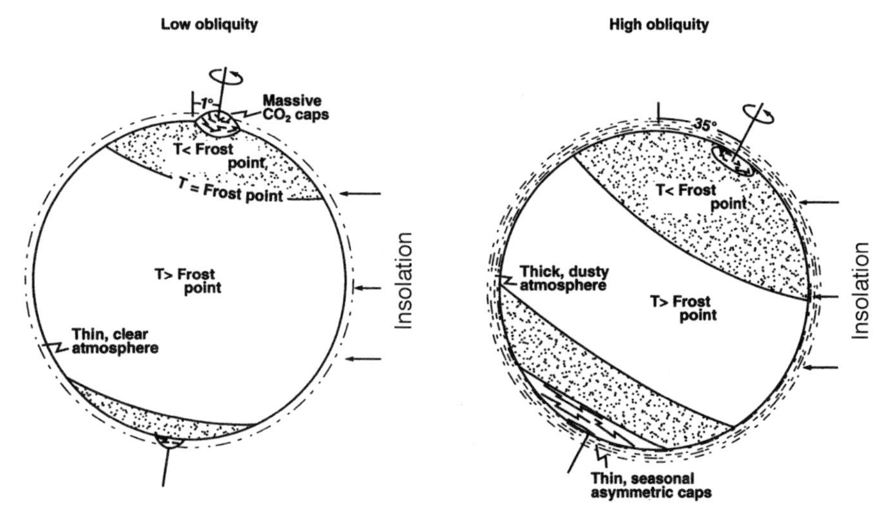

Figure 9.12. Schematic cartoons illustrating the main features of the two extreme climate states represented by conditions of (a) low and (b) high obliquity on Mars.
From Kieffer and Zent (1992).

will form at both poles, leading to substantial cold traps for water vapour. As a result, the atmosphere dessicates even further than at present, although the latitude range over which water in the regolith is unstable is likely to extend polewards somewhat from the present state. The net result is to incorporate more water ice into the permanent ice caps. The minimum sustained surface pressure reached during the cycle is somewhat sensitive to the minimum value of θ_o attained during the obliquity cycle, but for typical values believed to have been encountered during the past 10 Myr (around $\theta_o \sim 13°$) this could be as low as 0.1 hPa (e.g., Fanale and Jakosky 1982), strongly depending upon how the albedo of the polar ice evolves during the development of this phase of the obliquity cycle.

Long before this minimum in p_{s0} is reached, however, surface winds are likely to become insufficent to lift dust from the surface into the atmosphere. With a clearer atmosphere, the material systematically deposited onto the polar caps becomes almost pure ice, which has a significantly higher albedo than the present composition of the polar caps on Mars. This will further accelerate the cooling effects at the poles, leading to a rapid equilibration towards an extreme low surface pressure state characteristic of low obliquity conditions, which is therefore likely to be relatively clear of dust, with a much lower surface pressure than at present (see Figure 9.12(a)). Under these conditions, the supression of dust deposition provides at least one way in which the properties of the polar surface may be altered to contribute to the formation of the PLTs.

9.4.2 High obliquity conditions

As suggested from Figure 9.12(b), the high obliquity state of Mars is expected to be in many respects the antithesis of that at lower values of θ_o than at present. The

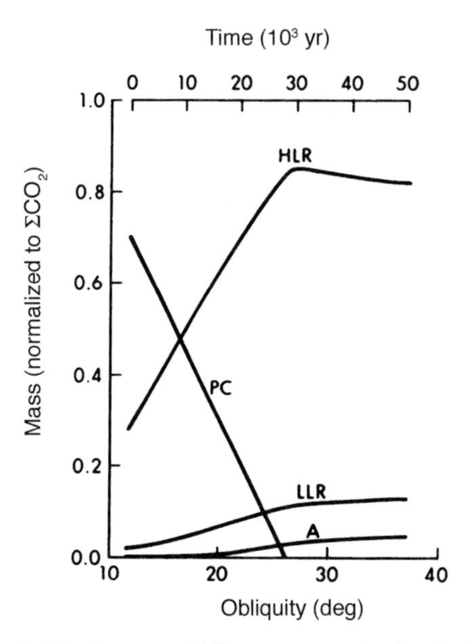

Figure 9.13. Exchange of CO_2 between different reservoirs in the Mars climate system as a function of obliquity, from the models of Fanale *et al.* (1982). 'PC' denotes the polar cap reservoir, 'A' denotes the atmosphere, 'LLR' is the low-latitude regolith and 'HLR' the high-latitude regolith, respectively.

details of the other extreme state are rather more uncertain, however, for reasons that will become clear in the following discussion.

Starting again from present conditions, as θ_o increases from the present state, changes to the climate are likely to be quite modest at first. As indicated in Figure 9.13 from the models of Fanale and Jakosky (1982), the permanent CO_2 ice cap in the south disappears around $\theta_o \sim 26°$. This is associated with a gradual increase in p_{s0}, although changes are relatively modest by this early stage in the cycle.

As p_{s0} increases, it is likely to become increasingly easy for low-altitude winds to raise dust into the atmosphere, so that dust storms may become more widespread and the atmosphere becomes more dusty, at least in the early stages of the cycle. All other things being equal, this would be expected to lead to an enhancement in the accumulation of dusty material in the seasonal polar caps, which would probably reduce the typical albedo of the polar caps over time. A decrease in A_c would be likely to accelerate the warming of the polar regions as solar insolation at high latitudes increases, allowing the steady increase in p_{s0} also to accelerate. This positive feedback might then lead to the atmosphere remaining relatively dusty for most of the Martian year, though the details here are somewhat uncertain since this may be strongly affected by the phase of the precessional and eccentricity oscillations, which control the amplitude and timing of perihelion during the year. The potential effects of changes in atmospheric humidity might also be very important (see below).

The increasing surface temperatures at high latitudes would allow both CO_2 and water ice systematically to desorb into the atmosphere from the polar caps. Somewhat counter to this trend, however, as p_{s0} increases the adsorption of CO_2 into the regolith also increases, leading to a systematic transfer of CO_2 ice from the polar ice caps into the high-latitude regolith (cf Figure 9.13). Adsorption also increases into the low-latitude regolith too, though at a somewhat slower rate. As the polar temperatures increase, however, eventually the CO_2 polar regolith starts to desorb some of its accumulated CO_2 ice back into the atmosphere, though at low latitudes the regolith continues to adsorb CO_2 monotonically until the maximum of θ_o is reached.

The maximum surface pressure attained during the obliquity cycle depends rather sensitively on the total amount of CO_2 ice, etc., which may be accessible to be released in the polar caps and (especially) the regolith on a timescale of 10^5 yr. The timescale is significant because this determines the depth to which the thermal 'wave' associated with diffusion of heat into the regolith will penetrate, and depends upon the thermal properties of the regolith itself. If we take the thermal conductivity K and density ρ of the soil to be constant with depth, then thermal conduction into the ground will satisfy the time-dependent diffusion equation:

$$\frac{\partial T}{\partial t} = \kappa \nabla^2 T \tag{9.3}$$

$$\simeq \kappa \frac{\partial^2 T}{\partial z^2} \tag{9.4}$$

where $\kappa = K/\rho c$ and c is the specific thermal capacity. Taking the surface temperature to vary with time as $T = T_{s0} + T_0 \exp i\omega t$, and taking $\tau_0 = 2\pi/\omega$ as the period of the oscillation, the solution to Eq. (9.4) takes the form:

$$T(z,t) = T_{s0} + T_0 \exp(z/s) \cos(k_S z - \omega t) \tag{9.5}$$

where $k_S = (\omega/2\kappa)^{1/2}$; $s = (\tau_0 \kappa/\pi)^{1/2}$ is the e-folding length scale (or 'skin depth') for the decrease in amplitude of the thermal wave with depth into the regolith; and T_{s0} and T_0 are constants. If we take a value for κ of around $4.75 \times 10^{-7}\,\mathrm{m^2\,s^{-1}} = 15\,\mathrm{m^2\,yr^{-1}}$, typical of frozen soils appropriate for polar regions on Mars, then the skin depth for an oscillation of period around 10^5 yr is somewhat less than 1 km. This may be compared with $s \simeq 3$ m for $\tau_0 = 1$ Mars year and $s \simeq 11$ cm for the diurnal period (cf Chapter 8). Thus, thermal variations associated with the cyclically varying obliquity on Mars will penetrate to a depth of 1–2 km, with the largest amplitude within 500–1,000 m of the surface, as illustrated in Figure 9.14 at a range of latitudes and obliquities. Hence, only the CO_2 adsorbed into the top several hundred metres of the regolith will be able to be released during the obliquity cycle.

The amount of CO_2 which can be stored in the rocky matrix of the regolith is strongly dependent on the mineralogy and grain size of the material, much as for water (discussed in the previous chapter), as well as on the temperature and ambient vapour pressure. But this is not known with any certainty for Mars, as is also the

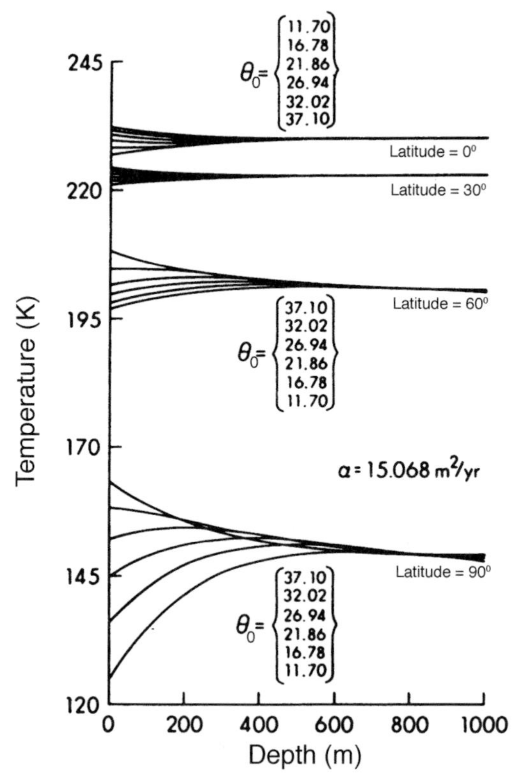

Figure 9.14. Temperature profiles in the Martian regolith as a function of depth at four representative latitudes during periods of increasing obliquity θ_o, from the models of Fanale *et al.* (1982).

case for the total amount of potentially exchangeable CO_2 available to be stored in the regolith. The absorptive capacity of a given soil tends to increase as the temperature decreases, so as the soil warms during the high-obliquity phase of the cycle (following the phase lag associated with the propagation of the thermal wave into the ground) CO_2 is released into the atmosphere. This is taken into account in energy balance models, such as those of Fanale and Jakosky (1982), though the amount of CO_2 available to be released is poorly constrained by observations. Table 9.1 summarizes some recent estimates of the capacities of the main reservoirs of CO_2 and water on Mars which are able to exchange mass on timescales comparable to (or smaller than) the obliquity cycle.

The amount of CO_2 stored in the regolith, however, is particularly uncertain. The amount quoted here is a moderate estimate, based on 'reasonable' assumptions concerning the adsorptive capacity of the soil at high latitudes (Carr, 1996), though it ignores the possibility of interstitial ice stored in voids within the soil and other potential forms of storage. Such an estimate places an upper limit on p_{s0} during the obliquity cycle of a few tens of hPa of CO_2. At the low extreme, assuming that

Table 9.1. Estimates of exchangeable volatile reservoirs.

	CO_2		H_2O	
	$kg\,m^{-2}$	hPa	$kg\,m^{-2}$	m
Atmosphere	150	5.6	0.01	10^{-5}
Seasonal caps	40	1.5	0.01	$>10^{-5}$
Perennial cap (N)	5×10^3	200	10^4	10
Perennial cap (S)	8	0.3	600	0.6
Polar layered deposits	23	850	$6,200–2.9 \times 10^4$	6.2–29
Regolith	1,000(?)	37(?)	>10(?)	$>10^{-2}$(?)

Estimates from Kieffer and Zent (1992), Carr (1996) and Jakosky *et al.* (1995).

the only CO_2 available for release is that stored in the vicinity of the southern permanent ice cap would increase p_{s0} by only a few hPa from its present value (perhaps leading to a maximum p_{s0} of 9–10 hPa). At the more generous extreme, an assumption of a more adsorptive mineralogy (such as various forms of clay) could enhance the amount of CO_2 stored in the regolith. In addition, CO_2 could conceivably be laid down in the form of a CO_2–H_2O clathrate, in which a molecular complex of both molecules (in the ratio of 5.75 : 1 in favour of H_2O) forms a crystalline form of ice under the ground. This form of CO_2 is only stable at pressures greater than around 100 hPa, so could only reside at depths in excess of around 10 m under the soil. Jakosky *et al.* (1995) have estimated that this form of storage could account for enough CO_2 to raise p_{s0} to 200 hPa during the peak of the high-obliquity state. If further clathrate reserves were locked up in deeper polar deposits underground, then a p_{s0} of up to 850 hPa might even be possible (Jakosky *et al.*, 1995). Such an increase could have a major impact on the climatic state and circulation of the atmosphere, though by itself would still not produce a sufficient greenhouse warming to raise the mean atmospheric temperature by around 10–30 K. Thus, the surface temperature would still be too cold to melt water ice into the liquid state (and hence, perhaps, account for the recent gullies observed in MOC images (Malin and Edgett 2000), although see below). Even so, this particular suggestion is quite controversial, and no evidence has been found so far of the possible presence of clathrate reserves of CO_2 and water in any quantity on Mars.

Various additional complications might also affect the maximum value of p_{s0} attained, such as possible changes in the albedo of the ice caps, dust storms, and changes corresponding to changes in the atmospheric circulation and transport. Somewhat counterintuitively, the effect of increased obliquity on the water cycle is likely to result in evaporation of the polar water ice caps, a substantial increase in the humidity of the atmosphere, leading to increased formation of water clouds, and the possibility of precipitation of water ice onto the surface at lower latitudes. Figure 9.15, for example, shows the result of a recent calculation based on an energy balance model by Mellon and Jakosky (1995), which shows an estimate of the dependence of atmospheric water column abundance as a function of obliquity θ_o.

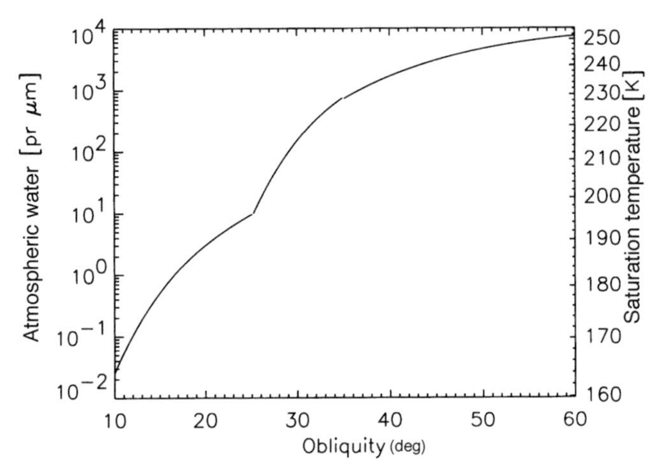

Figure 9.15. Average atmospheric water column abundance and surface saturation temperature as a function of obliquity θ_o, as computed in the energy balance model of Mellon and Jakosky (1995).

By the peak of the typical obliquity cycle around $\theta_o \sim 40°$, water abundances are expected to reach several precipitable mm of water, several hundred times the humidity of the present atmosphere. The changing distribution of insolation is also expected to make water ice stable throughout most of the year down to quite low latitudes. With the general increase in atmospheric humidity, even soil quite close to the surface can become rich in water ice through both direct precipitation of ice and adsorption into the porous regolith. If water ice were to accumulate across the surface from high to low latitudes, this might well inhibit the release of CO_2 into the atmosphere, and limit the maximum value of p_{s0} attained during the obliquity cycle. As indicated above, the present concensus suggests a value around 15–30 hPa, but this expectation could change dramatically in the light of more evidence from new observations.

As indicated above, the general increase in atmospheric humidity might also have a strong influence on the dust content of the atmosphere, associated with increased scavenging rates in developing water clouds. This is a feedback which is not well quantified by any model at present, but could be a critical factor affecting the dust content of ice deposited at the poles during polar layer formation.

9.4.3 Obliquity cycles and the 'gullies'?

The limited penetration of the thermal signature associated with the amplitude and timescale of Mars' obliquity oscillations has also been suggested recently (Mellon and Phillips, 2001) to play a possible role in the formation of the gully features discovered by Malin and Edgett (2000) (see Figure 9.16 for some further examples from MOC images). As was mentioned briefly in the previous chapter, these are small-scale features, commonly found on poleward-facing slopes cut into cliffs or the

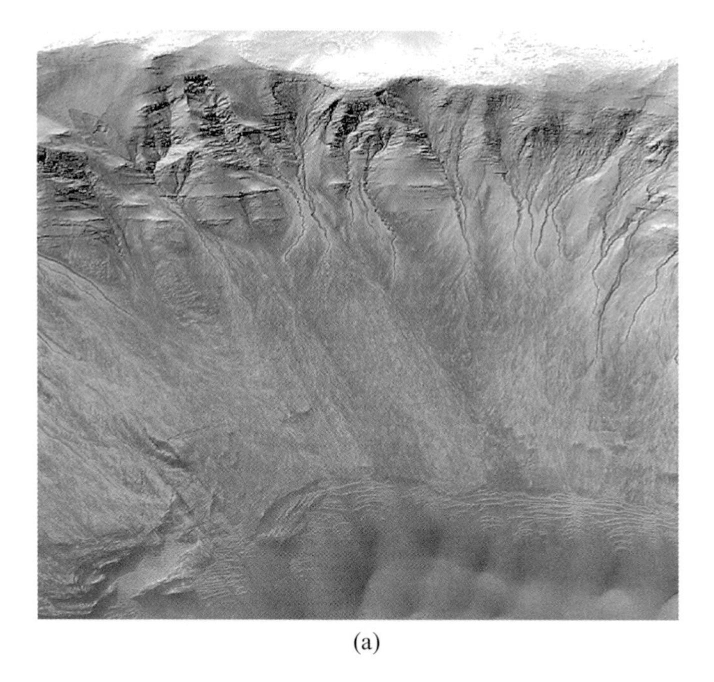

(a)

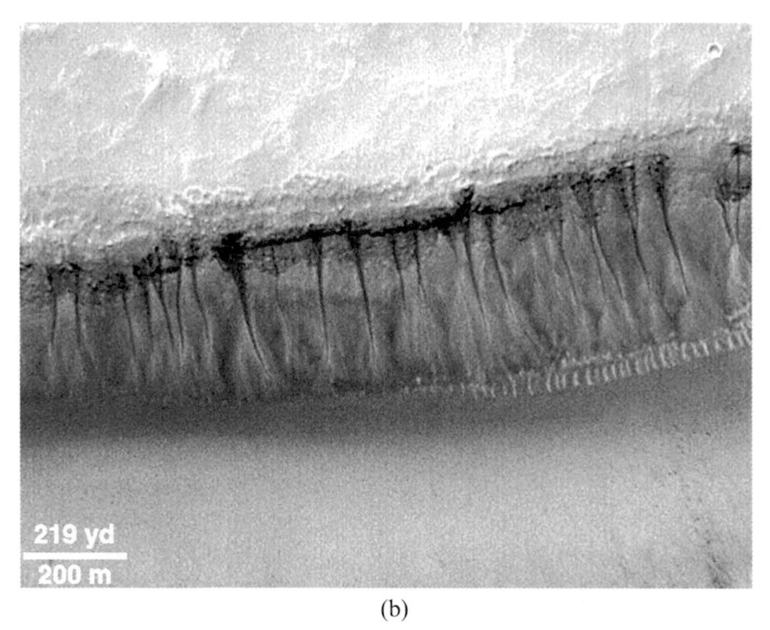

(b)

Figure 9.16. Two Mars Orbiter Camera (MOC) images of recently-formed gully features close to poleward-facing slopes on Mars (a) in the walls of a small crater and (b) 'weeping gully' features cut into the sides of a canyon.

From NASA/JPL and Malin Space Science Systems.

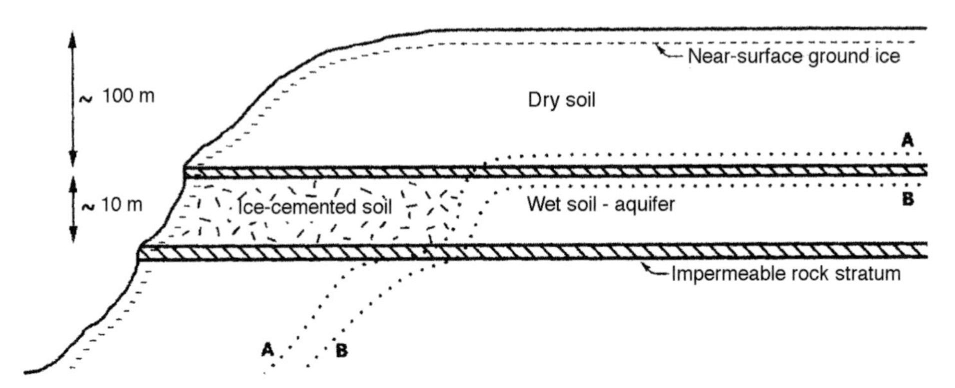

Figure 9.17. Conceptual model of the formation of gullies on Mars by the trapping and freezing of groundwater, followed by later release during the warm phase of the obliquity cycle, as suggested by Mellon and Phillips (2001). During this time, the trapped water in the wet layer between two impermeable rock strata may be warmed causing pressure to build up until the ice-cemented region close to a sloping surface fractures and allows the water to escape and flow down the slope.

sides of small craters, which appear to have been eroded by flowing water emerging from sites a few hundred metres below the top of the slope. Malin and Edgett (2000) estimate that around 2500 m^3 of water was necessary to carve the observed features. Moreover, the features are believed to be relatively young in age, perhaps having been formed sometime between 20 yr and 1 Myr ago (although the uncertainty is large).

Mellon and Phillips (2001) note that many of these features appear to form on poleward-facing slopes, away from direct sunlight during summer. This is precisely the configuration one might expect for water released from layers warmed by a thermal 'wave' propagating downwards into the regolith from above. Given the typical location of the head of the gullies, around 100–200 m from the top of the slope, the penetration depth of the thermal wave is suggestive of the 'skin depth' associated with the oscillation driven by the 10^5 yr obliquity cycle. Mellon and Phillips (2001) suggest that such gullies could form (as illustrated schematically in Figure 9.17) where a layer of water-rich soil is confined between two impermeable rock layers, all of which outcrop at a slope (e.g., at the edge of a crater or canyon). As the obliquity increases, the warm phase of the thermal wave will propagate into the regolith, warming the wet layer and allowing pressure to build up. At some point, the pressure is envisaged to reach a point where the ice-cemented soil close to the sloping surface fractures, allowing an outpouring of water roughly 100 m from the top of the slope. The water is likely to be in liquid form, because of the over-pressure due to the weight of soil above the trapped layer (which takes it well above the triple point). The result could be sporadic formation of eroded gullies during maxima of the obliquity cycle, on timescales of a few $\times 10^5$ yr.

This is by no means the only explanation for these gully features, however, and alternative explanations suggested to date include run-off either from direct melting of near-surface ice (perhaps enriched in salts to stabilize the liquid state) or from snow deposited during periods of high obliquity (Christensen, 2003).

9.5 MARS' VARIABLE CLIMATE

We have seen in this chapter that Mars' climate system is almost certainly still in a highly dynamic state, in response to substantial astronomically-forced variations in the external conditions affecting the distribution of solar insolation. The amplitude of such variations in association with the obliquity cycle appear to be very large and chaotic in character, although some cyclic timescales are identifiable. This almost certainly exerts a strong influence on the structure of the climate and circulation of the planet, changing its character substantially from those prevailing at present. At the present epoch, Mars appears to be roughly midway between its extreme states of high and low obliquity, so that its present obliquity coincidentally matches that of the Earth. At other times, however, Mars' atmospheric circulation is expected to be substantially different, ranging from either very clear and cold (at low obliquity) or much more dynamically active with enhanced dustiness and humidity (at high obliquity).

Our understanding of these variations is still very much in its infancy, but scientists are already bringing to bear the tools of full GCMs to examine possible states of extreme climates during the obliquity cycle. The details of these extreme climate states remain highly uncertain, however, not least because of the extreme sensitivity of the climate state and the volatile budgets to such factors as the albedo of the polar regions which are difficult and complex to model. The size and availability of the subsurface reservoirs of water and CO_2 ice are also a major source of uncertainty, which makes it difficult in particular to establish the full range of possible surface pressures during the extremes of the obliquity cycle. Further progress here will be highly dependent on obtaining new measurements of such reservoirs beneath the ground, to depths of at least 1–2 kilometres. The GRS instrument on Mars Odyssey has already made a valuable contribution to this problem in mapping subsurface water to depths of 1–2 metres, but much more remains to be done. At the time of writing, it is hoped that the availability of a ground-penetrating radar system on the Mars Express orbiter will be able to return information from much greater depths.

Even though erosion and deposition rates are presently very slow by terrestrial standards, the large amplitude of climate variability and its persistence over long periods of time almost certainly leaves a clear signature on the surface of the planet, in the form of various types of landform peculiar to Mars. This is one area of Mars research where close co-operation between atmospheric scientists, geologists, geochemists, and geomorphologists is going to be essential to achieve further progress.

10

Future climates: the human factor?

In this chapter we discuss the potential possibilities and pitfalls of human intervention in deliberately engineering future climate change on Mars. We conclude with a consideration of the ethics and morality of such an undertaking, were it ever to become feasible.

10.1 HUMAN EXPLORATION OF MARS IN THE SHORT TERM

Of the many scientific and technological goals of the current programme to explore Mars, one of the most ambitious is that of landing human explorers on the surface of the planet, with the aim (a) of studying the planet in more detail than may be possible with robotic spacecraft, and (b) ultimately of setting up a permanent presence there. While this may seem like science fantasy at the present time, many of the technological obstacles to transporting groups of people to the Martian surface from Earth, and sustaining at least small communities of scientists for lengthy periods on Mars, are already being overcome, at least in principle. The rate of progress towards this first, limited goal is determined as much by political and financial factors (the extent to which governments such as those of the USA, Europe, Russia, Japan, and China actually *want* to achieve it, against other competing priorities) as by the need to solve overwhelming technical issues.

The main technical problems are those associated with the transport of a group of explorers (most likely a team of between 5 and 10 professional astronauts and scientists in the first instance) from Earth to Mars and their subsequent safe return. While the payloads required to accommodate a human team to Mars would be significantly larger than that of any spacecraft which has so far been sent to Mars, even current generations of launch vehicles have sufficient capacity to enable payloads of several tonnes to reach Mars. The main difficulty is set by the celestial mechanics in that, in order to keep the energetic requirements of such a transfer to within reasonable limits, the journey can only sensibly be carried out when the Earth and Mars are suitably positioned relative to each other. For transfers in either direction (to or from Mars), this essentially means that launch opportunities only occur for a few weeks once every two Earth years or so. Thus, even though the transfer in either direction can be achieved with a travel time of 'only' around 6 months using current technology, a team would have to be prepared to camp out on Mars for more than a year before the next launch opportunity suitable for a return to Earth would occur.

This is because the route from one planet to the other can only be achieved economically if the spacecraft follows an approximation to the Hohmann transfer orbit, in which its trajectory consists of a segment of an ellipse which orbits the Sun in the same direction as the planets themselves, and which is tangential to the orbits of both the planet of origin and destination. This is illustrated schematically in Figure 10.1, which shows the required configurations of Earth and Mars for (a) the outward bound and (b) the inward bound trajectories. Note the timing of the launch for the return trajectory, assuming that the travel time for each leg of the journey lasts around 6 months. The return journey must occur in such a way that the spacecraft is launched from Mars in approximately the same direction as the orbit (to exploit the orbital kinetic energy of the planet itself) and arrives tangentially to the Earth's orbit just as the Earth itself arrives at the same location. By following this trajectory, not only is the expenditure of fuel by the propulsion system kept relatively small at launch, but also the spacecraft does not need to expend a large amount of fuel in slowing itself down to match the speed of either planet when it reaches its destination. The alignment of the two planets needed to allow this neat arrangement to work only occurs roughly once every Martian year or so. Hence, the party of astronauts on Mars would have to wait around for some 15 months after arriving on the planet before they can set off on their home journey.

For this reason, the logistics associated with even the shortest possible manned expedition to Mars will have to be able to deal with at least the following problems:

- Astronauts need to be able to survive periods of 6 months or more in deep space, without significant harm (e.g., from cosmic radiation, solar flares, or meteoritic impacts).
- Astronauts need to take sufficient resources with them to keep supplied with everything they need for a period of at least 15–18 months on the surface of Mars and to carry, or manufacture, sufficient fuel for the return journey.
- The team of people would need to be able to cope with large changes in environmental conditions (widely fluctuating temperatures, dust storms, etc.) during more than half of Mars' seasonal cycle (e.g., see Read *et al.*, 2003); this limits the choice of possible landing sites.
- To make the best use of the time which they need to spend on the surface, they need to be able to move around on the surface from place to place efficiently and easily.
- Individual astronauts also need to be able to perform a wide variety of tasks

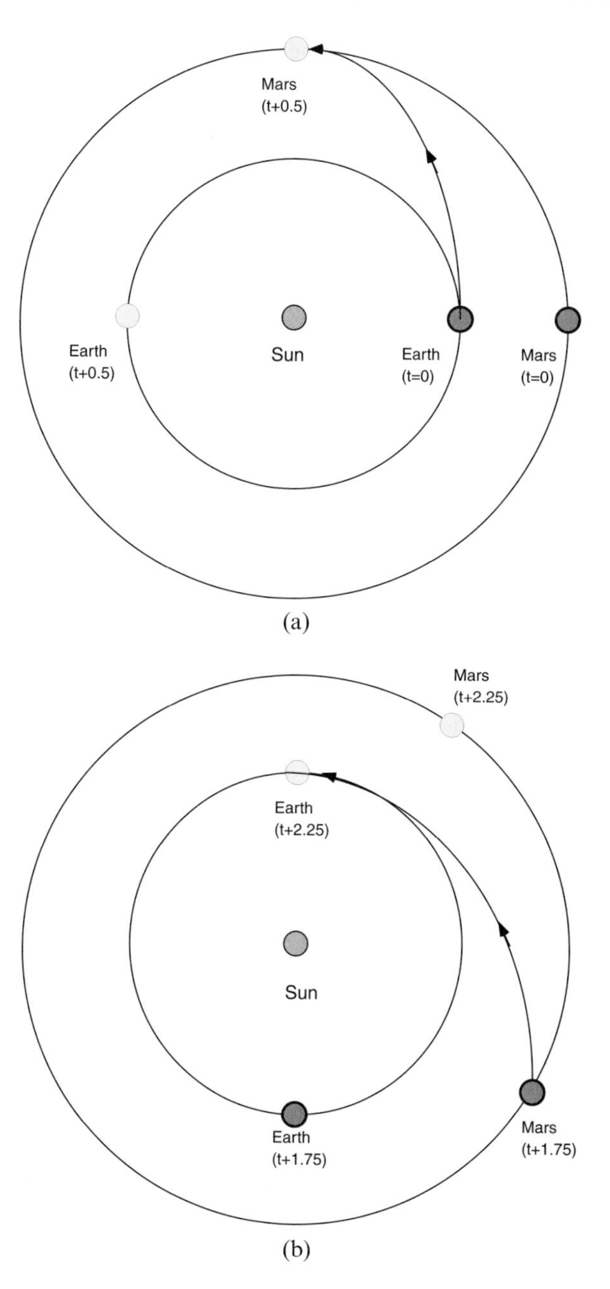

Figure 10.1. Schematic diagrams showing the relative positions of Earth and Mars (a) for Mars transfer from Earth, and (b) for the return Earth transfer from Mars, approximately following the Hohmann transfer orbit for optimum expenditure of propulsion energy. Note the time delay required for the appropriate planetary alignment for the return journey, which has been estimated assuming a journey time in either direction of approximately 6 months. Time is in units of Earth years.

outside space vehicles, despite having to wear protective clothing against the extreme environment, and to carry personal oxygen supplies – (e.g., see Figure 10.2, colour section).

The technologies needed for this kind of expedition are already fairly well developed, sometimes in other contexts. For example the survival techniques needed and medical issues associated with living in deep space for long periods of time have been well studied in long-duration sojourns of astronauts in space vehicles kept in Earth orbit, such as the Mir space station. But there is no doubt that the costs involved in planning such expeditions are considerable, and greatly exceed those associated with even relatively sophisticated unmanned, robotic missions to Mars.

10.2 LONG-TERM EXPLORATION AND COLONIZATION?

If the decision is eventually taken to begin the human-led exploration of Mars, the first missions are likely to stay the minimum possible time on the surface of the planet. Even so, a mission lasting more than one Earth year at the surface would enable a great deal of scientific investigation to be carried out, especially if the problems of mobility and extravehicular working on Mars can be solved effectively.

Eventually, however, one can envisage that humans will wish to establish a more permanent presence on Mars, perhaps even involving the establishment of self-sustaining communities or colonies. But in this case, one of the most serious obstacles to the large-scale colonization and ultimate exploitation of the Mars environment for human use would be the extremely hostile nature of that environment itself, not only to humans, but to any other form of terrestrial life – at least above the ground. The thin, CO_2-rich atmosphere, extremes of temperature (especially cold), and the harsh radiation environment in its present form would be fatal for virtually all known living organisms on Earth, except perhaps for a few hardy viruses or other micro-organisms capable of shutting down into a dormant form. Yet without some sort of indigenous Martian agriculture, it is unlikely that a human community of any substantial size could be sustained at the Martian surface.

From the previous chapter and elsewhere in this book, however, we have seen that Mars' climate is not a static system, but has sustained conditions in its past which have probably been sufficient to allow at least the transient presence of large amounts of liquid water, which is at least one critical element for sustaining any form of terrestrial life. While it is most likely that these kinds of hospitable conditions, which approach those experienced in certain locations on present-day Earth (e.g., in the Arctic or Antarctic), probably occurred only during the very early stages of Mars' evolution, it is evident from the previous chapter that cyclic variations in Mars' orbit and rotation can lead to significant changes in the climate, and that these changes are ongoing. In particular, there are certain aspects of the present climate at high obliquity which appear to push (at least certain regions of) Mars in the direction of increasing surface temperature and atmospheric pressure, towards the point where liquid water might be capable of existing for significant periods of time at the surface. This potential on Mars for substantial change in the surface climate and the possible release of stored volatiles has led some scientists seriously to consider the possibility of bringing about similar, or even more extreme, changes in the climate by artificial means. Indeed, the most ambitious of proposals for this strategy have even suggested the possibility of producing highly Earth-like conditions on Mars, with a breathable, oxygen-rich atmosphere and a freely existing biosphere covering much of the planet. Perhaps more realistically, however, proposals have tended to focus on an objective to enable the development of an environment in which some plants can function unaided, though humans would still need some protective clothing and breathing apparatus to work and live at the Martian surface (e.g., see Figure 10.3, colour section).

While such a possibility may sound enticing (at least to some people), it is important to keep the practical (and moral and political) issues firmly based on sound scientific and ethical principles. In the remaining part of this chapter, therefore, we give a scientific overview of the basic issues involved in 'terraforming' Mars. In what follows, we draw heavily on the recent work of McKay *et al.* (1991), subsequent studies, as reviewed extensively by McKay (1999), and on the excellent website maintained by Martyn Fogg (http://www.users.globalnet.co.uk/~mfogg/index.htm – hereafter referred to as Fogg) to which readers are referred for further details. Such a discussion of the questions 'how?' and 'what is possible?' in this context would not be complete without some consideration of the issues concerned with 'why?' and 'should we?' be terraforming a unique and virgin wilderness elsewhere in the Solar System. We conclude this chapter, therefore, with a few brief thoughts on these moral and ethical questions, in order to stimulate the reader to make his or her own judgment on the matter.

10.3 BASIC REQUIREMENTS FOR TERRAFORMING MARS

It is widely understood that several crucial aspects of the Earth's environment (e.g., the free oxygen content of the atmosphere, the surface albedo, and degree of greenhouse warming) are either the direct result of, or at least significantly influenced by, the presence of living organisms at the Earth's surface or in the oceans. Given the presence of relatively abundant CO_2 in the Martian atmosphere ($\sim$150 kg m^{-2} on Mars, see Table 9.1 – around 30 times the column density of CO_2 in the Earth's atmosphere), therefore, one might expect that it would be sufficient to place some form of plant life (which uses CO_2 as fuel and processes it into oxygen and carbohydrates) onto the surface of Mars and the process of terraforming would be self-sustaining. Unfortunately, in order for even the most basic plant-like organism to function successfully, it also needs water (preferably in liquid form) and a reasonable temperature range. Some protection from ionizing radiation would also be necessary to prevent biological damage and eventual sterilization and death. Other chemical elements (such as nitrogen) would also be needed to sustain a basic plant biochemistry.

As a minimum requirement, therefore, it is likely to be necessary to achieve the following changes to the surface climate and conditions on Mars (McKay *et al.*, 1991; McKay, 1999):

- substantially increase the mean global temperature (by around 60 K?);
- significantly increase the atmospheric mass, and consequent greenhouse warming;
- enable large amounts of liquid water to be available at the surface;
- provide protection from extremes of UV and other ionizing radiation;
- modify the atmospheric composition to provide significant amounts of N_2 and O_2.

This is still insufficient to allow higher organisms, such as oxygen-breathing animals or even humans, to function, but would be enough to be compatible with certain kinds of anaerobic organisms and some simple kinds of plants. Such a minimal change to Mars' environment is often referred to as 'ecopoiesis' (Haynes, 1990), from the Greek *οικος* (*oikos*, meaning 'house' or 'abode') and *ποιησις* (*poiesis*, meaning 'doing' or 'creating').

The increase in global temperature is important both to produce conditions in which biochemical reactions can proceed at a reasonable rate, and also to keep liquid water more or less permanently at least at some locations on Mars. The increase in atmospheric mass is needed not so much to provide more CO_2 for anaerobic organisms (the atmospheric mass of CO_2 is already much greater than on Earth and abundantly available for photosynthesis) but to provide additional protection to organisms at the surface from UV and other ionizing radiation. Note that, because of the reduced gravitational acceleration at Mars' surface, a surface pressure of around 300 hPa would provide as much atmospheric mass between the surface and space as on Earth. A greatly increased mass of atmosphere in the form of CO_2 would also help to increase the greenhouse warming of the surface, though a surface pressure of some 2,000 hPa of CO_2 would be needed to produce a warming of around 60 K by CO_2 alone (Pollack *et al.*, 1987). Amounts of liquid water required are somewhat uncertain, though they would have to correspond perhaps at least to several metres of water covering a significant portion of the planet's surface. Finally, it is a requirement of even relatively elementary anaerobic organisms that they need small amounts of N_2 in the atmosphere to function, and some O_2 is also necessary if plants are to survive. Estimated minimum amounts depend on the kind of organism, but some kinds of terrestrial organism can survive with just a few hPa of N_2, while plants would need a few hPa of O_2 as well.

Requirements for producing a more hospitable climate and conditions under which animals and humans could survive without specialized protection are much more ambitious (e.g., requiring at least 200 hPa of O_2 in the atmosphere, corresponding amounts of an inert buffer gas (such as nitrogen), and even warmer temperatures). As we will see below, even the technological demands for producing the minimal changes for ecopoiesis are daunting enough, and so we concentrate on those here. Further quantitative details of the requirements for producing a warm, breathable atmosphere can be found in the papers by McKay *et al.* (1991), McKay (1999), and in references listed on the website by Fogg.

10.4 FEEDBACKS AND THE 'RUNAWAY GREENHOUSE'?

Despite these cautionary remarks, there is a suggestion from some of the energy balance studies discussed here (cf. McKay *et al.*, 1991) and in the previous chapter that certain inbuilt feedbacks in the Martian climate system might actually help to bring about many of the changes listed above if just one or two of them are begun artificially. We have already seen in the previous chapter the suggestion that as Mars' obliquity increases, stored volatiles in the form of water and CO_2 might be released into the atmosphere from the polar caps and deep regolith as surface temperature at high latitudes are warmed. Such changes could lead to significant increases in surface pressure of CO_2, provided the subsurface reservoirs adsorb large amounts of CO_2 during cooler periods of low obliquity and can be persuaded to desorb these reserves by increasing the soil temperature.

The relationship between the mass of CO_2 desorbed from the soil (and consequent increase in atmospheric pressure) and the atmospheric and soil temperatures is not well known, but one possibility based on the van t'Hoff law for changes in chemical equilibrium with temperature, might be of the form:

$$p_d \simeq C_d M_a \exp(T/T_d) \tag{10.1}$$

where p_d is the desorbed pressure of CO_2; M_a is the total mass of adsorbed CO_2 in the regolith (expressed in units of pressure); and C_d is a constant. T_d is another constant expressed in units of temperature, which measures the characteristic energy associated with the release of gas molecules from adsorption within the soil matrix. Neither of these constants are at all well known for Mars at present, and may not be known for some time to come. But by making some educated guesses for plausible numbers, McKay *et al.* (1991) and Zubrin and McKay (1997) have made some simple calculations based on Eq. (10.1) which illustrate some potentially important qualitative features.

They further estimate the variation of the global mean T_{mean} and polar T_{pole} atmospheric temperatures on Mars as a function of CO_2 atmospheric pressure using highly simplified formulae of the forms:

$$T_{\text{mean}} \simeq S^{1/4} T_B + 20(1+S)p^{1/2} \tag{10.2}$$

$$T_{\text{pole}} \simeq T_{\text{mean}} + \frac{\Delta T}{(1+5p)} \tag{10.3}$$

where S represents the solar constant, normalized by the present value (so $S = 1$ if the Sun has the same luminosity as today); T_B is the black body equilibrium temperature of Mars (at present $T_B \simeq 213.5$ K); and ΔT is the temperature difference between equator and poles which would occur in the absence of an atmosphere (approximately 70 K for $S = 1$). Figure 10.4 shows plots of T_{CO_2} (the vapour

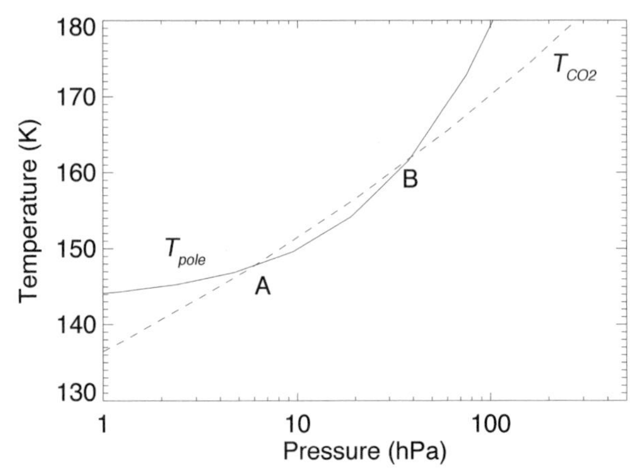

Figure 10.4. Variation of CO_2 equilibrium vapour pressure (dashed line) and estimated temperature of the pole (solid line) from Eq. (10.3), shown as a function of either temperature or atmospheric pressure.

equilibrium temperature of CO_2), and T_{pole} as a function of atmospheric CO_2 pressure. From this plot it can be seen that T_{pole} is equal to T_v at two points, A and B, representing possible equilibria. Point A, however, is stable whilst B is unstable. This is because wherever $T_{pole} < T_{CO_2}$, rapid condensation of CO_2 will take place, reducing the atmospheric pressure and moving the system towards the left. In the region between A and B, therefore, the atmosphere will tend to move away from B and cool towards A in a 'runaway icebox'. If $T_{pole} > T_{CO_2}$, however, CO_2 will sublime into the atmosphere from surface ice, increasing the pressure and moving the system towards the right. Thus, for surface pressures lower than at A (which is at around 6 hPa), the system will tend to sublime CO_2 into the atmosphere until A is reached. If the equilibrium polar temperature can be made to exceed that at point B, however, then $T_{pole} > T_{CO_2}$, and the system will continue to move towards the right, away from B and towards ever increasing T_{pole} and p, until the entire CO_2 reservoir has been evaporated. Thus, we could imagine bringing about a relatively modest initial rise in global mean temperature at first, which in turn would allow the release of some CO_2, thereby increasing the atmospheric mass and consequent greenhouse warming. This enables the surface temperature to rise a little further, releasing yet more stored CO_2, and so on, until much of the stored reservoir is released. The final equilibrium pressure would be set largely by the size of the stored reservoir, partly in the poles, but most likely in the top few 100 m or so of the regolith. This kind of positive feedback, in which a modest (though finite) perturbation to the surface temperature causes trends which reinforce the original perturbation, is known as a 'runaway greenhouse'.

The same kinds of feedback would eventually affect the reservoirs of water, though water would not appear in large quantities in liquid form until the latter stages of the process, and then only provided the temperature can be raised sufficiently. A significant uncertainty, however, is how large the initial artificial perturbation needs to be to push the present climate into an unstable track towards the other metastable equilibrium. Simple models such as those of McKay *et al.* (1991) suggest that an initial perturbation of between 5 and 20 K is likely to be enough to get the process started, though it is far from clear whether other factors may also be important. The extraction of the available CO_2 from the regolith may also have an implicit finite efficiency suggested from Eq. (10.1) in which, depending upon the unknown parameter T_d, only a certain fraction of the stored CO_2 in the regolith may be accessible to increase atmospheric pressure. McKay *et al.* (1991) considered a range of possible values of T_d from 10 to 60 K. The corresponding equilibrium atmospheric pressure at a constant surface temperature turned out to be very sensitive to the assumed values for T_d, with $p \sim 300$ hPa for $T_d \simeq 20$ K, for example, but this decreases to only 15–30 hPa if T_d is increased to 25–30 K. Such a low pressure would provide little more greenhouse warming than at present, leaving the atmosphere well short of the state needed for ecopoiesis.

The effect of releasing significant amounts of water vapour into the atmosphere, however, might also help the process along considerably. This is because water vapour is at least as effective a greenhouse gas as CO_2, and actually contributes as much or more to the Earth's greenhouse warming as does atmospheric CO_2. Thus, provided it proves possible to raise temperatures sufficiently to increase substantially amounts of atmospheric water vapour, the latter is likely to help accelerate the process towards a new equilibrium state which is much warmer and wetter than at present.

10.5 ENERGY REQUIREMENTS AND TIMESCALES

While this all sounds quite encouraging, in suggesting that relatively modest changes (arguably comparable with what has already been achieved by anthropogenic activity on Earth, for example) can initiate the process, it is still important to keep even these modest requirements in some perspective. The energetic requirements, for example, to heat solid CO_2 at 150 K to an amount of CO_2 gas equivalent to an atmospheric pressure of 200 hPa at 288 K would need around 3.7×10^9 J m^{-2}. Integrated over large tracts of the planet, this amounts to a prodigious amount of energy which has to be provided, and which would be quite impractical to supply by direct artificial means. It is almost inevitable, therefore, that these kinds of changes would require some ingenuity in diverting natural sources of energy, such as sunlight, towards the task of modifying the thermodynamic state of the volatile inventory on Mars. One possibility might be to divert an asteroid to impact directly on Mars, thereby also depositing new reserves of volatiles as well as providing a transient input of huge amounts of energy associated with the kinetic energy of the incoming projectile. Such an approach is also likely to be extremely destructive, however, and methods which provide the required energy at a gentler rate should perhaps be preferred. The other obvious source of energy is that of sunlight itself, though the energy densities achievable at the orbit of Mars are relatively modest.

As an illustration, assuming complete conversion of incoming sunlight, heating our CO_2 reservoir from 150 K in solid form to gas at 288 K would require about a year's worth of solar energy at the orbital distance of Mars, but it is most unlikely that anything like this level of efficiency in the use of solar energy could be achieved. Warming large quantities of water (say 5 m equivalent of liquid water) from ice at 215 K to liquid at 288 K would require about the same amount of energy. Allowing for realistic efficiencies of (say) 1–10% for the exploitation of solar energy would suggest a minimum timescale of several tens of years to make a significant impact on the present climate.

Moreover, if the CO_2 needed to be released from deep in the regolith, the intrinsic thermal inertia of the soil would introduce a diffusive time delay of many years, even to penetrate a few metres into the ground. If it was necessary to warm the soil a few hundred metres deep to bring about these changes, however, the timescales would extend to $\sim 10^5$ yr instead, unless other factors were to apply, such as a substantial pressure difference between air in the deep regolith and the atmosphere (enabling more rapid desorption of CO_2). Such a very long timescale would make the whole process pretty unattractive, notwithstanding the relatively low material cost of the energy required.

10.6 PRACTICAL METHODS OF ACHIEVING GLOBAL WARMING

We have seen from the above that, provided we take a fairly optimistic view concerning the size of the available reservoir of CO_2 on Mars, either locked up currently in polar caps or in the regolith, there is the potential for bringing about major changes to the Martian climate. The main trigger for such changes, however, must be somehow to raise the temperature of the surface for sustained periods of time sufficiently to put the planet into the 'runaway greenhouse' mode. The achievement of this mode requires the provision of some additional source of energy, especially in the polar regions, in order to get the process started. Suggestions for how to achieve this have tended to fall into three or four main categories:

(1) increase solar heating at the poles, by providing external reflectors to direct additional sunlight to these regions;
(2) modify the albedo of the polar regions on Mars (e.g., by sprinkling dark, absorbant material onto the ice);
(3) deflect volatile-rich meteors and asteroids to impact on Mars, releasing large quantities of greenhouse-active volatiles (such as ammonia) into the atmosphere; or
(4) introduce so-called 'super-greenhouse gases' such as chlorofluorocarbon (CFC) and perfluorocarbon (PFC) into the atmosphere, which are capable of causing strong changes in the radiation budget at relatively small concentrations (much as CFCs have been invoked as a source of recent anthropogenic greenhouse warming on Earth).

10.6.1 Polar reflectors

The idea of placing a large set of reflectors in orbit around Mars to reflect sunlight onto the polar regions is an attractive one in some respects. The basic idea would be to place a large reflective surface into orbit in such a way as to reflect sunlight onto the polar regions from the opposite direction to that of normal sunlight, by effectively positioning the reflector 'behind' Mars. This would be most effective during polar night, when there is little or no direct sunlight and any additional source of insolation could be very efficient at keeping the polar winter temperature from dropping too far.

The major drawback, however, is that the size and mass of such a reflector would have to be immense by current engineering standards to make a significant global difference on Mars. For warming the entire polar region, it would require a reflective surface as large as a major European country. More realistically, the use of smaller mirrors (perhaps 'only' 100–200 km in diameter) would allow the possibility of heating more localized regions by 5–10 K. Even allowing for the possibility of using extremely thin sheets of material, however, this would still require huge quantities of material (on the order of 10^5 tonnes) and the ability to deploy it and control its orientation in orbit.

This approach would almost certainly have to await the development of orbital manufacturing methods (e.g., using materials from asteroids). But the control of the device could probably use the techniques of solar sails in using radiation pressure from sunlight to maintain an orbit in the appropriate position relative to Mars.

10.6.2 Polar albedo

The suggestion to modify the albedo and emissivity of the polar regions relies on reducing the amount of available sunlight which is reflected from the bright, ice-covered surface. We have already seen in the previous chapter that small changes in ice albedo (as perhaps occurs at other obliquities) may have substantial effects on the equilibrium climate, not only in the polar regions but perhaps globally. Thus, a method of depositing highly absorbent material onto the ice caps might well have a substantial impact on increased warming of the climate. The practicalities of depositing material at sufficient densities to make a difference are far from trivial. Given the annual renewal and deposition of seasonal ice, the covering of the poles with an absorber would presumably have to be repeated every year in order to produce a sustained effect.

10.6.3 Artificially-enhanced greenhouse warming

The latter two categories above both rely primarily on changing the composition of the Martian atmosphere in order to enhance its ability to absorb outgoing infrared radiation from the surface. Ammonia is an efficient absorber in the infrared which occurs naturally in large quantities elsewhere in the Solar System (though not significantly on Mars itself), leading to one suggestion of harnessing and deflecting volatile-rich asteroids to impact onto the surface of Mars. Such a series of impacts could well produce both an immediate direct warming in the locality of the impact, and a transient increase in the concentration of ammonia in the atmosphere – perhaps sufficient to begin the process of the runaway greenhouse discussed above. The main source of such asteroids would almost certainly have to come from the outer Solar System, well beyond the orbit of Jupiter. Again the energetic and technological requirements for capturing and deflecting asteroid-sized objects in the outer Solar System are probably well into the realm of science fiction, at least for the foreseeable future. This would also have to be a long-term undertaking, even given the technology to modify the orbit of such large objects, since even the travel time from the outer Solar System to the orbit of Mars is likely to be on the order of years or even decades.

Perhaps the most realistic approach to this problem comes from the suggestion of James Lovelock and his associates (Lovelock and Allaby, 1984) to introduce a mixture of CFC and PFC gases and other strongly active greenhouse agents into the atmosphere. Such an approach would require the release of relatively small quantities of material compared with other methods, though still sufficient to require an industrial-scale operation to produce such materials on Mars itself from indigenous materials. CFCs have absorption coefficients per mole in the infrared which are at least 10^4 times larger than CO_2, are chemically inert with very long residence times in the atmosphere, and are non-toxic to any living organisms which humans might wish to plant in the modified environment. Decomposition by the action of UV sunlight may greatly reduce the lifetime of CFCs on Mars, however, because there is little else to protect the atmosphere from the action of UV, though the resistance of PFCs to photolysis may be significantly higher, making them more attractive as potential greenhouse enhancers.

Taken alone, this approach is perhaps capable of raising surface temperatures by up to 30 K. This is perhaps not enough by itself to achieve ecopoiesis, but if taken in combination with other techniques outlined above, then it may be that a sustained warming of the Martian environment could be achievable, at least in principle provided that the runaway greenhouse paradigm is applicable to Mars itself – which is far from certain at the present time. The practicalities of achieving this goal are also far from trivial, and it is clear that successive generations of Martian terraformers will have to be both highly committed to their ultimate objective and very patient.

10.7 HOW LONG WOULD IT LAST?

Having achieved the objective of ecopoiesis, it is legitimate to ask how stable and durable this environmental state might be expected to be. On Earth, there is a complex carbon cycle which involves the very long-term interaction between the biosphere and CO_2 in the atmosphere, the oceans, and the tectonic activity of the surface itself and the mantle. This is because in the presence of liquid water, CO_2 in the atmosphere will dissolve to form carbonic acid and (via weathering effects) carbonate salts, at least some of which will precipitate out onto the ocean bed in the form of limestone sedimentary rocks. If the Earth was not tectonically active, this would result in a permament sink for CO_2 from the atmosphere, since limestone will only release its stored carbon if it is heated to a temperature approaching 1,000 K. On Earth, however, the subduction of carboniferous limestone at the edges of major continental tectonic plates leads to its chemical breakdown at the high temperatures encountered deep below the Earth's surface into CaO and CO_2, the latter of which may be released back into the atmosphere through volcanic outgassing. Thus, on geological timescales the Earth comes to a dynamic equilibrium in which the solution of CO_2 into the oceans and its 'fixation' into limestone sediments is balanced in the long run by the release of CO_2 in volcanic eruptions.

On a terraformed Mars, however, enhanced levels of CO_2 would have been produced in the presence of liquid water, leading to the same kind of sink for CO_2 as occurs in the Earth's oceans. But Mars is not (so far as we know) tectonically active, so there is no mechanism to release CO_2 from the carbonates deposited on the beds of Martian lakes. Thus, unless an alternative method can be devised to reprocess these carbonate deposits, the terraformed environment is likely to have a finite 'shelf-life' of perhaps a few tens of millions of years. This is still a long time by human standards, but does indicate a long term need to continue to 'manage' the Martian environment, once it has been terraformed.

10.8 WE CAN – BUT SHOULD WE?

The task of converting Mars to an environment in which it becomes possible to grow plants and allow terrestrial organisms to roam freely on the Martian surface is viewed by many scientists as a noble and exciting enterprise which follows naturally from the scientific study of the interaction of the climate with various other components of the planetary system. Such a branch of study in the context of the Earth is increasingly being referred to as 'Earth System Science', and has already begun to spawn new research institutes and journals to facilitate the interdisciplinary communication necessary for its progress.

In the case of the Earth, however, the problems of human intervention and terraforming in the environment are more concerned with understanding and responding to what mankind has already been doing inadvertently through industrial and agricultural practices over several centuries. In this case, the challenge is to understand what changes have already taken place as the result of human activity, and to seek the means of managing these changes in the future.

For Mars, however, we are faced with a pristine and ancient environment which might, perhaps, have nurtured an indigenous biology in its distant past, but which has survived intact for perhaps 3 Gyr or more (at least in places). As such, therefore, it is crucially important to explore and understand its environment and its climatic and biological history while the evidence on the Martian surface is still uncontaminated by human or other terrestrial biological activity. It is crucial, therefore, that scientists are given every opportunity to explore and investigate the surface geology, climate, and biological potential of Mars before steps are taken to modify that environment and begin the task of transplanting terrestrial flora/fauna to begin an unconstrained Martian agriculture.

The reasons in favour of initiating active terraforming activities on Mars would seem to derive from:

(1) a desire to see whether such active control of a planetary environment is actually possible;
(2) a perceived need eventually to create an alternative planetary home for terrestrial life as a potential refuge should the Earth ever become uninhabitable; and
(3) an ultimate wish to exploit the Martian environment for human occupation and commercial gain.

The first is primarily a scientific and technological goal as an end in itself, and is perhaps ethically neutral if taken in isolation. As such, however, it is also unlikely to command a sufficiently high priority with the rest of human society to be commensurate with the resources needed to achieve it in practice – especially given the more pressing needs back on Earth. The second objective seems ethically defensible as a last resort for our species, but is not currently perceived as a high priority in the absence of a clear and imminent threat to the Earth's viability as our home planet.

The latter, however, with its prospects of commercial gain for some, though possibly at the expense of others, is more contentious, not least because the prospect of gaining a sufficient return on even a substantial capital investment might justify it (at least to some governments or multinational companies) in practice. In terms of the commercial potential for exploitation (say) of Mars' mineral resources, the question of whether we should adapt Mars for human occupation has direct parallels with the issues associated with the Antarctic continent and regions of the Arctic wilderness (e.g., in Canada and Alaska). In those regions, intranational and international agreements restrict the degrees of development, mining, drilling, and logging activities in order to protect the integrity of the environment, because of its rarity, fragility, and value for scientific research. The position of Antarctica, especially, has many resonances with the situation of Mars, for which international agreement would be necessary to ensure that planetary protection measures are enforced at the expense of commercial priorities, for the sake of continued and unimpeded scientific investigation. Speaking as scientists, we would hope that the widespread acceptance of the pre-eminence of the scientific imperative over the desire for commercial exploitation would continue at least until such time as the main questions over the nature of Mars' early history and its biological potential can be resolved to everyone's satisfaction.

Of course, as indicated in (2) above, it may eventually become necessary to consider the possibility that the inhabitability of Earth may become difficult for humankind to manage (e.g., through the increasing luminosity of the Sun itself). In that case, the survival of terrestrial life itself would be at stake and mankind would be forced to look elsewhere in the Solar System for a new home. Under those circumstances, mankind's right to adapt the environment of other planets to suit its needs surely rests on much firmer ground. But such circumstances are extreme, and unlikely to occur for a very long time to come (unless there were to be some other kind of environmental catastrophe on Earth which would render it hostile to human occupation).

In the meantime, it seems to us that the scientific imperative should take precedence in decisions concerning the deliberate modification of the Martian environment, and that all necessary steps should be taken to ensure planetary protection measures are fully implemented in all future missions to Mars. This is not to say that scientists should not be free to perform 'thought experiments' on issues relating to terraforming Mars. But for the moment these 'experiments' should be confined to models, theories, and speculations, for our edification and entertainment. In any event, as we have seen in this chapter, the theoretical basis for practical terraforming is still extremely crude, and scarcely provides a secure foundation for the controlled management of an environment as complex as that of Mars.

Appendix A

A climate database for Mars

The results of Mars GCM simulations are potentially of great use for a variety of applications other than direct atmospheric investigations. One example is in spacecraft mission planning (e.g., providing upper atmosphere density profiles for aerobreaking or aerocapture, estimating entry profiles for landers, and assessing the near-surface atmospheric conditions a lander may experience, both in terms of the climatological mean and a likely range of variability). Another might be in providing trial profiles for testing algorithms for retrieving remotely-sounded atmospheric observations, or in providing a 'first guess' profile for retrievals. One immediate problem with such studies is that they often require many runs of a GCM, sampled at a particular location or time, or run under particular dust distribution assumptions. If the particular requirements have not been anticipated, or all the relevant variables have not been stored at the right sampling frequency, it is then a major undertaking to conduct a series of GCM experiments on request, especially at short notice. It is equally impractical to supply a full GCM to the group requiring the output, since the code and initialization procedure is often very complex and considerable experience is required to run such a numerical model correctly. Few, if any, Mars GCM teams are in the position to offer much regular support or detailed documentation.

The LMD and Oxford GCM groups have collaborated, supported partly by contracts from the European Space Agency (ESA) and Centre National d'Etudes Spatiales (CNES), with the aim of making some of their GCM data more widely available and easily accessible in the form of a climate database, with access software which can be run on a small computer. The Mars Climate Database (MCD) (Lewis *et al.*, 1999) was formulated for use both as an engineering resource and as a scientific database and has been taken up for a wide range of studies, from spacecraft engineering to more casual interest in Martian weather and likely conditions facing future astronauts. One particular strength of using a GCM to compile such a database is that it provides a physically consistent estimate of the environmental

Table A.1. Mars Climate Database MCD 'months'.

Month	L_S	Sol
1	0°–30°	0–61
2	30°–60°	61–127
3	60°–90°	127–193
4	90°–120°	193–258
5	120°–150°	258–318
6	150°–180°	318–372
7	180°–210°	372–422
8	210°–240°	422–468
9	240°–270°	468–515
10	270°–300°	515–562
11	300°–330°	562–613
12	330°–360°	613–669

conditions on Mars for times of year and dust loadings which are not covered by past observations.

A previous set of environmental statistics for Mars were collected in the Mars Global Reference Atmosphere Model (MarsGRAM, Justus *et al.*, 1996). MarsGRAM originally used a blend of observational data and simple physical models (e.g., to produce realizations of environmental conditions). For example temperatures could be generated by fitting a series of functions to the Mariner 9 and Viking thermal measurements, and winds derived from temperatures by the use of thermal wind balance. Since 2001, more recent versions of MarsGRAM have included mean data from some Ames Mars GCM experiments, for similar reasons as the MCD.

All climate modellers, whether for the Earth or Mars, are faced with the formidable task of summarizing the results of their model simulations and usually resort to the compilation of spatially and temporally averaged statistics to reduce the data to a manageable level. The result is often in the form of a 'climate atlas' comprising maps or tables of the distribution of various averaged atmospheric fields (e.g., Hoskins *et al.*, 1989). The MCD has paid particular attention to the development of a strategy which enables the retention of scientifically useful information from the original high-resolution model simulations on both diurnal and seasonal timescales.

Seasonal variations are taken into account by dividing the Martian year into 12 periods, or 'months', each of 30° areocentric longitude, as shown in Table A.1 and illustrated in Figure A.1. Within each month, the diurnal cycle is also represented by storing mean values of all atmospheric and surface fields from the model at 12 different times of day. A simple measure of variability in the model is provided by also storing the variance of these fields over the month at the same local time of day.

The breakdown of the Martian year into twelve months, and how this may be used to reconstruct the seasonal cycle, is illustrated by a reconstruction of the surface

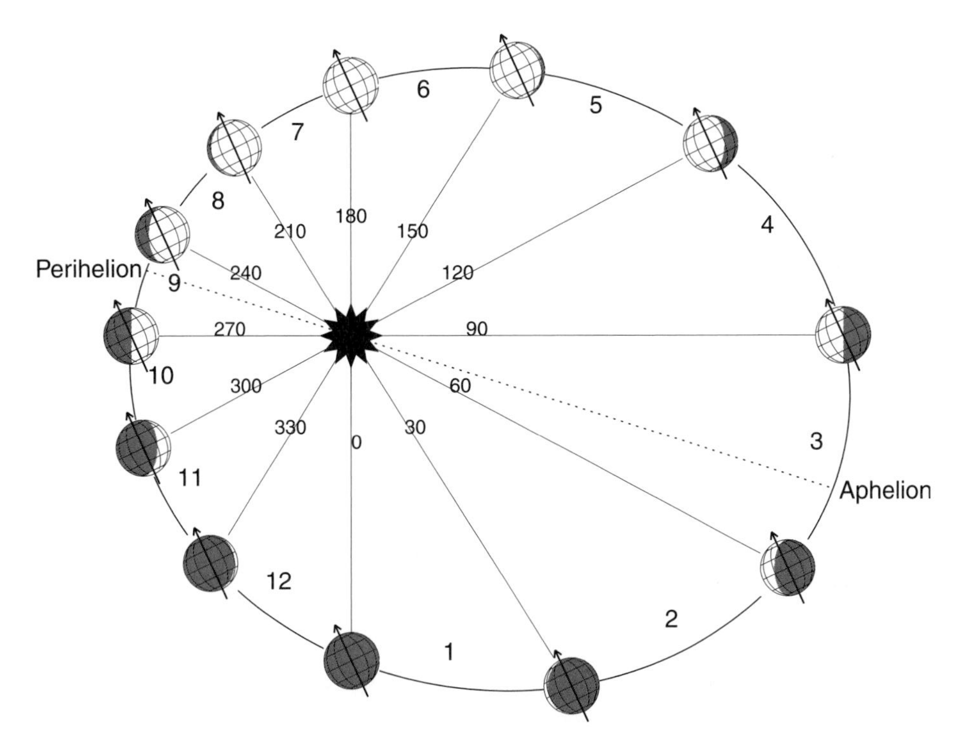

Figure A.1. A schematic diagram showing the orbit of Mars about the Sun. The eccentricity of the orbit (actually 0.093) has been exaggerated for clarity. The position of aphelion and perihelion (at $L_S = 251°$) are shown by the dashed line. The twelve 'months' used for the Mars climate database are numbered around the orbit, dividing the Martian year into twelve equal segments of 30° areocentric longitude; the lines are labelled with values of L_S. Northern hemisphere spring equinox ($L_S = 0°$) is directly below the Sun as drawn, summer solstice ($L_S = 90°$) to the right, autumn equinox ($L_S = 180°$) above and winter solstice ($L_S = 270°$) to the right. The shading indicates the region of day and night on the planet, as seen looking toward the Sun along the $L_S = 0°$ axis. The small vector indicates the rotation axis of the planet, inclined at 25.19° to the normal to the plane of the orbit.

pressure records at the two VL sites from the MCD. The real surface pressure data, with periods of 1 sol or less, for one Martian year is shown in Figure A.2 and the type of reconstruction possible from the MCD is shown in Figure A.3. In this case, no special vertical corrections were made to interpolate MCD data from the smooth GCM surface to the actual VL altitudes, so there is some systematic error in the mean pressures, but a realistic variation with time of year is obtained by a simple cubic spline fit to the mean pressure at the centre of each month. The non-diurnal component of the surface pressure variance is also stored in the climate database and, after removing that component of the variance which can be accounted for by the seasonal trend within a month, an estimate of the size of the day-to-day variability was also obtained and shaded in Figure A.3. As with the real weather

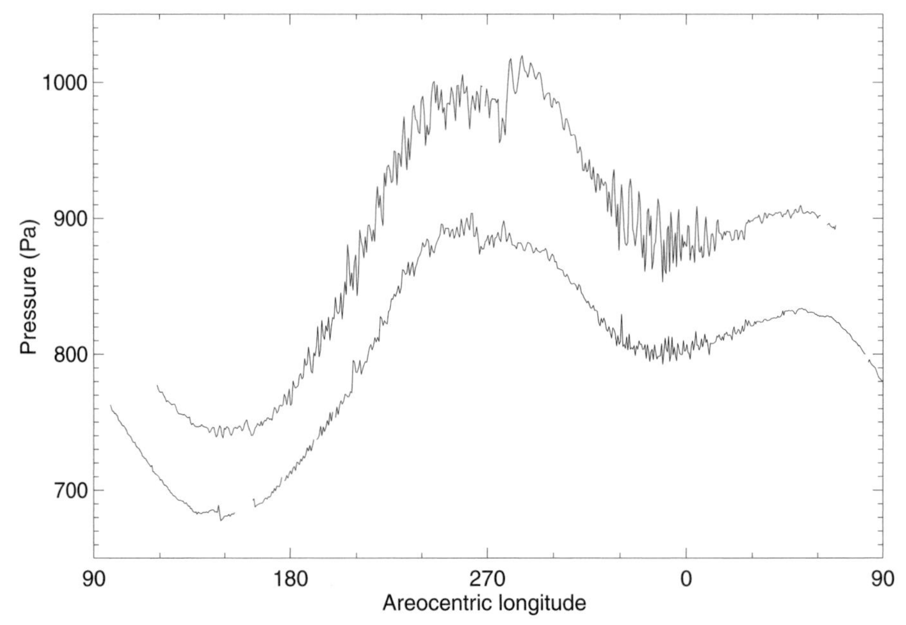

Figure A.2. Observed Viking Lander 1 (lower curve) and 2 (upper curve) surface pressure for the first Martian year of observations, filtered to remove diurnal and higher frequency periods. The remaining high-frequency oscillations seen are caused by weather systems passing over the landers. The jump in pressure, and modulation in the amplitude of the weather signal, close to $L_S = 280°$, was caused by a global dust storm.

variability, seen in Figure A.2, there is evidence for more day-to-day variability in the winter and spring and very little in the summer.

As previously described, the major source of interannual variability on Mars is related to the atmospheric dust loading. The total model variability running under prescribed dust conditions, which are identical from year-to-year, will, of course, be smaller than the total possible range of variability in reality. For this reason, the MCD provides annual dust scenarios run under default, 'best guess', dust conditions, and other annual scenarios which represent a cold, clear atmosphere and a warmer, dustier annual cycle, similar to that seen at the Viking lander sites (Lewis *et al.*, 1999). Global dust storms at moderate ($\tau = 2$) and heavy ($\tau = 5$) dust loadings are provided for the northern hemisphere winter months, when such storms are most likely to occur. Figure A.3 shows a single example from the heavy global dust storm at $L_S = 285°$, which shows how the mean surface pressure at the site is higher, and the variability lower, under global dust storm conditions, for comparison with observations at that time, Figure A.2.

It is often desirable to provide a large set of profiles as if sampled from the full GCM under identical conditions and times of day, from many different Mars years, in order to gain some understanding of day-to-day variability. Simple measures of variance are not enough since variability in one model field at any

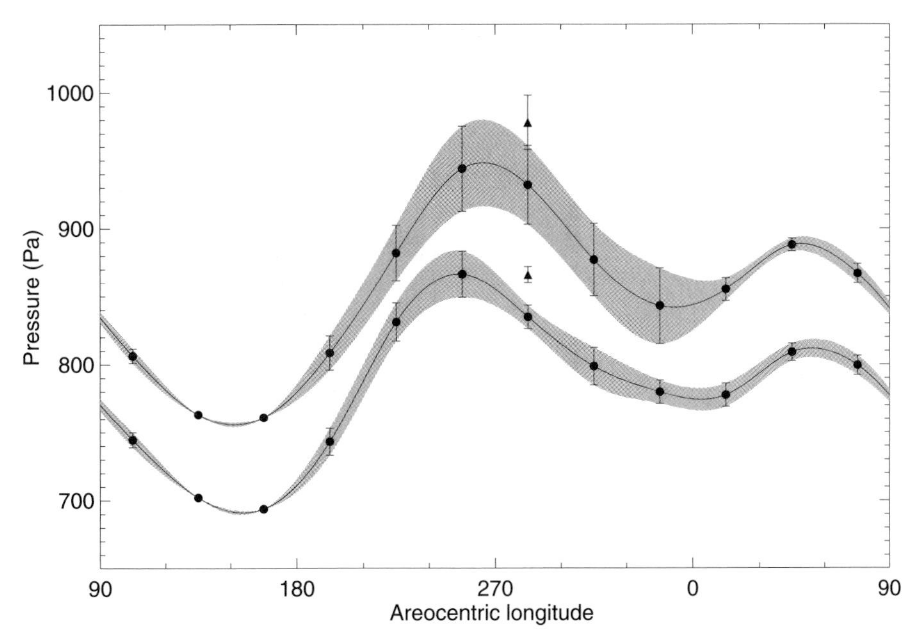

Figure A.3. Pressure at the Viking Lander 1 (bottom curve) and 2 (top curve) sites from the Mars Climate Database (MCD). This data was taken from a previous version of the database which did not have a mechanism for an accurate correction to the correct local topographic height; hence the errors in the absolute values of the mean surface pressure for these point measurements compared to the data in Figure A.2. The solid circles show the diurnally averaged, seasonal values at intervals of $L_S = 30°$, and the smooth curve is a cubic spline fit to the mean pressure. The error bars and shaded region indicate the non-diurnal variability (at two standard deviations from the mean) of the surface pressure with the seasonal trend component removed from the total variance. The triangles and error bars at $L_S = 285°$ show the pressure and non-diurnal variance at this time in a global dust storm scenario, to demonstrate the jump in pressure and reduction in variability when the dust loading is high.

point is generally correlated with variability at other locations and in other fields. For example, a synoptic-scale weather system has both horizontal and vertical correlations between pressure, temperature, and winds. It is desirable to be able to reconstruct this variability in some way that is faster than simply rerunning the Mars GCM many times, and considerable attention is paid to allow this to be done.

One means by which this problem was tackled was to represent large-scale variability in the database using an approach based on empirical orthogonal function (EOF) analysis (North, 1984; Mo and Ghil, 1987; Preisendorfer, 1988), a technique that is widely used in meteorological data analysis, as well as in other subject areas. The eigenvalues and eigenvectors are found for a large covariance matrix of a set of atmospheric fields stored at regular intervals. This data set can be formed by storing GCM output at frequent intervals throughout an annual integration. The eigenvectors, or EOFs, form a linear basis for the data such that those with the largest eigenvalues account for the most variance. The 'variance compression' feature of the EOFs means that it is possible to reconstruct the main characteristics of the original data from a limited set of EOFs and corresponding time coefficients, sometimes called principal components (PCs). In the case of data contaminated with uncorrelated noise, the EOFs separate the 'signal' from the 'noise'.

Commonly in meteorology, EOFs are computed for single variables only. For the present application, however, multivariate EOFs were formed of the main atmospheric variables: surface pressure, atmospheric temperature, zonal and meridional wind, and density. There is significant cross correlation between dynamical variables because the processes which generate such variability are generally large-scale features, such as baroclinic waves with wavenumbers in the longitudinal direction of typically 1, 2, or 3.

If $\bar{\mathbf{D}}$ is a vector representing the database mean field of a meteorological variable (stored on a horizontal and vertical grid), it is possible to form a new **D** by adding a series of vectors to the mean field:

$$\mathbf{D} = \bar{\mathbf{D}} + \sum_{i=1}^{M} p_i \mathbf{e}_i \tag{A.1}$$

where the vectors $\mathbf{e}_i$ are the EOFs of the covariance matrix computed from profiles extracted from the full GCM history and the scalars p_i are the principal components, which can be thought of as the projection of the EOFs onto the original model history fields. M is the truncation, which must be small enough to limit storage requirements while being large enough to capture a significant portion of the profile variance; at present the 72 most significant EOFs are retained in the MCD. The principal components themselves are a function of time of year, for example, if the EOF represents a baroclinic wave, consisting of a train of high- and low-pressure systems propagating around the winter pole, it is likely that its principal component will have a high mean amplitude in the winter and almost zero mean amplitude in the summer.

An illustration of the way in which EOFs can be rapidly used to build an ensemble of realistic profiles is given in Figure A.4. The profiles shown here were generated by the EOF technique, but are very similar in form and statistical properties to a real ensemble generated from a long run of the Mars GCM.

Account needs to be taken also of possible variability which is not modelled explicitly by the GCM, generally because of limits in resolution. The gravity wave drag parameterization for the Mars GCM has already been discussed in Section 2.5.2.7, and it is possible to add a set of small-scale waves of this type back onto simulated profiles, with random phase. The model is based on the parameterization scheme used in the GCM.

The surface stress exerted by a vertically propagating, stationary gravity wave can be written, as in Eq. (2.24):

$$\tau_{\mathrm{GW}} = \kappa \rho_s N_s U_s \mu^2 \tag{A.2}$$

where κ is a characteristic gravity wave horizontal wave number; ρ_s is the surface

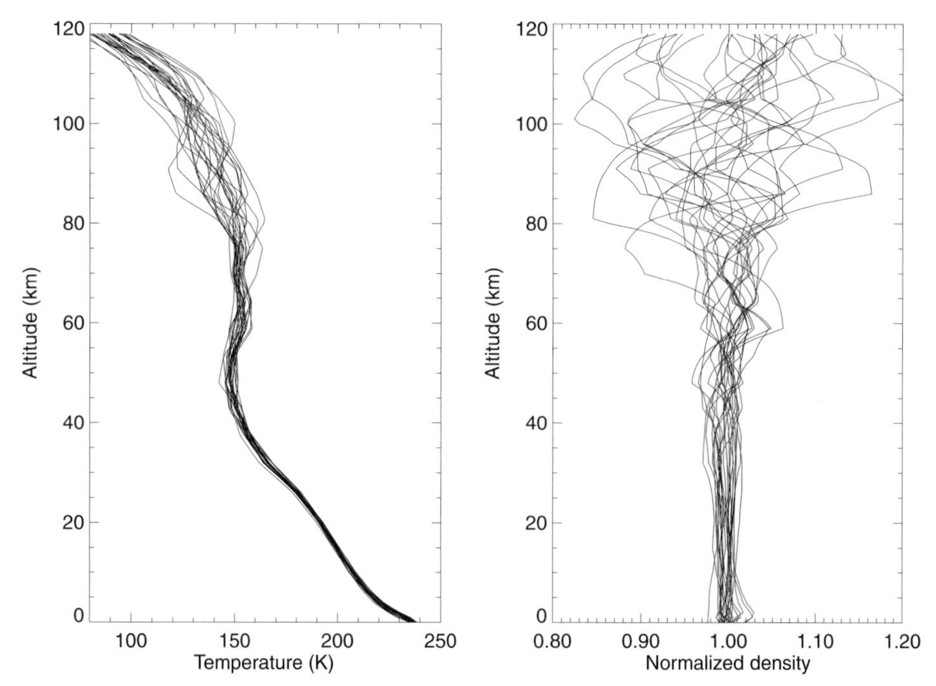

Figure A.4. An ensemble of profiles simulated using the Mars Climate Database (MCD) empirical orthogonal function model, at the same location and time as the Viking Lander 1 entry. The left panel shows the temperature profile and the right the density profile, normalized by the mean density profile for clarity.

density; N_s is the surface Brunt–Väisälä frequency; U_s is the surface wind speed; and μ^2 is a measure of the model subgrid-scale topographic vertical variance. Above the surface, the stress can be related to the gravity wave vertical material surface displacement δh by:

$$\tau_{\mathrm{GW}} = \kappa \rho N U \delta h^2 \tag{A.3}$$

This leads to an expression for δh at height z:

$$\delta h = \sqrt{\frac{\rho_s N_s U_s}{\rho N U}} \mu \tag{A.4}$$

where ρ, N, and U are the density, Brunt–Väisälä frequency, and wind speed at height z, respectively. Perturbations to variables in a mean vertical profile can then be applied by considering vertical displacements:

$$\delta z = \delta h \cos\left(\frac{2\pi z}{\lambda_z} + \phi\right) \tag{A.5}$$

where λ_z is a vertical wavelength for the gravity wave (typical values could be in the range 1–20 km) and ϕ a surface phase, which can be chosen at random.

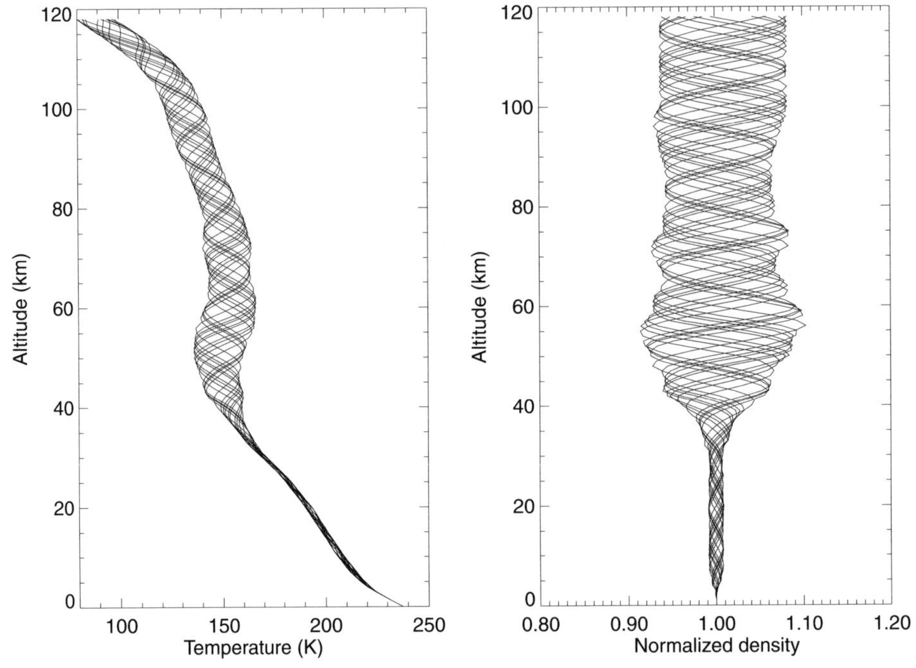

Figure A.5. An ensemble of perturbations generated by the small-scale variability model added to mean profiles of temperature and density from the Mars Climate Database (MCD); these correspond to the Viking Lander 1 entry location and time as in Figure 2.21. Gravity waves of the same vertical wavelength (16 km) and random phase have been added to the mean profiles and allowed to propagate upwards. They can be seen to saturate as the waves begin to break at about 50 km altitude.

An illustration of a simple ensemble of profiles with a gravity wave perturbation of vertical wavelength 16 km and random phase is shown in Figure A.5, all based on the single mean profile used in Figure A.4. The waves can be seen to grow with height and to saturate at about 50-km altitude as they begin to break. Of course, in reality inertia gravity waves will not necessarily be monochromatic and the contribution of both small- and large-scale variability needs to be assessed together. This can be done with the MCD by applying both variability models together and propagating small-scale perturbations upward along each of the large-scale perturbed profiles.

Lewis *et al.* (1999) described an earlier version of the MCD; this has now undergone three generations of major revisions, although the underlying structure and motivation remain the same. Among the improvements since the publication of Lewis *et al.* (1999) are an extension to the vertical range of the GCMs to 32 levels reaching to over 120 km altitude, the incorporation of more variables including upward and downward radiative fluxes in two wavebands, and the extension of the variability model to include both horizontal and vertical correlations between variables. In addition to technical improvements in the model and database, the model parameters and physical schemes have been updated with the advent of new spacecraft data from missions such as MGS and new dust distribution scenarios are provided.

A full distribution of the MCD requires several gigabytes of disk space, and the installation of some software to read the portable binary format, but it is also possible to access data and graphical plots via a WWW interface at `http://www.lmd.jussieu.fr/mars.html`, though there may be some restrictions on the variability models available over the internet and it is a less suitable route for intensive use.

Afterword

In this book we have attempted to provide a reasonably comprehensive overview of a physicist's understanding of the climate, meteorology, and environment of Mars. This subject has made considerable progress in the past 20 years, to the extent that we can address quite sophisticated questions relating to how the atmosphere of Mars interacts with the surface, how the different cycles of volatiles, dust, and atmospheric transport operate and evolve over geological timescales, and how the climate responds to astronomically induced variations in insolation. While many of these issues require a highly complex and interdisciplinary approach, bringing together climatologists, meteorologists, geologists, geophysicists, and biologists (amongst others), we have seen in this book that many of the key phenomena and components of the Martian climate and circulation systems can be understood using relatively simple physical models, many of which have direct counterparts on Earth. This is highly satisfying, and gives cause for hope that this body of knowledge can be further developed to improve our understanding of other planetary systems, and ultimately to improve our knowledge and understanding of the Earth itself and its potential for change in the future.

In concluding this book, we can scarcely do better than to leave our readers with the following quotation from the Nobel Prize-winning physicist Richard P. Feynman, taken from Volume II of his famous *Lectures on Physics* (Feynman *et al.*, 1970). These words express an uplifting view of the ultimate 'mission' of physical planetary science with which we can readily concur:

> There are those who are going to be disappointed when no life is found on other planets. Not I – I want to be reminded and delighted and surprised once again, through interplanetary exploration, with the infinite variety and novelty of phenomena that can be generated from such simple principles. The test of science is its ability to predict. Had you never visited the Earth, could you predict the thunderstorms, the volcanoes, the ocean waves, the

auroras, and the colorful sunset? A salutary lesson it will be when we learn of all that goes on on each of those dead planets – those eight or ten balls, each agglomerated from the same dust cloud and each obeying exactly the same laws of physics.

References

Allison, M., J. D. Ross, and N. Solomon (1999). Mapping the Martian meteorology. In The Fifth International Conference on Mars, Abstract #6102, LPI Contribution No. 972, Lunar and Planetary Institute, Houston (CD-ROM).

Anderson, D. L. T. (1976). The low-level jet as a western boundary current. *Mon. Weather Rev.*, **104**, 907–921.

Anderson, E. and C. Leovy (1978). Mariner 9 television limb observations of dust and ice hazes on Mars. *J. Atmos. Sci.*, **35**, 723–734.

Andrews, D. G. (2001). *An Introduction to Atmospheric Physics*. Cambridge, UK: Cambridge University Press.

Andrews, D. G., J. R. Holton, and C. B. Leovy (1987). *Middle Atmosphere Dynamics*. Orlando, Fl: Academic Press.

Angelats i Coll, M., F. Forget, F. Hourdin, Y. Wanherdrick, M. A. López-Valverde, F. González-Galindo, P. L. Read, and S. R. Lewis (2003). Towards a global model of the Martian atmosphere. *Workshop on Mars Atmosphere Modelling and Observations* (Abstract 11-5). Granada, Spain: ESA/CNES.

Atkinson, B. W. (1981). *Meso-Scale Atmospheric Circulations*. New York: Academic Press.

Bagnold, R. A. (1954). *The Physics of Blown Sand and Desert Dunes*. London: Methuen.

Baines, P. G. (1995). *Topographic Effects in Stratified Flows*. Cambridge, UK: Cambridge University Press.

Baker, V. R. (2001). Water and the Martian landscape. *Nature*, **412**, 228–236.

Baker, V. R., R. G. Strom, R. G. Gulick, J. S. Kargel, G. Komatsu, and V. S. Kale (1991). Ancient oceans, ice sheets and the hydrological cycle on Mars. *Nature*, **352**, 589–594.

Baker, V. R., M. H. Carr, V. C. Gulick, C. R. Williams, and M. S. Marley (1992). Channels and valley networks. In H. H. Kieffer, B. M. Jakosky, C. W. Snyder, and M. S. Matthews (eds.), *Mars* (pp. 493–522). Tucson, AZ: University of Arizona Press.

Bandfield, J. L., V. E. Hamilton, and P. R. Christensen (2000). A global view of Martian surface compositions from MGS-TES. *Science*, **287**, 1626–1630.

Banfield, D., A. Ingersoll, and C. Keppenne (1995). A steady-state Kalman filter for assimilating data from a single polar-orbiting satellite. *J. Atmos. Sci.*, **52**, 737–753.

Banfield, D., A. D. Toigo, A. P. Ingersoll, and D. A. Paige (1996). Martian weather correlation length scales. *Icarus*, **119**, 130–143.

Banfield, D., B. J. Conrath, J. C. Pearl, M. D. Smith, and P. Christensen (2000). Thermal tides and stationary waves on Mars as revealed by Mars Global Surveyor Thermal Emission Spectrometer. *J. Geophys. Res.*, **105**, 9521–9537.

Banin, A. (1992). The mineralogy and formation processes of Mars soil. *MSATT Workshop on Chemical Weathering on Mars* (Technical Report 92-01, 1-2). Houston, TX: LPI.

Barnes, J. R. (1980). Time spectral analysis of mid-latitude disturbances in the Martian atmosphere. *J. Atmos. Sci.*, **37**, 2002–2015.

Barnes, J. R. (1981). Midlatitude disturbances in the Martian atmosphere: A second Mars year. *J. Atmos. Sci.*, **38**, 225–234.

Barnes, J. R. (1984). Linear baroclinic instability in the Martian atmosphere. *J. Atmos. Sci.*, **41**, 1536–1550.

Barnes, J. R. (1990). Possible effects of breaking gravity waves on the circulation of the middle atmosphere of Mars. *J. Geophys. Res.*, **95**, 1401–1421.

Barnes, J. R. (2001). Asynoptic fourier transform analyses of MGS TES data: Transient baroclinic eddies. *Bull. Am. Met. Soc.*, **33**, 1067.

Barnes, J. R. and R. M. Haberle (1996). The Martian zonal-mean circulation: Angular momentum and potential vorticity structures in GCM simulations. *J. Atmos. Sci.*, **53**, 3143–3156.

Barnes, J. R. and J. L. Hollingsworth (1987). Dynamical modelling of a planetary wave mechanism for a Martian polar warming. *Icarus*, **71**, 313–334.

Barnes, J. R., J. B. Pollack, R. M. Haberle, R. W. Zurek, C. B. Leovy, H. Lee, and J. Schaeffer (1993). Mars atmospheric dynamics as simulated by the NASA/Ames general circulation model, 2, Transient baroclinic eddies. *J. Geophys. Res.*, **98**/(E2), 3125–3148.

Barnes, J. R., R. M. Haberle, J. B. Pollack, H. Lee, and J. Schaeffer (1996). Mars atmospheric dynamics as simulated by the NASA/Ames general circulation model, 3, Winter quasi-stationary eddies. *J. Geophys. Res.*, **101**, 12753–12776.

Blackadar, A. K. (1957). Boundary layer wind maxima and their significance for the growth of nocturnal inversions. *Bull. Am. Met. Soc.*, **38**, 283–290.

Blackadar, A. K. (1962). The vertical distribution of wind and turbulent exchange in a neutral atmosphere. *J. Geophys. Res.*, **67**, 3095–3102.

Blamont, J. E. and E. Chassefière (1993). First detection of ozone in the middle atmosphere of Mars from solar occultation measurements. *Icarus*, **104**, 324–336.

Blasius, K. R., S. J. Ferry, S. H. Silverman, D. F. Murphy, P. R. Christensen, and G. L. Mehall (2002). THEMIS makes a thermal Odyssey to Mars. *Laser Focus World*, **38**(4), 47.

Blumsack, J. L. (1971). On the effects of topography on planetary atmospheric circulation. *J. Atmos. Sci.*, **28**, 1134–1143.

Blumsack, J. L. and P. J. Gierasch (1972). Mars: The effects of topography on baroclinic instability. *J. Atmos. Sci.*, **29**, 1081–1089.

Böttger, H. (2003). *Modelling water transport in the Martian atmosphere*. Ph.D. thesis, Oxford University.

Böttger, H., S. R. Lewis, P. L. Read, and F. Forget (2003). GCM simulations of the Martian water cycle. *Workshop on Mars Atmosphere Modelling and Observations* (Abstract 8-4). Granada, Spain: ESA/CNES.

Bougher, S. W., R. G. Roble, E. C. Ridley, and R. E. Dickinson (1990). The Mars thermosphere. II: General circulation with coupled dynamics and composition. *J. Geophys. Res.*, **95**, 14811–14827.

Bougher, S. W., S. Engel, R. G. Roble, and B. Foster (1999). Comparative terrestrial planet thermospheres. 2: Solar cycle variation of global structure and winds at equinox. *J. Geophys. Res.*, **104**, 16591–16611.

Bougher, S. W., S. Engel, R. G. Roble, and B. Foster (2000). Comparative terrestrial planet thermospheres. 3: Solar cycle variation of global structure and winds at solstices. *J. Geophys. Res.*, **105**, 17669–17689.

Bougher, S. W., D. P. Hinson, J. M. Forbes, and S. Engel (2001). MGS radio science electron density profiles and implications for the neutral atmosphere. *Geophys. Res. Lett.*, **28**, 3091–3094.

Bougher, S. W., J. R. Murphy, and S. Engel (2002). Coupling processes and model simulations linking the Mars lower and upper atmospheres. *34th COSPAR Scientific Assembly, Houston, TX.*

Boynton, W. V., W. C. Feldman, S. W. Squyres, T. Prettyman, J. Bruckner, L. G. Evans *et al.* (2002). Distribution of hydrogen in the near-surface of Mars: Evidence for sub-surface ice deposits. *Science*, **297**, 81–85.

Branscombe, L. E. (1983). The Charney baroclinic instability problem: Approximate solutions and modal structures. *J. Atmos. Sci.*, **40**, 1393–1409.

Briggs, G. and C. B. Leovy (1974). Mariner 9 observations of the Mars north polar hood. *Bull. Am. Met. Soc.*, **55**, 278–296.

Byrne, S. and A. P. Ingersoll (2003). A sublimation model for Martian south polar ice features. *Science*, **299**, 1051–1053.

Cantor, B., P. B. James, M. Caplinger, and M. J. Wolff (2001). Martian dust storms: 1999 Mars Orbiter Camera observations. *J. Geophys. Res.*, **106**, 23653–23687.

Carr, M. H. (1996). *Water on Mars*. Oxford, UK: Oxford University Press.

Chapman, S. and R. S. Lindzen (1970). *Atmospheric Tides*. Dordrecht, Netherlands: Reidel.

Charney, J. G. (1947). The dynamics of long waves in a baroclinic westerly current. *J. Meteor.*, **4**, 135–162.

Charney, J. G. and J. G. DeVore (1979). Multiple flow equilibria in the atmosphere and blocking. *J. Atmos. Sci.*, **36**, 1205–1216.

Charney, J. G. and M. E. Stern (1962). On the stability of internal baroclinic jets in a rotating atmosphere. *J. Atmos. Sci.*, **19**, 159–172.

Chen, S. and K. E. Trenberth (1988). Orographically forced planetary waves in the northern hemisphere winter: Steady state model with wave-coupled lower boundary formulation. *J. Atmos. Sci.*, **45**, 657–680.

Christensen, P. R. (2003). Formation of recent Martian gullies through melting of extensive water-rich snow deposits. *Nature*, **422**, 45–48.

Christensen, P. R., D. L. Anderson, S. C. Chase, R. N. Clark, H. H. Kieffer, M. C. Malin, *et al.* (1992). Thermal Emission Spectrometer experiment – Mars Observer mission. *J. Geophys. Res.*, **97**, 7719–7734.

Clancy, R. T. and S. W. Lee (1991). A new look at dust and clouds in the Mars atmosphere: Analysis of emission phase function sequences from global Viking IRTM observations. *Icarus*, **93**, 135–158.

Clancy, R. T., S. W. Lee, G. R. Gladstone, W. W. McMillan, and T. Rousch (1995). A new model for Mars atmospheric dust based upon analysis of ultraviolet through infrared observations from Mariner 9, Viking, and Phobos. *J. Geophys. Res.*, **100**/(E3), 5251–5263.

Clancy, R. T., A. W. Grossman, M. J. Wolff, P. B. James, D. J. Rudy, Y. N. Billawala *et al.* (1996). Water vapor saturation at low altitudes around Mars aphelion: A key to Mars climate? *Icarus*, **122**, 32–62.

Clancy, R. T., B. J. Sandor, M. J. Wolff, P. R. Christensen, M. D. Smith, J. C. Pearl *et al.* (2000). An intercomparison of ground-based millimeter, MGS TES and Viking atmospheric temperature measurements: Seasonal and interannual variability of temperatures and dust loading in the global Mars atmosphere. *J. Geophys. Res.*, **105**, 9553–9572.

Colaprete, A. and O. B. Toon (2002). Carbon dioxide snow storms during the polar night on Mars. *J. Geophys. Res.*, **107**, doi:10.1029/2001JE001758.

Colburn, D. S., J. B. Pollack, and R. M. Haberle (1989). Diurnal variations in optical depth at Mars. *Icarus*, **79**, 159–189.

Collins, M. and I. N. James (1995). Regular baroclinic transient waves in a simplified global circulation model of the Martian atmosphere. *J. Geophys. Res.*, **100**/(E7), 14421–14432.

Collins, M., S. R. Lewis, P. L. Read, and F. Hourdin (1996). Baroclinic wave transitions in the Martian atmosphere. *Icarus*, **120**, 344–357.

Collins, M., S. R. Lewis, and P. L. Read (1997). Gravity wave drag in a global circulation model of the Martian atmosphere: Parameterisation and validation. *Adv. Space Res.*, **19**(8), 1245–1254.

Conrath, B. J. (1975). Thermal structure of the Martian atmosphere during the dissipation of the dust storm of 1971. *Icarus*, **24**, 36–46.

Conrath, B. J. (1981). Planetary-scale wave structure in the Martian atmosphere. *Icarus*, **48**, 246–255.

Conrath, B. J., J. C. Pearl, M. D. Smith, W. C. Maguire, S. Dason, M. S. Kaelberer *et al.* (2000). Mars Global Surveyor Thermal Emission Spectrometer (TES) observations: Atmospheric temperatures during aerobraking and science phasing. *J. Geophys. Res.*, **105**/(E4), 9509–9519.

Cook, K. H. and I. M. Held (1992). The stationary response to large scale orography in a general circulation model and a linear model. *J. Atmos. Sci.*, **49**, 525–539.

Cooley, J. W. and J. W. Tukey (1965). An algorithm for the machine calculation of complex Fourier series. *Math. Comp.*, **19**, 297–301.

Crapper, G. D. (1959). A three-dimensional solution for waves in the lee of mountains. *J. Fluid Mech.*, **6**, 51–76.

Crapper, G. D. (1962). Waves in the lee of a mountain with elliptical contours. *Phil. Trans. Roy. Soc. London*, **A354**, 601–623.

Daley, R. (1991). *Atmospheric Data Analysis.* Cambridge, UK: Cambridge University Press.

Davies, D. W. (1981). The Mars water cycle. *Icarus*, **45**, 398–414.

Deming, D. J., F. Mumma, F. Espenak, T. Kostius, and D. Zipoy (1986). Polar warming in the atmosphere of Mars. *Icarus*, **66**, 366–379.

Drazin, P. G. (1992). *Nonlinear Systems.* Cambridge, UK: Cambridge University Press.

Drazin, P. G. and W. H. Reid (1981). *Hydrodynamic Stability.* Cambridge, UK: Cambridge University Press.

Eady, E. T. (1949). Long waves and cyclone waves. *Tellus*, **1**, 33–52.

Emanuel, K. A. (1994). *Atmospheric Convection.* Oxford, UK: Oxford University Press.

Fanale, F. P. and B. M. Jakosky (1982). Regolith-atmosphere exchanges of water and carbon dioxide on Mars: Effects on atmospheric history and climate change. *Plan. Space Sci.*, **30**, 819–831.

Fanale, F. P., J. R. Salvail, W. B. Banerdt and R. S. Saunders (1982). Mars: The regolith–atmosphere–cap system and climate change. *Icarus*, **50**, 381–407.

Farmer, C. B. and P. E. Doms (1979). Global and seasonal variation of water vapor on Mars and the implications for permafrost. *J. Geophys. Res.*, **84**, 2881–2888.

Fels, S. B. and R. S. Lindzen (1974). The interaction of thermally excited gravity waves with mean flows. *Geophys. Fluid Dyn.*, **6**, 149–191.

Fenton, L. K. and M. I. Richardson (2001). Martian surface winds: insensitivity to orbital changes and implications for aeolian processes. *J. Geophys. Res.*, **106**, 32885–32902.

Feynman, R. P., R. Leighton, and M. Sands (1970). *The Feynman Lectures on Physics, Vol. II.* New York: Addison Wesley.

Findlater, J. (1969). A major low-level air current near the Indian Ocean during the northern summer. *Quart. J. R. Meteor. Soc.*, **95**, 362–380.

Findlater, J. (1977). Observational aspects of the low-level cross-equatorial jet stream system of the western Indian Ocean. *Pure Appl. Geophys.*, **115**, 1251–1262.

Fogg, M. http://www.users.globalnet.co.uk/~mfogg/index.htm

Forbes, J. M. and M. E. Hagan (2000). Diurnal Kelvin wave in the atmosphere of Mars: Towards an understanding of 'stationary' density structures observed by the MGS accelerometer. *Geophys. Res. Lett.*, **27**, 3563–3566.

Forget, F., F. Hourdin, and O. Talagrand (1998). CO_2 snow fall on Mars: Simulation with a general circulation model. *Icarus*, **131**, 302–316.

Forget, F., F. Hourdin, R. Fournier, C. Hourdin, O. Talagrand, M. Collins *et al.* (1999). Improved general circulation models of the Martian atmosphere from the surface to above 80 km. *J. Geophys. Res.*, **104**/(E10), 24155–24176.

Fouquart, Y. and B. Bonnel (1980). Computations of solar heating of the Earth's atmosphere: A new parametrization. *Contrib. Atmos. Phys.*, **53**, 35–62.

Fritts, D. C. (1984). Gravity wave saturation in the middle atmosphere: A review of theory and observations. *Rev. Geophys.*, **22**, 275–308.

Fritts, D. C. and W. Lu (1993). Spectral estimates of gravity wave energy and momentum fluxes. Part II: Parameterization of wave forcing and variability. *J. Atmos. Sci.*, **50**, 3093–3124.

Früh, W. G. and P. L. Read (1997). Wave interactions and the transition to chaos of baroclinic waves in a thermally driven rotating annulus. *Phil. Trans. Roy. Soc. London*, **A355**, 101–153.

Gadian, A. M. (1978). The dynamics of and the heat transfer by baroclinic eddies and large-scale stationary topographically-forced long waves in the Martian atmosphere. *Icarus*, **33**, 454–465.

Garratt, J. R. (1994). *The Atmospheric Boundary Layer.* Cambridge, UK: Cambridge University Press.

Gelb, A. (ed.) (1974). *Applied Optimal Estimation.* Cambridge, MA: MIT Press.

Ghil, M. (1989). Meteorological data assimilation for oceanographers. Part I: Description and theoretical framework. *Dyn. Atmos. Oceans*, **13**, 171–218.

Ghil, M. and P. Malanotte-Rizzoli (1991). Data assimilation in meteorology and oceanography. *Adv. Geophys.*, **33**, 141–266.

Gierasch, P. J. and R. Goody (1968). A study of the thermal and dynamical structure of the Martian lower atmosphere. *Plan. Space Sci.*, **16**, 615–646.

Gierasch, P. J. and R. M. Goody (1973). A model of a Martian great dust storm. *J. Atmos. Sci.*, **30**, 169–179.

Gierasch, P. J., P. Thomas, R. French, and J. Veverka (1979). Spiral clouds on Mars: A new atmospheric phenomenon. *Geophys. Res. Lett.*, **6**, 405–408.

Gill, A. E. (1982). *Atmosphere–Ocean Dynamics.* New York: Academic Press.

Goody, R. and M. J. S. Belton (1967). Radiative relaxation times for Mars. *Plan. Space Sci.*, **15**, 247–256.

Greeley, R. and J. D. Iversen (1985). *Wind as a Geological Process on Earth, Mars, Venus and Titan.* Cambridge, UK: Cambridge University Press.

Greeley, R., N. Lancaster, S. Lee, and P. Thomas (1992). Martian aeolian processes, sediments and features. In H. H. Kieffer, B. M. Jakosky, C. W. Snyder, and M. S. Matthews (eds), *Mars* (pp. 730–766). Tucson, AZ: University of Arizona Press.

Green, J. S. A. (1960). A problem in baroclinic instability. *Quart. J. R. Meteor, Soc.*, **86**, 237–251.

Haberle, R. M. and B. M. Jakosky (1990). Sublimation and transport of water from the north residual polar cap on Mars. *J. Geophys. Res.*, **95**, 1423–1437.

Haberle, R. M., H. C. Houben, R. Hertenstein, and T. Herdtle (1993). A boundary layer model for Mars: Comparison with Viking Lander and entry data. *J. Atmos. Sci.*, **50**, 1544–1559.

Haberle, R. M., J. B. Pollack, J. R. Barnes, R. W. Zurek, C. B. Leovy, J. R. Murphy *et al.* (1993). Mars atmospheric dynamics as simulated by the NASA/Ames general circulation model. 1: The zonal-mean circulation. *J. Geophys. Res.*, **98**, 3093–3123.

Haberle, R. M., C. P. McKay, J. Schaeffer, N. A. Cabrol, E. A. Grin, and A. P. Zent (2001). On the possibility of liquid water on present-day Mars. *J. Geophys. Res.*, **106**, 23317–23326.

Haberle, R. M., J. R. Murphy, and J. Schaeffer (2003). Orbital change experiments with a Mars general circulation model. *Icarus*, **161**, 66–89.

Hamilton, K. (1982). The effect of Solar tides on the general circulation of the Martian atmosphere. *J. Atmos. Sci.*, **39**, 481–485.

Hamm, A., T. Tél, and R. Graham (1994). Noise induced attractor explosions near tangent bifurcations. *Phys. Lett.*, **A185**, 313–320.

Hanel, R. A., B. J. Conrath, D. E. Jennings, and R. E. Samuelson (1992). *Exploration of the Solar System by Infrared Remote Sensing*. Cambridge, UK: Cambridge University Press.

Haynes, R. H. (1990). Ecce ecopoiesis: Playing God on Mars. In D. MacNiven (ed.), *Moral Expertise* (pp. 161–183). London: Routledge.

Hays, J. D., J. Imbrie, and N. J. Shackleton (1976). Variations in the Earth's orbit: Pacemaker of the ice ages. *Science*, **194**, 1121–1132.

Head, J. W., H. Hiesinger, M. A. Ivanov, M. A. Kreslavsky, S. Pratt, and B. J. Thompson (1999). Possible ancient oceans on Mars: Evidence from Mars Orbiter Laser Altimeter data. *Science*, **286**, 2134–2137.

Held, I. M. (1978). The vertical scale of an unstable baroclinic wave and its importance for eddy heat flux parameterizations. *J. Atmos. Sci.*, **35**, 572–576.

Held, I. M. and A. Y. Hou (1980). Nonlinear axially symmetric circulations in a nearly inviscid atmosphere. *J. Atmos. Sci.*, **37**, 515–533.

Hess, S. L. (1950). Some aspects of the meteorology of Mars. *J. Meteor.*, **7**, 1–13.

Hide, R. and P. J. Mason (1975). Sloping convection in a rotating fluid. *Adv. Phys.*, **24**, 47–100.

Hignett, P., A. A. White, R. S. Carter, W. D. N. Jackson, and R. M. Small (1985). A comparison of laboratory measurements and numerical simulations of baroclinic wave flows in a rotating cylindrical annulus. *Quart. J. R. Met. Soc.*, **111**, 131–154.

Hines, C. O. (1960). Internal gravity waves at ionospheric heights. *Canadian J. Phys.*, **38**, 1441–1481.

Hinson, D. P. and R. J. Wilson (2002). Transient eddies in the southern hemisphere of Mars. *Geophys. Res. Lett.*, **29**, doi:10.1029/2001GL014103.

Hinson, D. P., R. A. Simpson, J. D. Twicken, G. L. Tyler, and F. M. Flasar (1999). Initial results from radio occultation measurements with Mars Global Surveyor. *J. Geophys. Res.*, **104**, 26997–27012.

Hinson, D. P., G. L. Tyler, J. L. Hollingsworth, and R. J. Wilson (2001). Radio occultation measurements of forced atmospheric waves on Mars. *J. Geophys. Res.*, **106**, 1463–1480.

Hoffman, N. (2000). White Mars: A new model for Mars' surface and atmosphere based on CO_2. *Icarus*, **146**, 326–342.

Hollingsworth, J. L. and J. R. Barnes (1996). Forced stationary planetary waves in Mars's winter atmosphere. *J. Atmos. Sci.*, **53**, 428–448.

Hollingsworth, J. L., R. M. Haberle, J. R. Barnes, A. F. C. Bridger, J. B. Pollack, H. Lee *et al.* (1996). Orographic control of storm zones on Mars. *Nature*, **380**, 413–416.

Holopainen, E. O. (1983). Transient eddies in mid-latitudes: Observations and interpretation. In B. J. Hoskins and R. P. Pearce (eds), *Large-scale Dynamical Processes in the Atmosphere* (pp. 201–233). San Diego, CA: Academic Press.

Holton, J. R. (1982). The role of gravity wave induced drag and diffusion in the momentum budget of the mesosphere. *J. Atmos. Sci.*, **39**, 791–799.

Holton, J. R. (1983). The influence of gravity wave breaking on the general circulation of the middle atmosphere. *J. Atmos. Sci.*, **40**, 2497–2507.

Holton, J. R. (1992). *An Introduction to Dynamic Meteorology*. New York: Academic Press.

Hoskins, B. J. and A. J. Simmons (1975). A multi-layer spectral model and the semi-implicit method. *Quart. J. R. Meteor. Soc.*, **101**, 637–655.

Hoskins, B. J., H. H. Hsu, I. N. James, M. Masutani, P. D. Sardeshmukh, and G. H. White (1989). Diagnostics of the global atmospheric circulation, based on ECMWF analyses 1979–1989. Technical Report WMO/TD No. 326, World Meteorol. Organ., Geneva.

Houben, H. (1981). A global Martian dust storm model. In *Third International Colloquium on Mars, LPI Contrib. 441*, Houston, Texas, pp. 117. Lunar and Planetary Institute.

Houben, H. (1999). Assimilation of Mars Global Surveyor meteorological data. *Adv. Space Res.*, **23**/(11), 1899–1902.

Houben, H., R. M. Haberle, R. E. Young, and A. P. Zent (1997). Modeling the Martian seasonal water cycle. *J. Geophys. Res.*, **102**, 9069–9083.

Houghton, J. T. (1978). The stratosphere and mesosphere. *Quart. J. R. Meteor. Soc.*, **104**, 1–29.

Hourdin, F. (1992). A new representation of the CO_2 15 μm band for a Martian general circulation model. *J. Geophys. Res.*, **97**/(E11), 18319–18335.

Hourdin, F., P. Le Van, F. Forget, and O. Talagrand (1993). Meteorological variability and the annual surface pressure cycle on Mars. *J. Atmos. Sci.*, **50**, 3625–3640.

Hourdin, F., F. Forget, and O. Talagrand (1995). The sensitivity of the Martian surface pressure to various parameters: A comparison between numerical simulations and Viking observations. *J. Geophys. Res.*, **100**, 5501–5523.

Huck, F. O., D. J. Jobson, S. K. Park, S. D. Wall, R. E. Arvidson, W. R. Patterson *et al.* (1977). Spectrophotometric and color estimates of the Viking lander sites. *J. Geophys. Res.*, **82**, 4401–4411.

Hunt, G. E. and P. B. James (1979). Martian extratropical cyclones. *Nature*, **278**, 531–532.

Hvidberg, C. S. and H. J. Zwally (2003). Sublimation of water from the north polar cap on Mars. *Workshop on Mars Atmosphere Modelling and Observations* (Abstract 8-8). Granada, Spain: ESA/CNES.

Imbrie, J. (1982). Astronomical theory of the Pleistocene ice ages: A brief historical review. *Icarus*, **50**, 408–422.

Imbrie, J. and J. Z. Imbrie (1980). Modeling the climatic response to orbital variations. *Science*, **207**, 943–953.

Ingersoll, A. P. and J. R. Lyons (1993). Mars dust storms: Interannual variability and chaos. *J. Geophys. Res.*, **98**, 10951–10961.

Jakosky, B. M. (1983). The role of seasonal reservoirs in the Mars water cycle. II: Coupled models of the regolith, polar caps, and atmospheric water transport. *Icarus*, **55**, 19–39.

Jakosky, B. M. (1985). The seasonal cycle of water on Mars. *Space Sci. Rev.*, **41**, 131–200.

Jakosky, B. M. and C. B. Farmer (1982). The seasonal and global behavior of water vapor in the Martian atmosphere: Complete global results of the Viking Atmospheric Water Detector. *J. Geophys. Res.*, **87**, 2999–3019.

Jakosky, B. M. and R. M. Haberle (1992). The seasonal behaviour of water on Mars. In H. H. Kieffer, B. M. Jakosky, C. W. Snyder, and M. S. Matthews (eds), *Mars* (pp. 969–1016). Tucson, AZ: University of Arizona Press.

Jakosky, B. M. and T. Z. Martin (1987). Mars: North-polar atmospheric warming during dust storms. *Icarus*, **72**, 528–534.

Jakosky, B. M., B. G. Henderson, and M. T. Mellon (1995). Chaotic obliquity and the nature of the Martian climate. *J. Geophys. Res.*, **100**, 1579–1584.

James, I. N. (1994). *Introduction to Circulating Atmospheres.* Cambridge, UK: Cambridge University Press.

James, P. B. (1985). The Martian hydrologic cycle: Effects of CO_2 mass flux on global water distribution. *Icarus*, **64**, 249–264.

James, P. B. (1990). The role of water ice clouds in the Martian hydrologic cycle. *J. Geophys. Res.*, **95**, 1439–1445.

James, P. B., J. L. Hollingsworth, M. J. Wolff, and S. W. Lee (1999). North polar dust storms in early spring on Mars. *Icarus*, **138**, 64–73.

Jin, F. F., J. D. Neelin, and M. Ghil (1994). El Niño on the devil's staircase: Annual subharmonic steps to chaos. *Science*, **269**, 70–72.

Jones, K. L., R. E. Arvidson, E. A. Guiness, S. L. Bragg, S. D. Wall, C. E. Carlston *et al.* (1979). One Mars year: Viking Lander imaging observations. *Science*, **204**, 799–806.

Joseph, J. H., W. J. Wiscombe, and J. A. Weinman (1976). The delta-Eddington approximation for radiative flux transfer. *J. Atmos. Sci.*, **33**, 2452–2459.

Joshi, M. M., S. R. Lewis, P. L. Read, and D. C. Catling (1994). Western boundary currents in the atmosphere of Mars. *Nature*, **367**, 548–551.

Joshi, M. M., B. N. Lawrence, and S. R. Lewis (1995a). Gravity wave drag in three-dimensional atmospheric models of Mars. *J. Geophys. Res.*, **100**, 21235–21245.

Joshi, M. M., S. R. Lewis, P. L. Read, and D. C. Catling (1995b). Western boundary currents in the Martian atmosphere: Numerical simulations and observational evidence. *J. Geophys. Res.*, **100**, 5485–5500.

Joshi, M. M., B. N. Lawrence, and S. R. Lewis (1996). The effect of spatial variations in unresolved topography on gravity wave drag in the Martian atmosphere. *Geophys. Res. Lett.*, **23**, 2927–2930.

Joshi, M. M., R. M. Haberle, J. R. Barnes, J. R. Murphy, and J. Schaeffer (1997). Low-level jets in the NASA Ames Mars general circulation model. *J. Geophys. Res.*, **102**, 6511–6523.

Joshi, M. M., J. L. Hollingsworth, R. M. Haberle, and A. F. C. Bridger (2000). An interpretation of Martian thermospheric waves based on an analysis of a general circulation model. *Geophys. Res. Lett.*, **27**, 613–616.

Jossaume, S. (1990). Three-dimensional simulations of the atmospheric cycle of desert dust particles using a general circulation model. *J. Geophys. Res.*, **95**, 1909–1941.

Justus, C. G., B. F. James, and D. L. Johnson (1996). Mars Global Reference Atmospheric Model (Mars-GRAM 3.34): Programmer's guide. Technical Report NASA TM-108509, NASA, Washington, DC.

Kahn, R. (1980). *Some Properties of the Martian Atmosphere Obtained from the Viking Experiment.* Ph.D. thesis, Harvard University.

Kalman, R. E. (1960). A new approach to linear filtering and prediction problems. *Trans. ASME*, **82**D, 35–45.

Keating, G. M., S. W. Bougher, R. W. Zurek, R. H. Tolson, G. J. Cancro, S. N. Noll *et al.* (1998). The structure of the upper atmosphere of Mars: In situ accelerometer measurements from Mars Global Surveyor. *Science*, **279**, 1672–1676.

Kieffer, H. H. (1990). H_2O grain size and the amount of dust in Mars' residual north polar cap. *J. Geophys. Res.*, **95**, 1481–1493.

Kieffer, H. H. and A. P. Zent (1992). Quasi-periodic climate change on Mars. In H. H. Kieffer, B. M. Jakosky, C. W. Snyder, and M. S. Matthews (eds), *Mars* (pp. 1180–1218), Tucson: University of Arizona Press.

Kieffer, H. H., B. M. Jakosky, C. W. Snyder, and M. S. Mathews (eds) (1992). *Mars*. Tucson, AZ: University of Arizona Press.

Killworth, P. D. and M. E. McIntyre (1985). Do Rossby-wave critical layers absorb, reflect or over-reflect? *J. Fluid Mech.*, **161**, 449–492.

Kuiper, G. P. (1952). *The Atmospheres of the Earth and Planets*. Chicago: University of Chicago Press.

Kuo, H. L. (1949). Dynamic instability of two-dimensional, non-divergent flow in a barotropic atmosphere. *J. Meteor.*, **6**, 105–122.

Laskar, J. and P. Robutel (1993). The chaotic obliquity of the planets. *Nature*, **361**, 608–612.

Laskar, J., B. Levrard, and J. F. Mustard (2002). Orbital forcing of the martian polar layered deposits. *Nature*, **419**, 375–377.

Lee, S. W. (1987). Regional sources and sinks of dust on Mars: Viking observations of Cerberus, Solis Planum and Syrtis Major. In V. Baker, M. Carr, F. Fanale, R. Greeley, R. Haberle, C. Leovy and T. Maxwell (eds) *MECA Symposium on Mars: Evolution of its Climate and Atmosphere* (LPI Tech. Rept. 87-01, pp. 57–58 (abstract)). Tucson, AZ: Lunar and Planetary Institute.

Leighton, R. B. and B. C. Murray (1966). Behavior of carbon dioxide and other volatiles on Mars. *Science*, **153**, 136–144.

Leovy, C. B. (1969). Mars: Theoretical aspects of meteorology. *App. Optics*, **8**, 1279–1286.

Leovy, C. B. (1981). Observations of Martian tides over two annual cycles. *J. Atmos. Sci.*, **38**, 30–39.

Leovy, C. B. (1985). The general circulation of Mars: Models and observations. *Adv. Geophys.*, **28a**, 327–346.

Leovy, C. B. (1999). Wind and climate on mars. *Science*, **284**, 1891a.

Leovy, C. B. and Y. Mintz (1969). Numerical simulations of the atmospheric circulation and climate of Mars. *J. Atmos. Sci.*, **26**(6), 1167–1190.

Leovy, C. B., R. W. Zurek, and J. B. Pollack (1973). Mechanisms for Mars dust storms. *J. Atmos. Sci.*, **30**, 749–762.

Lewis, S. R. and P. L. Read (1995). An operational data assimilation scheme for the Martian atmosphere. *Adv. Space Res.*, **16**(6), 9–13.

Lewis, S. R. and P. L. Read (2003). Equatorial jets in the dusty Martian atmosphere. *J. Geophys. Res.*, **108**/(E4), doi:10.1029/2002JE001933.

Lewis, S. R., M. Collins, and P. L. Read (1996). Martian atmospheric data assimilation with a simplified general circulation model: Orbiter and lander networks. *Plan. Space Sci.*, **44**, 1395–1409.

Lewis, S. R., M. Collins, and P. L. Read (1997). Data assimilation with a Martian atmospheric GCM: An example using thermal data. *Adv. Space Res.*, **19**(8), 1267–1270.

Lewis, S. R., M. Collins, P. L. Read, F. Forget, F. Hourdin, R. Fournier *et al.* (1999). A climate database for Mars. *J. Geophys. Res.*, **104**(E10), 24177–24194.

Lewis, S. R., P. L. Read, B. J. Conrath, J. C. Pearl, and M. D. Smith (2003). Assimilation of Thermal Emission Spectrometer atmospheric data during the Mars Global Surveyor aerobraking period. *J. Geophys. Res.*, submitted.

Lindzen, R. S. (1970). The application and applicability of terrestrial atmospheric tidal theory to Venus and Mars. *J. Atmos. Sci.*, **27**, 536–549.

Lindzen, R. S. (1981). Turbulence and stress owing to gravity wave and tidal breakdown. *J. Geophys. Res.*, **86**, 9707–9714.

Lindzen, R. S. and A. Y. Hou (1988). Hadley circulations for zonally-averaged heating centered off the equator. *J. Atmos. Sci.*, **45**, 2416–2427.

Longuet-Higgins, M. S. (1968). The eigenfunctions of Laplace's tidal equations over a sphere. *Phil. Trans. Roy. Soc. London*, **A262**, 511–607.

López–Puertas, M. and M. A. López–Valverde (1995). Radiative energy balance of CO_2 non-LTE infrared emissions in the Martian atmosphere. *Icarus*, **114**, 113–129.

López–Valverde, M. A. and M. López–Puertas (1994). A non-local thermodynamic equilibrium radiative transfer model for infrared emissions in the atmosphere of Mars. 1: Theoretical basis and nighttime populations of vibrational states. *J. Geophys. Res.*, **99**, 13093–13115.

López–Valverde, M. A., D. P. Edwards, M. López–Puertas, and C. Roldán (1998). Non-local thermodynamic equilibrium in general circulation models of the Martian atmosphere. 1: Effects of the local thermodynamic equilibrium approximation on thermal cooling and solar heating. *J. Geophys. Res.*, **103**(E7), 16799–16811.

Lorenc, A. C. (1986). Analysis methods for numerical weather prediction. *Quart. J. R. Meteor. Soc.*, **112**, 1177–1194.

Lorenc, A. C., R. S. Bell, and B. Macpherson (1991). The Meteorological Office analysis correction data assimilation scheme. *Quart. J. R. Meteor. Soc.*, **117**, 59–89.

Lorenz, E. N. (1963). Deterministic nonperiodic flow. *J. Atmos. Sci.*, **20**, 130–141.

Lorenz, R. D., J. I. Lunine, J. A. Grier, and M. A. Fisher (1995). Prediction of aeolian features on planets: Application to Titan palaeoclimatology. *J. Geophys. Res.*, **100**, 26377–26386.

Lott, F. and M. J. Miller (1997). A new subgrid-scale orographic drag parametrization: Its formulation and testing. *Quart. J. R. Meteor. Soc.*, **123**, 101–127.

Lovelock, J. E. and M. Allaby (1984). *The Greening of Mars.* New York: Warner Brothers Inc.

Lowell, P. (1895). *Mars.* Boston: Houghton Mifflin.

Lowell, P. (1906). *Mars and Its Canals.* New York: Macmillan.

Lowell, P. (1908). *Mars and the Abode of Life.* New York: Macmillan.

MacVean, M. K. (1983). The effect of horizontal diffusion on baroclinic development in a spectral model. *Quart. J. R. Meteor. Soc.*, **38**, 771–783.

Madsen, M. B. (1999). The magnetic properties experiments on Mars Pathfinder. *J. Geophys. Res.*, **104**, 8761–8779.

Maki, J. N., J. J. Lorre, P. H. Smith, R. D. Brandt, and D. J. Steinwand (1999). The color of Mars: Spectrophotometric measurements at the Pathfinder landing site. *J. Geophys. Res.*, **104**, 8781–8794.

Malin, M. C. and K. S. Edgett (2000). Evidence for recent groundwater seepage and surface runoff on Mars. *Science*, **288**, 2330–2334.

Malin, M. C., G. E. Danielson, A. P. Ingersoll, H. Masursky, J. Veverka, M. A. Ravine *et al.* (1992). Mars Observer camera. *J. Geophys. Res.*, **97**, 7699–7718.

Martin, T. Z. (1986). Thermal infrared opacity of the Mars atmosphere. *Icarus*, **66**, 2–21.

Martin, T. Z. and H. H. Kieffer (1979). Thermal infrared properties of the Martian atmosphere. 2: The 15 μm band measurements. *J. Geophys. Res.*, **84**, 2843–2852.

Martin, T. Z. and R. W. Zurek (1993). An analysis of the history of dust activity on Mars. *J. Geophys. Res.*, **98**, 3221–3246.

Mass, C. and C. Sagan (1976). A numerical circulation model with topography for the Martian southern hemisphere. *J. Atmos. Sci.*, **33**, 1418–1430.

McCleese, D. J., R. D. Haskins, J. T. Schofield, R. W. Zurek, C. B. Leovy, D. A. Paige *et al.* (1992). Atmosphere and climate studies of Mars using the Mars Observer Pressure Modulator Infrared Radiometer. *J. Geophys. Res.*, **97**, 7735–7757.

McIntyre, M. E. and T. N. Palmer (1983). Breaking planetary waves in the stratosphere. *Nature*, **305**, 593–600.

McKay, C. P. (1999). Bringing life to Mars. In *The Future of Space Exploration. Scientific American Quarterly*, **10**(1), 52–57.

McKay, C. P., O. B. Toon, and J. F. Kasting (1991). Making Mars habitable. *Nature*, **352**, 489–496.

McKim, R. J. (1999). Telescopic Martian dust storms: A narrative and catalogue. *Mem. Br. Astron. Assoc.*, **44**, 1–200.

Mellon, M. T. and B. M. Jakosky (1995). The distribution and behavior of Martian ground ice during past and present epochs. *J. Geophys. Res.*, **100**, 11781–11799.

Mellon, M. T. and R. J. Phillips (2001). Recent gullies on Mars and the source of liquid water. *J. Geophys. Res.*, **106**, 23165–23179.

Mellon, M. T., B. M. Jakosky, H. H. Kieffer, and P. R. Christensen (2000). High resolution thermal inertia mapping from the Mars Global Surveyor Thermal Emission Spectrometer. *Icarus*, **148**, 437–455.

Mellor, G. L. and T. Yamada (1974). A hierarchy of turbulence closure models for planetary boundary layers. *J. Atmos. Sci.*, **31**, 1791–1806.

Mellor, G. L. and T. Yamada (1982). Development of a turbulence closure model for geophysical fluid problems. *Rev. Geophys. Space Phys.*, **20**, 851–875.

Metzger, S. M. and J. R. Carr (1999). Dust devil vortices seen by the Mars Pathfinder camera. *Geophys. Res. Lett.*, **26**, 2781–2784.

Michaels, T. I. and S. C. R. Rafkin (2003). Large eddy simulation of atmospheric convection on mars. *Quart. J. R. Meteor. Soc.*, in press.

Michelangeli, D. V., R. W. Zurek, and L. S. Elson (1987). Barotropic instability of midlatitude zonal jets on Mars, Earth and Venus. *J. Atmos. Sci.*, **44**, 2031–2041.

Michelangeli, D. V., O. B. Toon, R. M. Haberle, and J. B. Pollack (1993). Numerical simulations of the formation and evolution of water ice clouds in the Martian atmosphere. *Icarus*, **100**, 261–285.

Mintz, Y. (1961). The general circulation of planetary atmospheres. *The Atmospheres of Mars and Venus, NAS-NRC Publ.*, **944**, 107–146.

Mischna, M. A., M. I. Richardson, R. J. Wilson, and D. J. McCleese (2003). On the orbital forcing of Martian water and CO_2 cycles: A general circulation model study with simplified volatile schemes. *J. Geophys. Res.*, **108(E6)**, doi:10.1029/2003JE002051.

Mo, K. C. and M. Ghil (1987). Statistics and dynamics of persistent anomalies. *J. Atmos. Sci.*, **55**, 877–901.

Montmessin, F. and F. Forget (2003). Water-ice clouds in the LMD Martian general circulation model. *Workshop on Mars Atmosphere Modelling and Observations* (Abstract 8-8). Granada, Spain: ESA/CNES.

Montmessin, F., P. Rannou, and M. Cabane (2002). New insights into Martian dust distribution and water-ice clouds. *J. Geophys. Res.*, **107**, doi:10.1029/2001JE001520.

Moon, F. C. (1979). *Chaotic Vibrations*. New York: John Wiley & Sons.

Moriyama, S. and T. Iwashima (1980). A spectral model of the atmospheric general circulation of Mars: A numerical experiment including the effects of the suspended dust and the topography. *J. Geophys. Res.*, **85**(C5), 2847–2860.

Mouginis-Mark, P. J. (1990). Recent water release in the Tharsis region of Mars. *Icarus*, **84**, 362–373.

Murphy, J. R., J. B. Pollack, R. M. Haberle, C. B. Leovy, O. B. Toon, and J. Schaeffer (1995). 3-dimensional numerical simulation of Martian global dust storms. *J. Geophys. Res.*, **100**(E12), 26357–26376.

Nayvelt, L., P. J. Gierasch, and K. H. Cook (1997). Modeling and observations of Martian stationary waves. *J. Atmos. Sci.*, **54**, 986–1013.

Neumann, G. A., D. E. Smith and M. T. Zuber (2003). Two Mars years of clouds detected by the Mars Orbiter Laser Altinater. *J. Geophys. Res.*, **108(E4)**, doi:10.1029/2002JE001849.

Newman, C. E. (2001). *Modelling the Dust Cycle in the Martian Atmosphere.* Ph.D. thesis, Oxford University.

Newman, C. E., S. R. Lewis, P. L. Read, and F. Forget (2002a). Modeling the Martian dust cycle. 1: Representations of dust transport processes. *J. Geophys. Res.*, **107**(E12), doi:10.1029/2002JE001910.

Newman, C. E., S. R. Lewis, P. L. Read, and F. Forget (2002b). Modeling the Martian dust cycle. 2: Multiannual radiatively active dust transport simulations. *J. Geophys. Res.*, **107**(E12), doi:10.1029/2002JE001920.

Newman, C. E., P. L. Read, and S. R. Lewis (2003). Effects of the obliquity cycle on Mars' atmospheric climate and circulation as simulated in a Mars general circulation model. *Icarus*, submitted.

Niver, D. S. and S. L. Hess (1982). Band-pass filtering of one year of daily mean pressures on Mars. *J. Geophys. Res.*, **87**, 10191–10196.

North, G. R. (1984). Empirical orthogonal functions and normal modes. *J. Atmos. Sci.*, **41**, 879–887.

Ockert-Bell, M. E., J. F. Bell, C. McKay, J. Pollack, and F. Forget (1997). Absorption and scattering properties of Martian dust in the solar wavelengths. *J. Geophys. Res.*, **102**, 9039–9050.

Paige, D. A. (1985). *The Annual Heat Balance of the Martian Polar Caps from Viking Observations.* Ph.D. thesis, California Institute of Technology.

Paige, D. A. and K. D. Keegan (1994). Thermal and albedo mapping of the polar regions of Mars using Viking Thermal Mapper observations. 2: South Polar region. *J. Geophys. Res.*, **99**, 25993–26013.

Paige, D. A., J. E. Bachman and K. D. Keegan (1994). Thermal and albedo mapping of the polar regions of Mars using Viking Thermal Mapper observations. 1: North Polar region. *J. Geophys. Res.*, **99**, 25959–25991.

Palmer, T. (1981). Aspects of stratospheric sudden warmings studied from a transformed Eulerian-mean viewpoint. *J. Geophys. Res.*, **86**, 9679–9687.

Palmer, T. (1982). Properties of the Eliassen-Palm flux for planetary scale motion. *J. Atmos. Sci.*, **39**, 992–997.

Palmer, T. N., G. J. Shutts, and R. Swinbank (1986). Alleviation of a systematic westerly bias in general circulation and numerical weather prediction models through an orographic gravity wave drag parametrisation. *Quart. J. R. Meteor. Soc.*, **112**, 1001–1039.

Pankine, A. A. and A. P. Ingersoll (2001). Interannual variability of Martian global dust storms. *Icarus*, **155**, 299–323.

Parker, T. J., D. S. G. and R. S. Saunders, D. Pieri, and D. M. Schneeberger (1993). Coastal geomorphology of the Martian northern plains. *J. Geophys. Res.*, **98**, 11061–11078.

Pedlosky, J. (1987). *Geophysical Fluid Dynamics*. Berlin: Springer.
Pettengill, G. H. and P. G. Ford (2000). Winter clouds over the north Martian polar cap. *Geophys. Res. Lett.*, **27**, 609–612.
Phillips, N. A. (1956). A coordinate system having some special advantages for numerical forecasting. *J. Meteor.*, **14**, 184–185.
Phillips, S. P. (1984). Analytical surface pressure and drag for linear hydrostatic flow over three-dimensional elliptical mountains. *J. Atmos. Sci.*, **41**, 1073–1084.
Pickersgill, A. O. and G. E. Hunt (1979). The formation of Martian lee waves generated by a crater. *J. Geophys. Res.*, **84**, 8317–8331.
Pickersgill, A. O. and G. E. Hunt (1981). An examination of the formation of linear lee waves generated by giant Martian volcanoes. *J. Atmos. Sci.*, **38**, 40–51.
Pirraglia, J. A. (1976). Martian atmospheric lee waves. *Icarus*, **27**, 517–530.
Plaut, G. and R. Vautard (1994). Spells of low frequency oscillations and weather regimes in the northern hemisphere. *J. Atmos. Sci.*, **51**, 210–236.
Plaut, J. J., R. Kahn, E. A. Guinness, and R. E. Arvidson (1988). Accumulation of sedimentary debris in the south polar region of Mars and implications for climate history. *Icarus*, **75**, 357–377.
Pollack, J. B., D. S. Colburn, R. Kahn, J. Hunter, W. van Kamp, C. E. Carlston *et al.* (1977). Properties of aerosols in the Martian atmosphere as inferred from Viking Lander imaging data. *J. Geophys. Res.*, **82**, 4479–4496.
Pollack, J. B., D. S. Colburn, F. M. Flasar, R. Kahn, C. E. Carlston, and D. C. Pidek (1979). Properties and effects of dust suspended in the Martian atmosphere. *J. Geophys. Res.*, **84**, 2929–2945.
Pollack, J. B., C. B. Leovy, P. W. Greiman, and Y. Mintz (1981). A Martian General Circulation Model experiment with large topography. *J. Atmos. Sci.*, **38**, 3–29.
Pollack, J. B., J. F. Kasting, S. M. Richardson, and K. Poliakoff (1987). The case for a wet, warm climate on early Mars. *Icarus*, **71**, 203–224.
Pollack, J. B., R. M. Haberle, J. Schaeffer, and H. Lee (1990). Simulations of the general circulation of the Martian atmosphere. 1: Polar processes. *J. Geophys. Res.*, **95**, 1447–1473.
Pollack, J. B., R. M. Haberle, J. R. Murphy, H. Schaeffer, and J. Lee (1993). Simulations of the general circulation of the Martian atmosphere. 2: Seasonal pressure variations. *J. Geophys. Res.*, **98**, 3149–3181.
Pollack, J. B., M. E. Ockert-Bell, and M. K. Shepard (1995). Viking Lander image analysis of Martian atmospheric dust. *J. Geophys. Res.*, **100**, 5235–5250.
Preisendorfer, R. W. (1988). *Principal Component Analysis in Meteorology and Oceanography*. Amsterdam: Elsevier.
Rafkin, S. C. R., R. M. Haberle, and T. I. Michaels (2001). The Mars Regional Atmospheric Modelling System: Model description and selected simulations. *Icarus*, **151**, 228–256.
Rayleigh, L. (1880). On the stability, or instability, of certain fluid motions. *Scientific Papers (Cambridge University Press)*, **3**, 594–596.
Read, P. L. (1988). On the scale of baroclinic instability in deep, compressible atmospheres. *Quart. J. R. Meteor. Soc.*, **114**, 421–437.
Read, P. L., S. R. Lewis, and S. Bingham (2003). Predicting weather conditions and climate for Mars expeditions. *J. Brit. Interplan. Soc.*, in press.
Reader, M. C., I. Fung, and N. McFarlane (1999). The mineral dust aerosol cycle during the last glacial maximum. *J. Geophys. Res.*, **104**, 9381–9398.

Rennó, N. O., M. L. Burckett, and M. P. Larkin (1998). A simple thermodynamical theory for Dust Devils. *J. Atmos. Sci.*, **55**, 3244–3252.

Rennó, N. O., A. A. Nash, J. Lunine, and J. Murphy (2000). Martian and terrestrial Dust Devils: Test of a scaling theory using Pathfinder data. *J. Geophys. Res.*, **105**, 1859–1865.

Richardson, M. I. (1999). A general circulation model study of the Mars water cycle. PhD Thesis, University of California.

Richardson, M. I. and R. J. Wilson (2002a). Investigation of the nature and stability of the Martian seasonal water cycle with a general circulation model. *J. Geophys. Res.*, **107**(E5), doi:10.1029/2001JE001536.

Richardson, M. I. and R. J. Wilson (2002b). A topographically forced asymmetry in the Martian circulation and climate. *Nature*, **416**, 298–301.

Richardson, M. I., R. J. Wilson, and A. V. Rodin (2002). Water ice clouds in the Martian atmosphere: General circulation model experiments with a simple cloud scheme. *J. Geophys. Res.*, **107**(E9), doi:10.1029/2001JE001804.

Rossow, W. B. (1978). Cloud microphysics: analysis of the clouds of Earth, Venus, Mars and Jupiter. *Icarus*, **36**, 1–50.

Ryan, J. A. and R. M. Henry (1979). Mars atmospheric phenomena during two major dust storms as measured at the surface. *J. Geophys. Res.*, **84**, 2821–2829.

Ryan, J. A. and R. D. Lucich (1983). Possible Dust Devils vortices on Mars. *J. Geophys. Res.*, **88**, 11005–11011.

Ryan, J. A., R. M. Henry, S. L. Hess, C. B. Leovy, J. E. Tillman, and C. Walcek (1978). Mars meteorology: Three seasons at the surface. *Geophys. Res. Lett.*, **5**, 715–718.

Sadourny, R. (1975). The dynamics of finite-difference models of the shallow water equations. *J. Atmos. Sci.*, **32**, 680–689.

Salby, M. L. (1982). Sampling theory for asynoptic satellite observations. 2: Fast fourier synoptic mapping. *J. Atmos. Sci.*, **39**, 2601–2614.

Santee, M. and D. Crisp (1993). Thermal structure and dust loading of the Martian atmosphere during late southern summer: Mariner 9 revisited. *J. Geophys. Res.*, **98**, 3261–3279.

Sasaki, Y. (1970). Some basic formalisms in numerical variational analysis. *Mon. Weather Rev.*, **98**, 875–883.

Savijärvi, H. (1991). A model study of the PBL structure on Mars and the Earth. *Beitr. Physik Atmosph.*, **64**, 219–229.

Savijärvi, H. and T. Siili (1993). The Martian slope winds and the nocturnal PBL jet. *J. Atmos. Sci.*, **50**, 77–88.

Sawyer, J. S. (1962). Gravity waves in the atmosphere as a three-dimensional problem. *Quart. J. R. Meteor. Soc.*, **88**, 412–425.

Schneider, E. K. (1983). Martian great dust storms: Interpretive axially symmetric models. *Icarus*, **55**, 302–331.

Schofield, J. T., D. Crisp, J. R. Barnes, R. M. Haberle, J. A. Magalhães, J. R. Murphy *et al.* (1997). The Mars Pathfinder Atmospheric Structure Investigation/Meteorology (ASI/MET) experiment. *Science*, **278**, 1752–1757.

Scorer, R. S. (1956). Airflow over an isolated hill. *Quart. J. R. Meteor. Soc.*, **82**, 75–81.

Scorer, R. S. and M. Wilkinson (1956). Waves in the lee of an isolated hill. *Quart. J. R. Meteor. Soc.*, **82**, 419–427.

Segschneider, J., B. Grieger, H. U. Keller, F. Lunkeit, E. Kirk, K. Fraedrich *et al.* (2003). Towards an intermediate complexity Martian climate simulator. *Workshop on Mars Atmosphere Modelling and Observations* (Abstract 4-1). Granada, Spain: ESA/CNES.

Segura, T. L., O. B. Toon, A. Colaprete, and K. Zahnle (2002). Environmental effects of large impacts on Mars. *Science*, **298**, 1977–1980.

Sharman, R. D. and J. A. Ryan (1980). Mars atmospheric pressure periodicities from Viking observations. *J. Atmos. Sci.*, **37**, 1994–2001.

Siili, T. (1996). Modeling of albedo and thermal inertia induced mesoscale circulations in the midlatitude summertime Martian atmosphere. *J. Geophys. Res.*, **101**(E6), 14957–14968.

Simmons, A. J. and D. M. Burridge (1980). An energy and angular-momentum conserving vertical finite-difference scheme and hybrid vertical coordinates. *Mon. Weather Rev.*, **109**, 758–766.

Slipher, E. C. (1962). *The Photographic Story of Mars*. Flagstaff, AZ: Northland Press.

Slipher, E. C. (1964). *A Photographic Study of the Brighter Planets*. Flagstaff, AZ: Lowell Observatory.

Smith, M. D. (2002). The annual cycle of water vapor on Mars as observed by the Thermal Emission Spectrometer. *J. Geophys. Res.*, **107**, doi:10.1029/2001JE001522.

Smith, M. D. (2003). Interannual variability in TES atmospheric observations of Mars during 1999–2003. *Icarus*, in press.

Smith, D. E. and M. T. Zuber (1996). The shape of Mars and the topographic signature of the hemispheric dichotomy. *Science*, **217**, 184–188.

Smith, D. E., M. T. Zuber, H. V. Frey, J. B. Garvin, J. W. Head, D. O. Muhleman *et al.* (1998). Topography of the northern hemisphere of Mars from the Mars Orbiter Laser Altimeter. *Science*, **279**, 1686–1692.

Smith, D. E., M. T. Zuber, S. Solomon, R. Phillips, J. Head, J. Garvin *et al.* (1999). The global topography of Mars and implications for surface evolution. *Science*, **284**, 1495–1503.

Smith, M. D., J. C. Pearl, B. J. Conrath, and P. R. Christensen (2000). Mars Global Surveyor Thermal Emission Spectrometer (TES) observations of dust opacity during aerobraking and science phasing. *J. Geophys. Res.*, **105**(E4), 9539–9552.

Smith, M. D. and J. C. Pearl, B. J. Conrath, and P. R. Christensen (2001). Thermal Emission Spectrometer results: Mars atmospheric thermal structure and aerosol distribution. *J. Geophys. Res.*, **106**, 23929–23945.

Smith, M. D., B. J. Conrath, J. C. Pearl, and P. R. Christensen (2002). Thermal Emission Spectrometer observations of Martian planet-encircling dust storm 2001A. *Icarus*, **157**, 259–263.

Smith, M. D., B. J. Conrath, J. C. Pearl and P. R. Christensen (2003). TES instrument description and thermal structure observations. *Workshop on Mars Atmosphere Modelling and Observations* (Abstract 1-1). Granada, Spain: ESA/CNES.

Smith, R. B. (1979). The influence of mountains on the atmosphere. *Adv. in Geophys.*, **21**, 87–230.

Sommerer, J. C., W. L. Ditto, C. Grebogi, E. Ott, and M. L. Spano (1991). Experimental confirmation of a scaling theory for noise-induced crises. *Phys. Rev. Lett.*, **66**, 1947–1950.

Spinrad, H. G., L. D. Münch, and D. Kaplan (1963). The detection of water vapour on Mars. *Astrophys. J.*, **137**, 1319–1321.

Squyres, S. W., S. M. Clifford, R. O. Kuzmin, J. R. Zimbelman, and F. M. Costard (1992). Ice in the Martian regolith. In H. H. Kieffer, B. M. Jakosky, C. W. Snyder, and M. S. Matthews (eds), *Mars* (pp. 523–556). Tucson, AZ: University of Arizona Press.

Sutton, J. L., C. B. Leovy, and J. E. Tillman (1978). Diurnal variations of the Martian surface layer meteorological parameters during the first 45 sols at two Viking Lander sites. *J. Atmos. Sci.*, **35**, 2346–2355.

Takahashi, Y. O., H. Fujiwara, H. Fukunishi, M. Odaka, and Y.-Y. Hayashi (2003a). Zonal mean circulation obtained by a newly developed Martian atmospheric general circulation model. *Workshop on Mars Atmosphere Modelling and Observations* (Abstract 2-5). Granada, Spain: ESA/CNES.

Takahashi, Y. O., H. Fujiwara, H. Fukunishi, M. Odaka, Y.-Y. Hayashi, and S. Watanabe (2003b). Topographically induced north-south asymmetry of the meridional circulation in the Martian atmosphere. *J. Geophys. Res.*, **108**, doi:10.1029/2001JE001638.

Talagrand, O. and P. Courtier (1987). Variational assimilation of meteorological observations with the adjoint vorticity equation. I: Theory. *Quart. J. R. Meteor. Soc.*, **113**, 1311–1328.

Tanaka, H. L. and M. Arai (1999). Linear baroclinic instability in the Martian atmosphere: Primitive equation calculations. *Earth Planets & Space*, **51**, 225–232.

Tanaka, K. L., D. H. Scott, and R. Greeley (1992). Global stratigraphy. In H. H. Kieffer, B. M. Jakosky, C. W. Snyder, and M. S. Matthews (eds), *Mars* (pp. 545–582). Tucson, AZ: University of Arizona Press.

Théodore, B., E. Lellouch, E. Chassefière, and A. Hauchecorne (1993). Solsticial temperature inversions in the Martian middle atmosphere: Observational clues and 2-D modelling. *Icarus*, **105**, 512–528.

Thomas, P. and P. J. Gierasch (1985). Dust devils on mars. *Science*, **230**, 175–177.

Thomas, P. C., J. Veverka, D. Gineris, and L. Wong (1984). 'Dust' streaks on Mars. *Icarus*, **49**, 398–415.

Thomas, P., S. Squyres, K. Herkenhoff, A. Howard, and B. Murray (1992). Polar deposits of Mars. In H. H. Kieffer, B. M. Jakosky, C. W. Snyder, and M. S. Matthews (eds), *Mars* (pp. 767–795). Tucson, AZ: University of Arizona Press.

Thomas, N., W. J. Markiewicz, R. M. Sablotny, M. W. Wuttke, H. U. Keller, J. R. Johnson *et al.* (1999). The color of the Martian sky and its influence on the illumination of the Martian surface. *J. Geophys. Res.*, **104**, 8795–8808.

Thompson, J. M. T. and H. B. Stewart (1986). *Nonlinear Dynamics and Chaos* (376pp.). Chichester, UK: John Wiley & Sons.

Tillman, J. E. (1977). Dynamics of the boundary layer of Mars. In A. V. Jones (ed.), *Proc. Symp. on Planetary Atmospheres* (pp. 145–149), Ottawa, Canada: Royal Society of Canada.

Tillman, J. E. (1988). Mars global atmospheric oscillations: Annually synchronized, transient normal-mode oscillations and the triggering of global dust storms. *J. Geophys. Res.*, **93**, 9433–9451.

Tillman, J. E., R. M. Henry, and S. L. Hess (1979). Frontal systems during passage of the Martian north polar hood over the Viking Lander 2 site prior to the first 1977 dust storm. *J. Geophys. Res.*, **84**, 2947–2955.

Titus, T. N., H. H. Kieffer, and P. R. Christensen (2002). Exposed water ice discovered near the south pole of Mars. *Science*, **299**, 1049–1051.

Toigo, A. D. and M. I. Richardson (2002). A mesoscale model for the Martian atmosphere. *J. Geophys. Res.*, **107**, doi:10.1029/2001JE001489.

Toigo, A. D., M. I. Richardson, R. J. Wilson, H. Wang, and A. P. Ingersoll (2002). Dust lifting and dust storms near the South Pole of Mars. *J. Geophys. Res.*, **107**, doi:10.1029/2001JE001592.

Tomasko, M. G., L. R. Doose, M. Lemmon, P. H. Smith, and E. Wegryn (1999). Properties of dust in the Martian atmosphere from the imager on Mars Pathfinder. *J. Geophys. Res.*, **104**, 8987–9007.

Toon, O. B., J. B. Pollack, and C. Sagan (1977). Physical properties of the particles comprising the Martian dust storm of 1971–1972. *Icarus*, **30**, 663–696.

Touma, J. and J. Wisdom (1993). The chaotic obliquity of Mars. *Science*, **259**, 1294–1297.

Turco, R. P., O. B. Toon, T. P. Ackerman, J. B. Pollack, and C. Sagan (1983). Nuclear Winter – global consequences of multiple nuclear explosions. *Science*, **222**, 1283–1292.

Tyler, G. L., G. Balmino, D. P. Hinson, W. L. Sjogren, D. E. Smith, R. Woo *et al.* (1992). Radio science investigations with Mars Observer. *J. Geophys. Res.*, **97**, 7759–7779.

Tyler, D. and J. R. Barnes (2002). Simulation of surface meteorology at the Pathfinder and VL1 sites using a Mars mesoscale model. *J. Geophys. Res.*, **107**, doi:10.1029/2001JE001JE001618.

Valdes, P. J. and B. J. Hoskins (1991). Nonlinear orographically forced planetary waves. *J. Atmos. Sci.*, **48**, 2089–2106.

Vallis, G. K. (1986). El Niño: A chaotic dynamical system? *Science*, **232**, 243–245.

Wänke, H., J. Bruckner, G. Dreibus, R. Rieder and I. Ryabchikov (2001). Chemical composition of rocks and soils at the Pathfinder site. *Space Sci. Rev.*, **96**, 317–330.

Ward, W. R. (1974). Climatic variations of Mars. I: astronomical theory of insolation. *J. Geophys. Res.*, **79**, 3375–3386.

Ward, W. R. (1979). Present obliquity oscillations of Mars: Fourth order accuracy in orbital e and I. *J. Geophys. Res.*, **84**, 237–241.

Ward, W. R. (1982). Comments on the long term stability of the Earth's obliquity. *Icarus*, **50**, 444–448.

Ward, W. R. (1992). Long term orbital and spin dynamics of Mars. In H. H. Kieffer, B. M. Jakosky, C. W. Snyder, and M. S. Matthews (eds), *Mars* (pp. 298–320). Tucson, AZ: University of Arizona Press.

Webster, P. J. (1977). The low-latitude circulation of Mars. *Icarus*, **30**, 626–649.

Wells, R. A. (1984). *Geophysics of Mars.* Amsterdam: Elsevier.

White, A. A. (2003). Dynamic meteorology: Primitive equations. In J. Holton, J. Curry, and J. Pyle (eds), *Encyclopedia of Atmospheric Science* (pp. 694–702). New York: Academic Press.

White, B. R. (1979). Soil transport by winds on Mars. *J. Geophys. Res.*, **84**, 4643–4651.

Whitehouse, S. G. (1999). *POD-Galerkin Modelling of the Martian Atmosphere.* Ph.D. thesis, Oxford University.

Wilson, R. J. (1997). A general circulation model simulation of the Martian polar warming. *Geophys. Res. Lett.*, **24**, 123–126.

Wilson, R. J. (2000). Evidence for diurnal period Kelvin waves in the Martian atmosphere from MGS TES data. *Geophys. Res. Lett.*, **27**, 3889–3892.

Wilson, R. J. and K. Hamilton (1996). Comprehensive model simulation of thermal tides in the Martian atmosphere. *J. Atmos. Sci.*, **53**, 1290–1326.

Wilson, R. J. and M. I. Richardson (2000). The Martian atmosphere during the Viking mission. 1: Infrared measurements of atmospheric temperatures revisited. *Icarus*, **145**, 555–579.

Wilson, R. J., D. Banfield, B. J. Conrath, and M. D. Smith (2002). Travelling waves in the northern hemisphere of Mars. *Geophys. Res. Lett.*, **29**, doi:10.1029/2002GL014866.

Withers, P. A. and G. A. Neumann (2001). Enigmatic northern plains of Mars. *Nature*, **410**, 651.

Wu, S. S. C. (1981). A method of defining topographic datums of planetary bodies. *Adv. Geophys.*, **1**, 147–160.

Zent, A. P., R. M. Haberle, H. C. Houben, and B. M. Jakosky (1993). A coupled subsurface-boundary layer model of water on Mars. *J. Geophys. Res.*, **98**, 3319–3337.

Zhang, K. Q., A. P. Ingersoll, D. M. Kass, J. C. Pearl, M. D. Smith, B. J. Conrath *et al.* (2001). Assimilation of Mars Global Surveyor atmospheric temperature data into a general circulation model. *J. Geophys. Res.*, **106**(E12), 32863–32877.

Zuber, M. T. (2000). The crust and mantle of Mars. *Nature*, **412**, 220–227.

Zuber, M. T., D. E. Smith, S. C. Solomon, D. O. Muhleman, J. W. Head, J. B. Garvin *et al.* (1992). The Mars Observer Laser Altimeter investigation. *J. Geophys. Res.*, **97**, 7781–7797.

Zuber, M. T., D. E. Smith, S. C. Solomon, J. B. Abshire, R. S. Afzal, O. Aharonson *et al.* (1998). Observations of the north polar region of Mars from the Mars Observer Laser Altimeter. *Science*, **282**, 2053–2060.

Zubrin, R. and C. P. McKay (1997). Technological requirements for terraforming Mars. *J. Brit. Interplan. Soc.*, **50**, 83–92.

Zurek, R. W. (1976). Diurnal tide in the Martian atmosphere. *J. Atmos. Sci.*, **33**, 321–337.

Zurek, R. W. (1981). Inference of dust opacities for the 1977 Martian great dust storms from Viking Lander 1 pressure data. *Icarus*, **45**, 202–215.

Zurek, R. W. (1982). Martian great dust storm, an update. *Icarus*, **50**, 288–310.

Zurek, R. W. (1986). Atmospheric tidal forcing of the zonal-mean circulation: The Martian dusty atmosphere. *J. Atmos. Sci.*, **43**, 652–670.

Zurek, R. W. (1988). Free and forced modes in the Martian atmosphere. *J. Geophys. Res.*, **93**, 9452–9462.

Zurek, R. W., J. R. Barnes, R. M. Haberle, J. B. Pollack, J. E. Tillman, and C. B. Leovy (1992). Dynamics of the atmosphere of Mars. In H. H. Kieffer, B. M. Jakosky, C. W. Snyder, and M. S. Matthews (eds), *Mars* (pp. 835–933). Tucson, AZ: University of Arizona Press.

Index

Printing: Mercedes-Druck, Berlin
Binding: Stein + Lehmann, Berlin